Advances in Natural and Technological Hazards Research

Volume 50

More information about this series at http://www.springer.com/series/6362

S. P. Pradhan • V. Vishal • T. N. Singh
Editors

Landslides: Theory, Practice and Modelling

Springer

Editors
S. P. Pradhan
Department of Earth Sciences
Indian Institute of Technology Roorkee
Roorkee, India

V. Vishal
Department of Earth Sciences
Indian Institute of Technology Bombay
Mumbai, Maharashtra, India

T. N. Singh
Department of Earth Sciences
Indian Institute of Technology Bombay
Mumbai, Maharashtra, India

ISSN 1878-9897 ISSN 2213-6959 (electronic)
Advances in Natural and Technological Hazards Research
ISBN 978-3-030-08443-1 ISBN 978-3-319-77377-3 (eBook)
https://doi.org/10.1007/978-3-319-77377-3

Softcover re-print of the Hardcover 1st edition 2019

Printed on acid-free paper

This Springer imprint is published by the registered company Springer International Publishing AG part of Springer Nature.
The registered company address is: Gewerbestrasse 11, 6330 Cham, Switzerland

Foreword

I am happy to learn that Springer Publishing Company is bringing out a book on landslides under the NTHR series on natural hazards. The book is jointly edited by S. P. Pradhan, V. Vishal, and T. N. Singh who are eminent scholars and researchers in the field of geo-mechanics.

Landslides are associated with surface excavations in opencast mines, transportation routes (rail–road cuts), canals, tailing dams, urban cuts for buildings, power houses, and other civil constructions. With the advent of the larger excavation machineries, there is a tremendous increase in the size of the cuts particularly in surface mining. Therefore, stability of open pit slopes both in soil and rock must be properly ensured. These excavations should be designed and constructed to be absolutely safe at all times due to the great cost of failure.

As compared to soil, rock is a more complicated material. It is inhomogeneous, anisotropic, and inelastic. Strength parameters vary considerably and are difficult to determine precisely. Field conditions are complex and difficult to duplicate in a laboratory.

The book *Landslides: Theory, Practice and Modelling* is a compilation of fourteen chapters dealing with the different aspects of the subject. The chapters are grouped in five parts, based on the major themes of subject matter of the book. Each chapter is written by a group of authors who are eminent researchers in their own field of specialization.

Part I provides "Introduction to Landslides Dynamics". It deals with dump stability and shear strength parameters of rock mass and gives the details of a specific case of a landslide.

Part II discusses "Landslide Monitoring and Prediction". It also includes infrared thermography for the study of jointed and weathered rock mass and wireless instrumentation for landslide prediction.

Part III deals with landslides risk and hazard mitigation and gives the case study assessment in Iran. The development of early warning system in India is also described. A chapter on bio-engineering is included. The authors stress that it is an effective eco-friendly slope stabilization method. Soil nailing is described.

Part IV gives the details of various landslide numerical modeling techniques where as Part V deals with selected case studies in different countries.

It is a cohesive effort of a number of authors, researchers, and experts in the rock engineering across the country and other parts of the world. The editors have done exemplary job in collecting the papers, compiling, and editing them in a book form. I extend my warm greetings to all those associated with the publications and congratulate the Springer Publishing Company for the launch of the book.

D. P. Singh

Distinguished Professor
Rock Mechanics
Indian Institute of Technology
(Banaras Hindu University)
Varanasi, Uttar Pradesh, India

Former Vice Chancellor
Awadhesh Pratap Singh University
Rewa, Madhya Pradesh, India

University of Lucknow
Lucknow, Uttar Pradesh, India

U.P. Rajarshi Tandon Open University
Allahabad, Uttar Pradesh, India

Preface

The subject of landslides has always been fascinating from both theoretical and practical context that has gained attention of geologists, engineers, model developers, professionals and researchers, and students of geosciences worldwide. Landslides represent one of the most destructive geo-hazards, which can reach remarkably long distances and velocities, and are also susceptible of wiping out human communities and settlements. Landslides are more frequent nowadays due to extreme climatic conditions and intervention of human with the nature. The augmentation of human settlements and rapid developments that break the natural geomorphic equilibrium contribute to landslides. Even small slopes are prone to landslides and have potential to break the infrastructure, disrupt the traffic and bring loss to lives. The economic distress induced by landslides makes terrible impact on socio-economic conditions of human race.

This book is not an attempt to review the field of landslides, rather an attempt to have a better understanding of the subject, and the need was felt during our teaching exercises. In this book, the emphasis is placed on the dynamics of landslide research, its monitoring and prediction, landslide risk and hazard mitigation, landslide numerical modelling techniques and analysis of a few selected case studies. Each chapter is self-consistent, with questions and arguments introduced from the beginning.

The book is broadly categorised into five parts, which include fourteen chapters from contributors across the globe. The first part of the book introduces to landslide dynamics comprising four chapters that focus on basic understanding of slope stability, different types of mass movement, shear strength behaviour of jointed rock mass and introduces to rock fall and man-made slope instabilities. The second part of the book discourses about the modern monitoring and prediction techniques for landslides and consists of two chapters. These chapters highlight on infrared thermography and ground-based real time monitoring system using wireless instrumentation for prediction of slope instabilities. The third section of the book is dedicated to risk and mitigation of landslides. There are four chapters in this part, which discusses on landslide susceptibility assessment, mapping, vulnerability and risk assessment. This also includes innovative and eco-friendly stabilisation methods by using bioengineering and soil-nailing techniques. Soil-reinforcement by plant is an innovative way of stabilising the landslide prone slope area. Section four of the book comprises one chapter on review of optimisation techniques in slope stability. The concluding part consists of few case specific landslides from Italy, Turkey and permafrost regions of Russia in which the processes involving from slope instabilities, monitoring at length specific to site are discussed.

We thank all the contributing authors for supplementing interesting chapters to this book, and for contributing their expertise and knowledge as authors. We are also much indebted to reviewers for their valuable suggestions and comments.

And above all, we express our exuberant gratitude to Springer Publication Ltd. For believing in our work and enabling it for publication.

We are much beholden to Anasuya P. Pradhan for offering her editorial assistantship in the making of the book.

Roorkee, India — Sarada Prasad Pradhan
Mumbai, Maharashtra, India — Vikram Vishal
Mumbai, Maharashtra, India — Trilok Nath Singh

Contents

About the Editors and Authors

Editors

Sarada Prasad Pradhan is currently working as an Assistant Professor at the Indian Institute of Technology (IIT) Roorkee, India. He obtained his M.Sc. (Applied Geology) and Ph.D. from the IIT Bombay (India). He subsequently worked as a Reservoir Engineer for the Oil and Natural Gas Corporation Ltd. (ONGC) for 5 years, during which he was involved in many projects of national importance. He received the Young Scientist Award from CAFET INNOVA Technical Society, and the Asset Manager Award for his excellent performance in Gas Field Development from ONGC Ltd. His research findings have been published in leading national and international journals, book chapters and conference proceedings. His major research interests are Rock Mechanics, Engineering Geology, Reservoir Geomechanics, Petroleum Geo-science and Carbon Dioxide Sequestration, and he is currently participating in three major research projects on slope stability.

Vikram Vishal is an Assistant Professor at the Department of Earth Sciences, IIT Bombay, Mumbai. He completed his bachelor's degree at Presidency College, Kolkata, and his master's degree at the IIT Bombay. After a brief stint as a geologist at Tata Steel, he pursued his Ph.D. degree from 2010 to 2012, jointly at the IIT Bombay and Monash University (Australia). He worked as DST Inspire Faculty at the Department of Earth Sciences, Indian Institute of Technology (IIT) Roorkee from 2013 to 2015, then went on to pursue further research as a Fulbright-Nehru Postdoctoral Fellow at Stanford University, California, USA, in 2015. He is a recipient of the National Geosciences Award – Young Researcher by the Government of India, the Young Scientist Award by the Indian Science Congress Association, the Excellence in PhD Thesis Award by the IIT Bombay and Monash University, and the Best Undergraduate Gold Medal by Presidency College, Kolkata. He has produced more than 40 publications on various aspects of geomechanics, engineering geology and unconventional petrophysics in journals, book chapters and conference proceedings. Recently, he co-edited the book *Geologic Carbon Sequestration: Understanding Reservoir Concepts* for Springer. He also serves as an Associate Editor for the *Journal of Natural Gas Science and Engineering* (Elsevier) and is currently investigating three research projects funded by the Ministry of Science and Technology, Government of India.

Trilok Nath Singh is the Institute Geoscience Chair Professor at the Department of Earth Sciences, IIT Bombay, Mumbai, and an expert in the fields of rock mechanics, mining geology, and clean energy. He received his Ph.D. from the Institute of Technology BHU, Varanasi, in 1991, where he subsequently worked until 2003. He has been honored with many prestigious awards such as the National Mineral Award, the first P. N. Bose Mineral Award, the SEAGATE Excellence Award for Geo-Engineering, and the GSI Sesquicentennial Commemorative Award. He has nearly 28 years of experience in research and education and has authored more than 350 publications in various journals and for national and international conferences. He serves on the governing and advisory councils of several national institutes and universities and is currently leading projects of immense scientific and industrial importance on, e.g. coalbed methane, carbon sequestration, shale gas, nuclear waste repositories and mine slope stability.

Authors

Sayed Fakhreddin Afzali is an Assistant Professor in the Department of Natural Resource and Environmental Engineering at the University of Shiraz, where he has been a faculty member since 2008. He completed his PhD and undergraduate studies at Isfahan University of Technology (IUT) and his MSc studies at Shiraz University. His research interests lie in the area of soil management, especially in soil erosion and conservation, carbon sequestration and medicinal plants stress. In recent years, he has focused on wind erosion and carbon sequestration especially works with some mulches. He has collaborated actively with researchers in several other disciplines of environmental science, particularly sustainable management and pollution.

Ismail Colkesen received the PhD degree from the Istanbul Technical University, Istanbul, Turkey, in 2015. He is currently an Assistant Professor in the Department of Geomatics Engineering, Gebze Technical University, Gebze, Turkey. His major research areas are land use and land cover classification, image processing, machine learning techniques, landslide susceptibility mapping and application of artificial intelligence techniques in various engineering and environmental applications. He gives remote sensing courses at graduate levels and continues research about raster data analysis, digital image processing and natural hazard modelling.

Marina De Maio is working as an Associate Professor in the Department of Environmental Engineering, Land and Infrastructure at Politecnico di Torino, Italy. She received her master's degree in Geology from the University of Catania and PhD degree in Applied Hydrogeology from Politecnico di Torino, Italy. She has nearly 19 years of experience in research and teaching, and has authored more than 55 articles in the field of aquifer vulnerability, water pollution, geological risk assessment, hydro-geochemistry and remote sensing and GIS.

Gianpiero Amanzio was born on October 1, 1982, in Turin. Dr. Gianpiero holds a master's degree in Environmental Engineering in July 2009. During his studies, he has acquired skills in remote sensing, geophysics, geotechnical, geomatics and hydrogeology. He worked with engineering office dealing with risk analysis, CAD project, acquisition and processing of points clouds collected by laser scanner and drones (photogrammetry acquisition) in order to generate 3D models. He has acquired the necessary qualifications for the drone flight in noncritical areas, as required by the Civil Aviation Authority (ENAC). Since 2010, he is a research fellow at the Politecnico di Torino, and in April 2014, he received a PhD in Environmental Engineering in the field of Applied Geology. During his PhD, he developed some statistical method in order to study the groundwater vulnerability of mountain spring by the implementation of several algorithms in Matlab environment.

Tushar Gupta is a final year research scholar presently under the joint PhD programme of IITB-Monash Research Academy in IIT Bombay. His present research is in the field of utilisation of coal fly ash in haul roads and other applications in open-cast coal mines. He has completed his bachelors in technology (BTech) from Indian Institute of Technology (BHU), Varanasi, in mining engineering discipline, and worked on dump slope and coal rib stability problems in open-cast mines of NCL as his final year projects. Later, he has worked in underground coal mines of Coal India Limited for 3 years as assistant manager (mining) before joining the joint PhD programme at IIT Bombay.

Stanilovskaya Julia serves as a Research Associate of Permafrost Laboratory at the Sergeev Institute of Environmental Geoscience (IEG RAS) since 2007. During the last 10 years, she and her colleagues have established the long-term monitoring networks for permafrost studying in the Northern Transbaikalia (southern part of Siberian platform) about temperature regime of frozen ground and ground ice (ice wedges). In the meantime, their permafrost monitoring station in Chara provides data on permafrost diversity of cryogenic phenomena that are useful in scientific programmes and engineering projects. The permafrost-related hazard assessment along pipelines and roads is one of her research topics. She is interested in new techniques and devices that could be useful in permafrost research. She has participated in mountain expeditions to investigate permafrost on Kilimanjaro (5895 m), Qinghai Plateau on Tibet (4900 m), Elbrus (4200 m), Kodar and Udokan (2000 m) and Verkhoyan'e (2000 m). She has about more than 50 research articles in reputed journals and conference proceedings.

D. P. Kanungo born in 1967, is a Senior Principal Scientist in Geotechnical Engineering Group at CSIR-Central Building Research Institute, India. He is also a Professor of Academy of Scientific and Innovative Research (AcSIR), India, in the Faculty of Physical Sciences. He did his Masters of Technology in Applied Geology in 1990 from the University of Roorkee, India (now IIT Roorkee), and PhD in Earth Sciences in 2006 from IIT Roorkee, India. He has professional experience of more than 23 years in the field of engineering geology and landslide disaster mitigation. He has contributed more than 90 research papers in books, journals and conferences. His research interests are engineering geological investigations, landslide hazard and risk assessment, landslide dynamics through instrumentation and monitoring, landslide modelling and remote sensing and GIS applications. He has carried out a large number of infrastructure and landslide disaster mitigation-related projects in India. He is the recipient of Raman Research Fellowship Award in the year 2010 from Council of Scientific and Industrial Research (CSIR), India, and carried out his post-doctoral research at Research Centre on Landslides, Disaster Prevention Research Institute, Kyoto University, Japan.

Taskin Kavzoglu received the MSc degree in Geographical Information Science and the PhD degree in Remote Sensing from the University of Nottingham, UK, in 1997 and 2001, respectively. In 2001, he joined the Geomatics Department of Gebze Institute of Technology in Turkey as a faculty member. He is currently a full professor in Gebze Technical University. He received best PhD Thesis Award from RSPSoc, UK. He is an internationally acknowledged researcher in remote sensing and computer vision with a particular focus on machine learning and their applications to various fields. He has active interdisciplinary collaborations with geologists, geomorphologists, foresters, geodesists and photogrammetrists.

Aiding Kornejady was born in Bandar Torkaman, Golestan Province, Iran, in 1989. He graduated in 2011 with his BSc degree in Rangeland and Watershed Management from Ferdowsi University of Mashhad, Iran. He earned his MSc degree in 2013 at the Gorgan University of Agricultural Sciences and Natural resources, with a focus on assessing landslide susceptibility and semi-quantitative risk. He continued to pursue the same concept for his future scientific career. At this meantime, he is studying for a PhD in Watershed Management Science and Engineering specialises in developing a landslide management strategic plan. He dedicated his scientific career to the landslide issue and presented some articles and projects devoted to this field.

Muriel Lavy (1987) is an environmental engineer, graduated in 2012 at the Polytechnic of Turin in the protection of the territory engineering. In 2013, she started her PhD in Environmental Engineering, focussing on GIS geospatial analysis and environmental modelling based on digital elevation models. During her PhD, she took a summer internship in Esri Inc. (2014) in California, working with ArcGIS and Esri water-oriented applications, where she was awarded at the Esri Young Scholar Award. In 2016, she won a fellowship for an Erasmus Mundus programme in

Australia where she worked with the Airborne Research Australia. In 2017, she earned her PhD and started working for a company focusing on ICT in the geospatial and GIS market.

Simone Mineo is a geologist working in cooperation with University of Catania and University of Naples Federico II (Italy). He recently accomplished a PhD course in Earth, Environmental and Resources Sciences, with a thesis on the study of unstable rock slopes, for the assessment of rockfall risk, through traditional and innovative approaches.

He is author and co-author of several scientific articles dealing with rock mechanics and engineering geology, with particular reference to the characterisation of physico-mechanical behaviour of rocks at different scales, also through infrared thermography. He is a member of the Italian Geological Society (SGI) and of the Italian Association of Applied and Environmental Geology (AIGA), as well as a reviewer for some of the most relevant journals in his scientific fields.

Koushik Pandit is presently working as a Scientist in Geotechnical Engineering Group at CSIR – Central Building Research Institute, Roorkee, and is pursuing PhD in Geotechnical Engineering from IIT Roorkee. He is also an Assistant Professor in Academy of Scientific and Innovative Research (AcSIR), New Delhi. He has obtained his BTech in Civil Engineering from Jalpaiguri Government Engineering College, West Bengal, and MTech in Engineering of Infrastructure and Disaster Mitigation from CSIR-CBRI Roorkee campus of AcSIR. His current research interests include rock mechanics, slope stability and foundation engineering.

Giovanna Pappalardo is a geologist researcher at University of Catania (Italy) since 2003, where she teaches hydrogeology, rock mechanics, environmental geology and applied geology. She is author and co-author of numerous scientific publications focused on the study of rock masses, intact rock characterisation, slope stability in cultural heritage sites, application of infrared thermography to the geomechanics and hydrogeological modelling. She is currently part of the board of directors of the Italian Association of Applied and Environmental Geology (AIGA) and member of the Italian Geological Society (SGI) and of the International Association for Engineering Geology and the Environment. She recently organised scientific meetings on the remote survey of rock slopes and rock mechanics principles, and reviewed several papers for the most relevant journals in her scientific field.

Hamid Reza Pourghasemi is an Assistant Professor of Watershed Management Engineering in the College of Agriculture, Shiraz University, Iran. He has a BSc in watershed management engineering of the University of Gorgan (2004), Iran; an MSc in watershed management engineering, from Tarbiat Modares University, Iran (2008); and a PhD in watershed management engineering from the same university (2014). His main research interests are GIS-based spatial modelling using machine-learning/data mining techniques in different fields, such as landslide susceptibility and hazard, flood, gully erosion, forest fire and groundwater. Also, Hamid Reza works on multi-criteria decision-making methods in Natural Resources and Environmental Engineering. He has published more of 55 research papers in different international journals environmental engineering.

Piyush Punetha holds MTech from Academy of Scientific and Innovative Research in Building Engineering and Disaster Mitigation and BTech from Kumaon Engineering College, Dwarahat, in Civil Engineering. He is a University Silver Medallist in Civil Engineering during his bachelor's degree. He currently works as a Senior Research Fellow at CSIR-Central Building Research Institute Roorkee. He has research publications in reputed international/national journals and conferences. He received the Best Paper Award (BPA) in 6th Indian Young Geotechnical Engineers Conference (6IYGEC-2017), organised by NIT, Trichy, India. His research interests include geotechnical earthquake engineering, disaster mitigation, soil-geosynthetic interaction, ground improvement, bio-engineering, image-based damage assessment and constitutive modelling.

Emrehan Kutlug Sahin is a Research Assistant of Geomatics Engineering at the Gebze Technical University. His research focuses on the landslide susceptibility mapping, operation of GIS/CAD, spatial analysis and statistics, mapping and cartography tools, particularly ArcGIS and other open source tools, and developing basic program tools and scripts for GIS applications. His recent work examines the effect of landslide-related factors and automatic selection to define the best factor for the landslide susceptibility modelling. He has a master's degree from Gebze Technical University and a PhD from İstanbul Technical University.

Manojit Samanta currently works as a Scientist in CSIR-Central Building Research Institute Roorkee and an Assistant Professor in Academy of Scientific and Innovative Research. He did his MTech from Indian Institute of Technology Roorkee (IITR) in Geotechnical Engineering and BE from Indian Institute of Engineering Science and Technology, Shibpur (formerly Bengal Engineering College), in Civil Engineering. His research interests include ground improvement, geotechnical earthquake engineering, deep foundation, slope-stabilisation, soil-structure interaction and bio-engineering. He is a lifetime member of many prestigious professional bodies. He has several research publications in well-known international/national journals and conferences. He has received several Best Paper Awards for publications in national and international conference.

Shantanu Sarkar is a Senior Principal Scientist and Group Leader of Geotechnical Engineering Group at CSIR-Central Building Research Institute, Roorkee. He is also a Professor in Academy of Scientific and Innovative Research (AcSIR). He did his PhD and MSc from the Indian Institute of Technology Roorkee (IITR) and MTech from Indian School of Mines, Dhanbad. He has 27 years of professional experience in the field of engineering geology and geotechnical engineering. He has carried out very extensive geological and geotechnical studies in Western and Eastern Himalayas in various projects for landslide disaster mitigation. His research interests include landslide investigation, slope stability assessment, hazard and risk assessment, slope-stabilisation and remote sensing and GIS application. He has published more than 80 research publications in journals and conferences. He has been awarded prestigious National Geoscience Award in the area of Disaster Management by the Government of India and BOYSCAST Fellow by DST for post-doctoral research at EPFL, Switzerland.

Mahesh Sharma is currently working as a Trainee Scientist at CSIR-Central Building Research Institute, Roorkee, India. He holds MTech from Academy of Scientific and Innovative Research and BTech from Jaypee University of Information and Technology. He has research publications in reputed international/national journals and conferences. He has received the Best Paper Award in International Geotechnical Engineering Conference on Sustainability in Geotechnical Engineering Practices and Related Urban Issues, Mumbai, India, and also the Best Paper Award and Outstanding Paper Award in 6th Indian Young Geotechnical Engineers Conference (6IYGEC-2017), NIT, Trichy, India. His research interests include geotechnical earthquake engineering, slope stabilisation, disaster mitigation, ground improvement, constitutive/numerical modelling and foundation engineering.

Tariq Siddique is an Assistant Professor in Department of Geology, Aligarh Muslim University, India. He is also working as a part-time doctoral researcher in Department of Earth Sciences, Indian Institute of Technology, Roorkee, India. Presently, he is working on slope stability assessment of road cut slopes in the Himalayan terrain. He has four publications in international and national journals of repute. He has presented many research papers in national conferences. He has worked on a project entitled "Reservoir Characterization of Sandstone" during a summer internship at Oil and Natural Gas Corporation. He has attended three outreach programmes on different aspects of remote sensing and its applications conducted by Indian Space Research Organization (ISRO) under the Department of Space, Government of India. He has participated in industry-academia workshop on "Recent Trends and Challenges in Oil and Gas Drilling Technology" conducted by Oil and Natural Gas Corporation. He has been awarded gold medals for outstanding performance and securing highest grades during his post-graduation at Aligarh Muslim University, India.

Mahendra Singh is currently Professor of Civil Engineering at IIT Roorkee. He earned his Bachelor degree in Civil Engineering in 1983, MTech in Hydraulic and Water Resources Engineering in the year 1985 and PhD in the year 1997. Dr Singh is actively engaged in teaching, research and consultancy in the field of geotechnical engineering in general and in rock engineering in particular. His research interests include strength and deformation behaviour of jointed rocks, stability analysis of underground opening and rock slopes, and probabilistic analysis in rock engineering. He has supervised nine PhD Theses and more than fifty-five M Tech dissertations. He has published more than 100 papers in journals and conferences mainly in the area of rock mechanics. His research contribution is directly related to field problems in rock engineering projects, for example, bearing capacity of shallow foundation in rocks, squeezing potential of rocks in tunnels and nonlinear failure criteria for intact and jointed rocks. He has organised several conferences including prestigious Indian Geotechnical Conferences in various capacities. Some of these conferences are IGC-2002, IGC-2003, INDOROCK-2011 and IGC-2013.

Ashwani Kumar Tiwari obtained his MPhil and PhD degrees in Environmental Sciences. Currently, he is working as a Post-PhD Researcher at Politecnico di Torino, Italy. He was awarded Erasmus Mundus Scholarship, one of the meritorious and prestigious awards in Europe, and finished his research work as exchange doctoral researcher in Politecnico di Torino, Italy. He holds to his credit a total of 37 publications on different domains of hydro-geochemistry, aquifer vulnerability, water quality, health risk assessment, water resource management and remote sensing and GIS in various journals. He has contributed one research book and five chapters in the edited book volumes.

Dhananjay Verma is a Senior Geologist and is currently working in Geological Survey of India (GSI), Gandhinagar, India. He has completed his master's degree in Applied Geology from Allahabad University and MTech in Mineral Exploration from IIT Bombay, Mumbai. He has more than 8 years of experiences in the field of Engineering Geology, particularly in slope stability of cut slope as well as dumps slope. He is having more than 20 published papers in engineering geology and mineral exploration. He is also acting as associate editor and reviewer in number of international and national journals. His research interests are slope stability of hill slope, road cut and dump slope.

Sudesh Kumar Wadhawan is a former Director General, Geological Survey of India (GSI). He is a Gold Medallist MSc Applied Geology scholar from Gwyer Hall, University of Delhi. After serving as a Lecturer for a year in prestigious Delhi College of Engineering, Delhi, he joined GSI as a professional geologist and served in various capacities for 36 years with exceptional merit and commitment in different terrain conditions. Besides writing 40 technical reports in GSI, he has published over 50 scientific papers in peer-reviewed national and international journals and co-authored three books. He has played a crucial role in establishing an exclusive Geohazards Research and Management (GHRM) Cell in GSI CHQ with five regional units in India devoted to research in landslides and launched a National Program on Landslide Susceptibility Mapping on 1:50,000 scale on GIS platform utilising remote sensing and field survey data sets in 2014. Presently, he is a Visiting Faculty and nominated Geoscience Advisor and Expert Member of the Search-Cum-Selection Committee for Atomic Minerals Directorate for Exploration and Research, Department of Atomic Energy and other Government Departments/ Commissions.

Part I
Introduction to Landslide Dynamics

Chapter 1
Mass Wasting: An Overview

S. P. Pradhan and Tariq Siddique

Abstract Mass wasting is a natural phenomenon by which rock, soil and/or debris move downwards due to the action of gravity. It describes all the processes that act continuously with varied intensity on all type of slopes to lower the ground surface. The mass wasting process is controlled by the interaction of geological agents and processes with the geo-materials. The degree and type of movements depend upon a few aspects of geology, environment, geomorphology, hydrology, and some additional environmental stress factors, including biotic factors. It is more active in hilly regions like Himalayas, Western Ghats, Alps, and some other extensive mountain chains of the world. Sometimes it becomes disastrous to lives, property and economy. This chapter gives an overview of mass wasting processes and its classification. Some widely used mass movement classification schemes have been documented.

Keywords Mass wasting · Landslides · Slope failures · Natural hazards

1.1 Introduction

Mass wasting is also termed as "Mass Movement" and it is a downslope movement of rock, debris, soil, regolith etc. near the earth's surface under the action of gravity. The rate of mass wasting may be imperceptibly slow to rapid which involves the transfer of microscopic fractions as well as the mega-sized detritus down the slope. The triggering factors that assist mass movement are heavy precipitation, seismicity, volcanic activities, flooding and are often linked with slope failure at local and

S. P. Pradhan (✉)
Department of Earth Sciences, IIT Roorkee, Roorkee, India
e-mail: sppradhanfes@iitr.ac.in

T. Siddique
Department of Earth Sciences, IIT Roorkee, Roorkee, India

Department of Geology, Aligarh Muslim University, Aligarh, India

S. P. Pradhan et al. (eds.), *Landslides: Theory, Practice and Modelling*, Advances in Natural and Technological Hazards Research 50,
https://doi.org/10.1007/978-3-319-77377-3_1

regional scales. Mass wasting can also be defined as the continuum of erosional processes that exist between weathering and transportation. Sediment wasting process on continental slopes transfers significant volume of sediments in submarine environment leading to lowering of the ground surface [1, 2]. It is a significant part of the process of erosion as it transfers the material from higher elevation to lower elevation. A landslide may be defined as the downward movement of mass rock, earth, or debris [3]. According to Varnes [4] landslides include all types of mass movement on the slope which includes sliding, rockfalls, topples, debris flows etc. Brusden [5] defined landslide as a unique way of mass transport which does not require transportation agent for slope movement. According to Cozier [6] landslide is a downward and outward movement of slope forming material without the involvement of surface runoff as transporting medium under the influence of gravity. Hutchinson [7], Cruden [8] and Cruden and Varnes [9] described landslide as a rapid mass wasting process that causes down slope movement of the mass of rock, debris or earth which are induced by a variety of external stimulus. In the recent times, it has been noted that the term mass movement is used interchangeably with landslide.

Mass movements, floods, tsunamis, cyclones, earthquakes are the most common natural hazards and their impacts are very disastrous when they occur in clusters. In India, one such event was Kedarnath disaster of June 2013 in Uttarakhand. The Flash floods accompanied by mass movements in the various parts of the region devastated the economy, lives and ecosystem of the region. Such hazards significantly affect humans including loss of property and lives, cause severe injuries and sometimes fatalities as well. Large scale mass wasting events also affect certain elements of environment viz. topography of marine and terrestial landforms, quality of surface runoff, groundwater flow, forest cover and habitats, destruction of biotic components etc. In many countries, the number of casualties and economic losses occur because landslides are more frequent than the other commonly occurring hazards like earthquake, floods, windstorms [10–13].

Comprehensive understanding of mass movement in slopes require sound knowledge of factors associated with groundwater [14]. Scheidegger [15] reviewed surface-slips, deep seated soil creep, rock mass creep, surficial landslides, mudflows, Alpine debris flows and illustrated some recent developments in the field of mass movement. Mass movement occurs when the driving forces exceeds resisting forces to pull slope forming material down the slope [16]. This stability of a slope is quantified in terms of Factor of Safety (FoS) (Eq. 1.1). Prediction of mass movement in space and time is very difficult; the area may be categorized into homogeneous domains and can be ranked according to the degree of potential hazard.

In these years, landslides have grabbed the attention due to the growth of urbanization and its socio-economic consequences [17]. For monitoring and evaluation of mass movements, significant number of techniques are available and few widely used such techniques are Global Positioning System (GPS), Geographical Information System, aerial photography, inclinometers, acoustic emission [18]; ground based geodetic techniques, satellite based geodetic techniques, geophysical and geotechnical methods [19, 20]. The real time monitoring of landslides can be done by wireless sensor network (WSN) [21].

1.2 Factor of Safety (FoS)

Factor of safety is the capacity of a system beyond the expected or actual load. It may be defined as the ratio of resistive to driving forces.

$$\text{FoS} = \frac{\text{Resistive forces}}{\text{Driving forces}} \tag{1.1}$$

If FoS ≤ 1 slope is unstable
If FoS > 1 slope is stable

However, if FoS is slightly greater than 1, even small disequilibrium with the slope may cause slope failure. For example, FoS is 1.05, slope is marginally stable and it means that resistive forces are only 5% greater than driving forces. In such conditions, slight undercutting, heavy rainfall, seismicity etc. may cause the failure of the slope.

1.2.1 Resistive Forces

Resistive forces act opposite to the direction of the motion which tends to resist the movement. The resistive force can be defined in terms of shear strength of the material which is a function of cohesion and angle of internal friction. Cohesion is the ability of the particles to hold together. For example the clays and granites are cohesive, whereas the dry sand is non-cohesive. Angle of internal friction is the measure of frictional forces acting between constituent grains.

1.2.2 Driving Forces

Driving force acts in the direction of the motion and promotes the down slope movement. The major driving force is gravity, which plays a significant role in guiding or initiation of mass wasting phenomena. Gravitational force pulls every material or body downward, towards the centre of the earth. As depicted in Fig. 1.1, on flat surface gravitational force acts perpendicular to the ground. Hence, ground forming material will remain intact and will not move. In slopes gravitational force can be resolved into two components (Fig. 1.1):

(a) Acting perpendicular to the slope (g_p): Resistive force or shear strength which hold the object and resist movement.
(b) Acting tangential to the slope (g_s): Driving force or shear stress that promote down slope movement of the object.

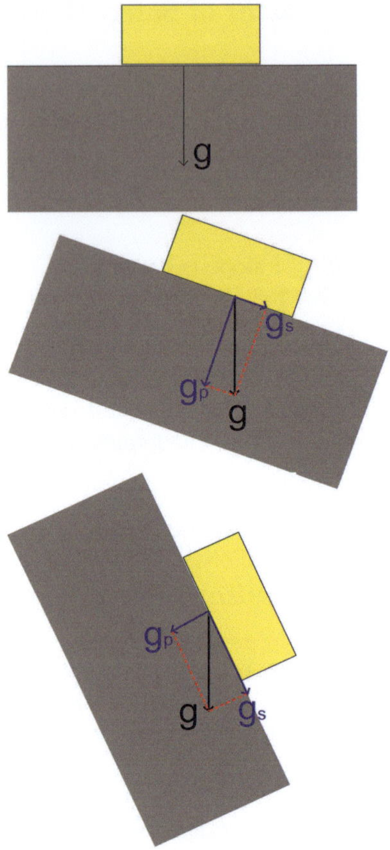

Fig. 1.1 Gravitational force and its components

When the shear stress (driving force) exceeds the shear strength (resistive force), the slope forming material will fail. Thus, on steeper slopes, the tangential component of gravity is greater than the resistive component and cause down slope movement of the mass. Slope angle, height of the slope, climatic conditions, types of slope materials, runoff and groundwater etc. are some other significant factors that affect the magnitude of the driving force. Water plays an important role in slope instability. When water is added to a slope, it causes slope failure due to the additional loading on the slope, accelerating erosion rates and increasing pore pressure that ultimately led to the reduction in shear strength of the slope forming material. Considering the above parameters FoS can be calculated from Eq. 1.2.

$$\mathrm{FoS} = \frac{c' + hg\ \cos^2\theta(\rho_r - \rho_w)\ \tan\varphi}{\rho_r\ hg\ \sin\theta\ \cos\theta} \tag{1.2}$$

where,

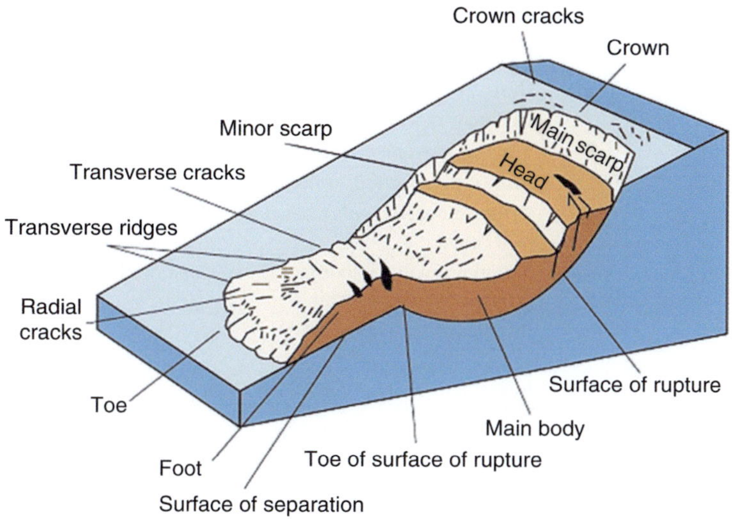

Fig. 1.2 Parts of slump-earth type landslide taken form Highland and Bobrowsky (Taken from [22])

c': Effective cohesion
h: Thickness of potential slide
g: Acceleration due to gravity
θ: Dip angle of potential sliding plane
ρ_r: Material density of potential sliding plane
ρ_r: Density of water
φ: Angle of internal friction

1.3 Parts of Landslide

In real practice, the different parts of a landslide depend upon the type of mass movement. Various parts of typical slump-earth flow type landslide have been depicted in Fig. 1.2.

1.4 Types of Mass Movement

Different classifications have been proposed worldwide, most of them consider different parameters viz. types of movement (falling, sliding, flows etc.), material involved (rock, soil or debris) and rate of movement (rapid or slow). Some other classification schemes incorporate certain additional variables such as the rate of movement and water, air and ice contained in wasted material. Some major

Table 1.1 Some significant contributions in framing classification of mass movements

Authors	Year	Parameters considered
Baltzer [23]	1875	Basic modes of motion: fall, slide, flow
Stini [24]	1910	Different types of debris movement
Sharpe [25]	1938	Type of movement, material involved and velocity of movement
Savage [26]	1951	Size of the material, water condition and speed of the movement
Varnes [27, 28]	1954, 1978	Extension of Sharpe's classification by considering type and rate of movement
Hutchinson [7, 29]	1968, 1988	Material involved, morphology of slope, water content, failure and propagation mechanism
Cruden and Varnes [9]	1996	Introduced velocity scale
Sassa [30]	1999	Liquefaction and pre-shearing clays

Table 1.2 Classification of landslide taken from Savage [26]

Kind of material	Size of material	Condition of material	Relative speed of movement	Term applied to phenomena
Blocky, angular rock	Cubic foot or larger	Wet, dry or frozen	Fast: free fall or slide	Rockfall or rockslide
Blocky, angular rock with snow	Cubic foot or larger	Wet, dry or frozen	Fast: free fall or slide	Debris fall avalanche Debris slide
Cobbles, pebbles and soil	Less than a cubic foot to rock flour	Wet, dry or frozen	Rapid slide	Slump slide
Cobbles, pebbles and soil	Less than a cubic foot to rock flour	Wet	Visible flow	Earthflow
Fine slit and clay	1/16 mm and less	Wet	Visible flow	Slumgullion flow, mudflow
Fine slit and clay	1/16 mm and less	Wet or semi-frozen	Imperceptible flow or slide	Solifluction, soil creep
Blocky, angular rocks	Usually cobbles and boulders	Wet or semi-frozen	Imperceptible flow or slide	Talus creep, rock creep
Cobbles, pebbles and soil	Less than a cubic foot to rock flour	Wet or semi-frozen	Rapid drop to imperceptible subsidence	Sink, wallow, slump depression, compaction depression

contributions in classifying varieties of mass movement have been illustrated in Tables 1.1, 1.2, 1.3, 1.4, 1.5, and 1.6.

In mass movement, material moves at different rate ranging from imperceptible as creep to extremely rapid as rock fall. On the basis of velocity of movement, Cruden and Varnes [9] classified landslides into seven classes (Table 1.4).

Classification schemes given by Varnes [28] and Cruden and Varnes [9] have been extensively acknowledged and widely accepted in varying scenario. Hungr

Table 1.3 Classification of landslides based on Varnes [28]

Type of movement	Bedrock		Type of material Engineering Soils Predominantly coarse	Predominantly fine
Falls	Rockfall		Debris fall	Earth fall
Topples	Rock topple		Debris topple	Earth topple
Slides	Rotational	Rock slump	Debris slump	Earth slump
	Translational	Rock block slide	Debris block slide	Earth block slide
Lateral spreads	Rock slide rock spread		Debris slide	Earth slide
			Debris spread	Earth spread
Flows	Rock flows		Debris flow	Earth flow
Complex	Combination of two or more principal types of movements			

Table 1.4 Classification of landslides based on Cruden and Varnes [9]

Velocity class	Description	Velocity (mm/s)	Probable destructive significance
7	Extremely Rapid	$>5 \times 10^3$	Catastrophe of major violence, building destroyed by impact of displaced material, many deaths, escape unlikely
6	Very Rapid	$<5 \times 10^3$	Some lives lost, velocity too great to permit all persons to escape
5	Rapid	$<5 \times 10^1$	Escape evacuation possible, structures, possessions and equipment destroyed
4	Moderate	$<5 \times 10^{-1}$	Some temporary and insensitive structures can be temporarily maintained
3	Slow	$<5 \times 10^{-3}$	Remedial construction can be undertaken during movement, insensitive can be maintained by frequent maintenance if total movement is not large during particular acceleration phase
2	Very slow	$<5 \times 10^{-5}$	Some permanent structures undamaged by movement
1	Extremely Slow	$<5 \times 10^{-7}$	Imperceptible without instruments, construction possible with precautions

[31] classified landslide forming materials (Table 1.5) and published an updated classification of landslides (Table 1.6) by modifying definitions of landslide forming materials and compatibility with geotechnical and geological terminologies of rock and soil.

British Geological Survey (BGS) also presented a comprehensive classification of mass movements by following classification scheme (Fig. 1.3) suggested by UNESCO [32, 33]. The classification criterion involves the type of movement and materials involved. Carson and Kirkby [34] proposed a ternary classification of mass movements on the basis of pure slide, heave and flow type of movement (Fig. 1.4).

Table 1.5 Types of landslide forming material from Hungr et al. [31]

Material	Character description if important	Specified field description for the purpose of classification	Corresponding unified soil classes	Laboratory indices (if available)
Rock	Strong	Strong-Broken with hammer	–	UCS > 25 MPa
	Weak	Weak-peeled with a knife	–	2 < UCS < 25 MPa
Clay	Stiff	Plastic can be molded into standard thread when moist, has dry strength	GC, SC, CL, MH, CH, OL and OH	$I_p > 0.05$
	Soft			
	Sensitive			
Mud	Liquid	Plastic, unsorted remolded and close to liquid limit	CL, CH, CM	$I_p > 0.05$
Silt, sand, gravel and boulders	Dry	Non-plastic (or very low plasticity), granular, sorted, silt particles cannot be seen by eye	ML	$I_p < 0.05$
	Saturated		SW, SP and SM	
	Partly saturated		GW,GP and GM	
Debris	Dry	Low plasticity, unsorted and mixed	SW-GW	$I_p < 0.05$
	Saturated		SM-GM	
	Partly saturated		CL, CH and CM	
Peat	–	Organic	–	–
Ice	–	Glacier	–	–

GC Clayey gravels, gravelsand-clay mixtures, *SC* Clayey sands, sand-clay mixtures, *CL* Inorganic clays of low to medium plasticity, gravelly/sandy/silty/lean clays, *MH* Inorganic silts, micaceous or diatomaceous fine sands or silts, elastic silts, *CH* Inorganic clays or high plasticity, fat clay, *OL* Organic silts and organic silty clays of low plasticity, *OH* Organic clays of medium to high plasticity, *CM* Silt-clay mixture, *ML* Inorganic silts, very fine sands, rock four, silty or clayey fine sands, *SW* Well-graded sands and gravelly sands, little or no fines, *SP* Poorly graded sands and gravelly sands, little or no fines, *SM* Silty sands, sand-silt mixtures, *GW* Well-graded gravels and gravel-sand mixtures, little or no fines, *GP* Poorly graded gravels and gravel-sand mixtures, little or no fines, *GM* Siltygravels, gravel-sand-silt mixtures, *UCS* Uniaxial compressive strength, Ip Point load index

The following description of some principal types of the mass movements are based on Varnes [28], Hutchinson [7], Hungr et al. [35], Cruden and Varnes [9], Highland et al. [22] and Hungr et al. [31].

Falls A fall occurs with the detachment of mass from a steep slope or cliff with no shear displacement. The movement involves free fall through air, bouncing and or rolling. Physical weathering, presence of interstitial water, seismic activity trigger falls.

Flows Flow is a type of mass movement which is characterised by downward and outward movement of ground/slope under water saturated condition in which material moves as a viscous fluid. Depending upon the type of material involved and degree of water saturation, variety of flows can be categorized as debris flow, mud flow, and avalanche flow. Flows type of mass movement has been widely noticed in slopes having finer material (silt and clay) at least 50% or more [28]. Such slopes become extremely susceptible to failure under saturated conditions. Failure is

Table 1.6 New version of the Varnes classification from Hungr et al. [31]

Type of movement	Rock	Soil
Fall	Rock/ice fall	Boulder/debris/silt fall
Topple	Rock block topple	Gravel/silt/sand topple
	Rock flexural topple	
Slide	Rock rotational slide	Clay/silt rotational slide
	Rock planar slide	Clay/silt planar slide
	Rock wedge slide	Gravel/sand/debris slide
	Rock compound slide	Clay/silt compound slide
	Rock irregular slide	
Spread	Rock slope spread	Sand/Silt liquefaction spread
		Sensitive clay spread
Flow	Rock/ice avalanche	Sand/silt/debris dry flow
		Sand/silt/debris dry flowslide
		Sensitive clay flowslide
		Debris flow
		Mudflow
		Debris flood
		Debris avalanche
		Earth flow
		Peat flow
Slope deformation	Mountain slope deformation	Soil slope deformation
	Rock slope deformation	Soil creep
		Solifluction

generally initiated from tension cracks at crown portion and forms nearly circular failure scarp.

Topples It is the movement of rock that involves forward rotation about a point or axis below the centre of gravity of the displaced mass. In jointed rock mass, closely spaced and steeply dipping discontinuity sets that dip opposite to the slope surface are necessary prerequisites for toppling failure.

Slides For landslides many technical terms have been proposed. However, slide is restricted where there is a distinct zone of weakness that separates the slide material from more stable underlying material. There are two types of slides i.e. rotational slide and translational slide or combination of both. In rotational slides, material slide outward and downward on one or more concave failure surfaces that impart a backward tilt to the slipping mass. In this type of mass movement, the slide movement is roughly rotational about an axis that is parallel to the ground surface and traverse across the slide. In translational slides, material moves along planar failure surface or discontinuities that may run more or less parallel to the slope.

Spreads Spread involves fracturing and lateral movement of rock or soil masses. This occurs due to the plastic flow or liquefaction of material. Usually they occur on gentle slopes that are triggered by rapid ground motion induced by earthquake or artificially induced seismic waves.

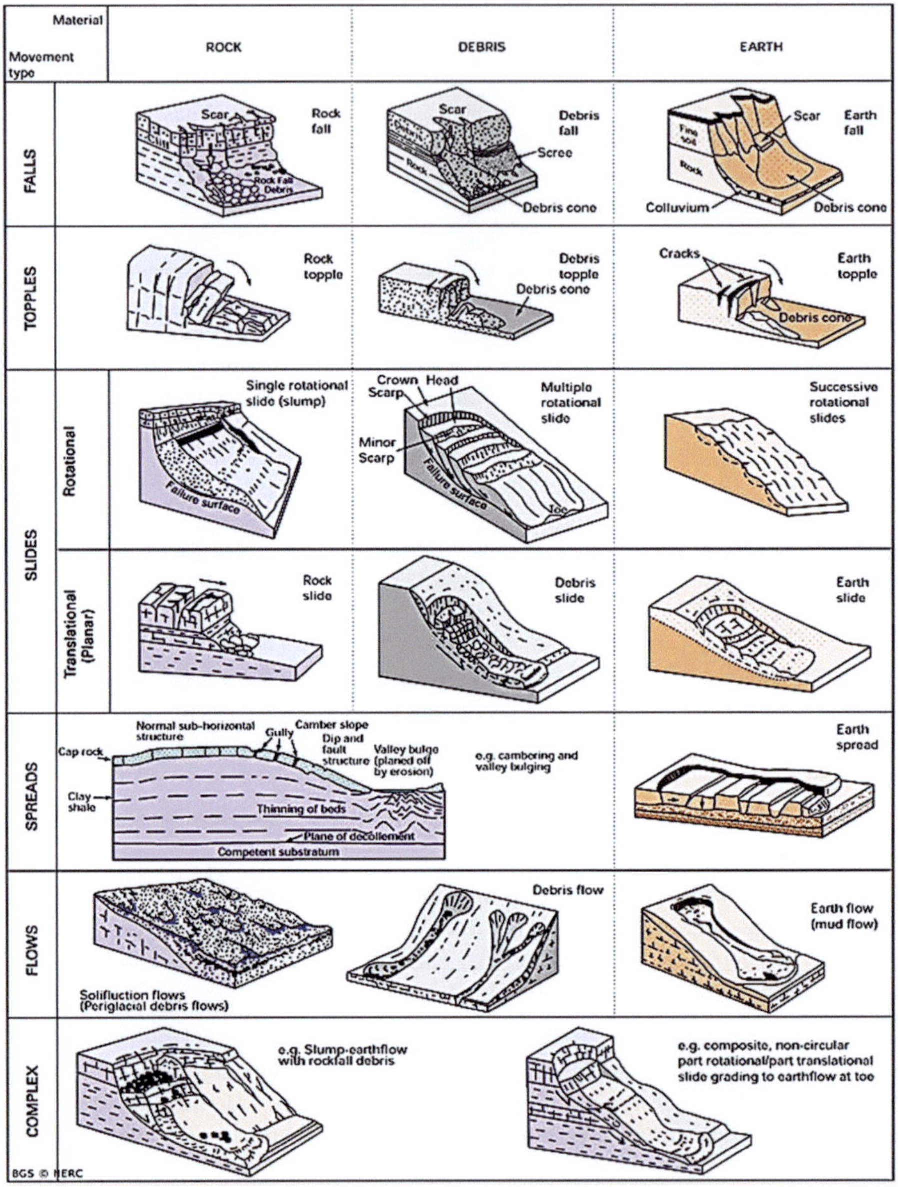

Fig. 1.3 Classification of landslides taken from British Geological Survey (BGS) http://www.bgs.ac.uk/landslides/how_does_bgs_classify_landslides.html

Creep Creep is imperceptibly slow, steady, downward and outward movement of slope forming material. Creep is a continuous deformation that continues under constant effective stress [9, 36]. According to Eberhardt et al. [37] creep is a time dependent deformation of material under a constant load. Shear stress acting on the slope promotes permanent deformation but they are too small to generate shear

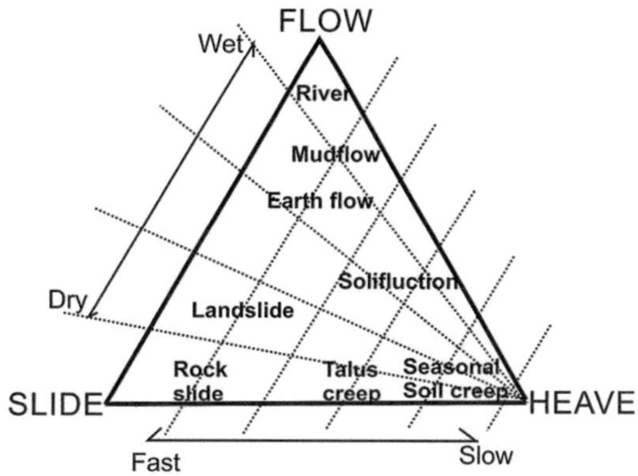

Fig. 1.4 Classification of mass movements in terms of pure Slide, Heave and Flow based on Carson and Kirkby [34]

failure in the ground. Creep can be seasonal, continuous or progressive in nature depending upon the local climatic conditions, geological and geotechnical parameters of the material involved. Major causative factors of creep are gravity, freeze and thaw cycles, shrinkage and swelling cycles. Chang et al. [38] conducted geomechanics based numerical modeling to quantitatively explain the cause and evolution of soil creep. As slope movement is very slow in creep and can be observed by certain geomorphological evidences, abnormal curvature of tree trunks, tilting of poles and walls and subsidence of roads etc. Global Positioning System (GPS) has capability to monitor sub-centimetre deformations of ground movement [39]. Creep in slopes can be investigated by extensometers, GPS, geodetic networks, aerial photographs, LiDAR and InSAR etc. [40].

Complex All the types of mass movements discussed above are mostly ideal cases. However, in real scenario such ideal conditions may not be found, changes may occur that could be gradational from one type of mass movement to the other. Complex mass movements are the combination of two or more principal type of mass movements.

1.5 Causes of Mass Movements

There are several factors that cause, trigger or promote mass movement. Terzaghi [41] classified such cause into:

(i) External causes (changes in geometry, loading of crest of the slope, unloading of toe of the slope, changes in water level, vibrations and shocks)
(ii) Internal causes (weathering, seepage, erosion, progressive failure)

Different types of mass movement reflect that there exist a diversity of causative factors. Many authors classified the causes of mass movements into natural and anthropogenic. Fundamental geological and geomorphological features of weathering related slope movements are common [42]. Impacts of chemical

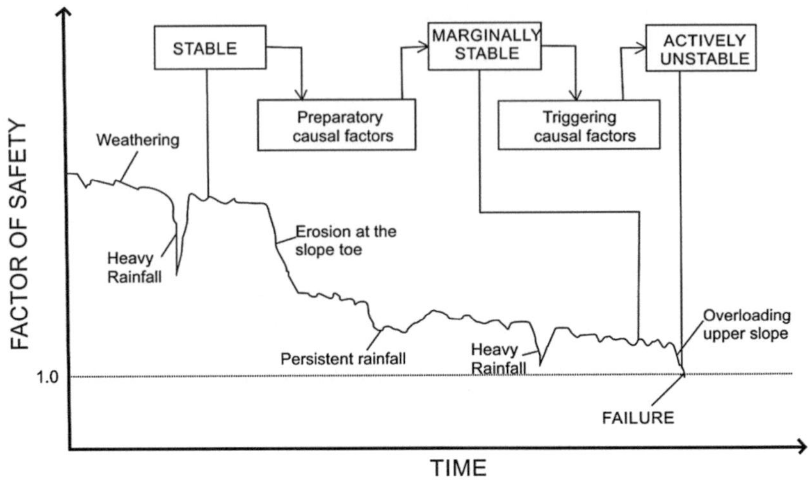

Fig. 1.5 Changes in factor of safety with time taken from Popescu [45]

weathering in rock slope stability have been illustrated by Jaboyedoff et al. [43]. Stoffel et al. [44] focused on the impact of climate change on mass movement activity in mountainous regions. Popescu [45] given a bi-variate plot and demonstrated that, for a given slope how factor of safety varies with time. He had also shown the effect of seasonal rainfall, evaporation erosion at toe of slope; persistent rainfall, heavy rainfall and overloading of crown portion affect FoS of a particular slope (Fig. 1.5).

Causes or factors influencing mass movement activities can be categorized as:

(A) Geological

(a) Weak and sensitive materials
(b) Weathered material
(c) Highly discontinuous, sheared, fractured, fissured material
(d) Adversely oriented discontinuities
(e) Contrast in permeability
(f) Competency or stiffness of the material
(g) Volcanic eruptions
(h) Composition of rock
(i) Geomechanical properties of rocks and soil forming slope material

(B) Geomorphological

(a) Tectonic and volcanic uplift
(b) Glacial rebound
(c) Fluvial, wave or glacial erosion at toe of the slope or at lateral margins
(d) Subterranean erosion
(e) Removal of vegetation by fire or drought
(f) Topography or geometry of the slope
(g) Gradient of slope

(C) Climatic

(a) Heavy rainfall
(b) Freeze and thaw cycles or frost wedging
(c) Dry and wet cycles
(d) Thawing
(e) Cloud bursts
(f) Loud claps of thunder

(D) Anthropogenic

(a) Excavation at toe of the slope
(b) Loading crest of the slope
(c) Drawdown of the reservoir
(d) Deforestation
(e) Excessive irrigation
(f) Mining
(g) Artificial vibrations
(h) Water leakage from utilities
(i) Landuse pattern

(E) Biotic

(a) Grazing
(b) Burrowing and boring

1.6 Monitoring and Remedial Measures for Arresting Mass Movements

Fort et al. [46] illustrated different research carried out on large scale landslides and suggested that field work for landslide monitoring should be well documented by triggers, predisposing factors to sort out better causative factors. He also suggested that the field data should be coupled with geotechnical survey (LIDAR, GPS and core logs) in order to give better concluding remarks on slope failure. Comprehensive evaluation of landslide hazards involve multi-dimensional problems, which require expertise from different disciplines viz. geomorphology, geology, mathematics, statistics, computer science, physics, chemistry, information technology, economics and sociology [47]. Wieczorek and Synder [48] suggested five basic elements or vital signs of monitoring slope movements and defined methods to evaluate each element, expertise required, technical needs, relative costs, personnel required and intensity of labour for each (Table 1.7).

(a) Determination of type of landslides
(b) Monitoring of landslide triggers and causes

Table 1.7 Summary of comparative landslide vital signs and monitoring methods taken from Wieczorek and Synder [48]

Vital signs	Methods	Expertise	Technical needs	Relative cost	Personnel	Labour intensity
Types of landslides	Identification	Volunteer	No	A	Individual	Medium
	Measurement	Volunteer	Yes	B	Group	Medium
	Imagery	Scientist	Yes	C	Individual	High
Landslide triggers and causes	Online data	Volunteer	No	A	Individual	Medium
	Climatic and seismic instruments	Volunteer	No	B	Individual	High
	Landslide instruments	Scientist	Yes	C	Group	High
Geologic material in landslides	Examination	Volunteer	No	A	Individual	Medium
	Surface sampling	Scientist	Yes	B	Group	High
	Surface sampling and testing	Scientist	Yes	C	Group	High
Measurement of landslide movement	Tapes and GPS	Volunteer	Yes	A	Individual	High
	Extensometers	Scientist	Yes	B	Group	High
	Aerial photos, LIDAR and InSAR	Scientist	Yes	C	Group	High
Assessing landslide hazard and risks	Inventory and susceptibility	Scientist	No	A	Individual	High
	Volume, velocity and travel distance	Scientist	Yes	B	Individual	High
	Modeling	Scientist	Yes	C	Individual	High

Relative cost in US$: A- up to $1000; B- $1000-$10,000; C- >$10,000
Labour intensity: Low = < few hours; Medium = < full day; High = > full day
GPS Global Positioning System, *LIDAR* Light detection and ranging, *InSAR* Interferometric synthetic aperture radar

(c) Determination of geological material involved in landslide
(d) Measurement and assessment of landslide movements
(e) Regional assessment of hazards and risks

In landslide prone terrains, for safer operations of transportation routes, power generations and transmission sites, mining activities, residential and commercial areas, long term stable slopes are prerequisite. Inadequate considerations of different landslide causative factors cause large number of fatalities, injuries and economic losses. A number of slope stabilization techniques have been proposed and adopted to attain better safety, economics and engineering design. For cost effective remedial measures for slopes certain factors are to be considered:

(a) Purpose of stabilization
(b) Significance of slope
(c) Availability of time
(d) Cost of project
(e) Accessibility of the site
(f) Construction material and equipments required
(g) Geotechnical problems
(h) Socio-economic issues
(i) Political aspects
(j) Available time during traffic closure
(k) Availability of disposal site
(l) Controlling and triggering factors

Slope stabilization methods can be broadly categorized in three components:

(A) Removal or excavation

   (a) Removal of material from crown portion
   (b) Removal of overhang by trim blasting
   (c) Removal of trees having big roots and growing along discontinuities in the rock mass
   (d) Hand scaling of loose portions of the slope

(B) Repair

   (a) Benching
   (b) Grading and serrating of slope to attain more stable configuration
   (c) Fill light weight material on slope to reduce impact of gravity
   (d) Add material at toe portion of the slope

(C) Protection

   (a) Ditches
   (b) Wide shoulders
   (c) Catchment by engineered benches
   (d) Berms
   (e) Steel barriers
   (f) Retaining walls
   (g) Nets
   (h) Fences
   (i) Concrete walls
   (j) Piers
   (k) Caissons

(D) Drainage of water (by improving surface and sub-surface drainage system)

(a) Reshaping of ditches on slope to allow staged water to flow
(b) Trenches
(c) Horizontal drains
(d) Relief wells
(e) Drain wells
(f) Well points
(g) Deep wells
(h) Drainage galleries

(E) Reinforcement

(a) Rock bolting
(b) Shear keys
(c) Steel reinforcement
(d) Anchoring
(e) Rock dowels
(f) Soil nailing
(g) Piles
(h) Shotcrete
(i) Guniting
(j) Cable anchorage
(k) Grouting
(l) Gabions
(m) Growing vegetation on debris or soil slopes
(n) Shear pins

References

1. Burda J, Zizka L, Dohnal J (2011) Monitoring of recent mass movement activity in anthropogenic slopes of the Kruˇsn'eHory Mountains (Czech Republic). Nat Haz Ear Syst Sci 11:1463–1473
2. ElverhØi A, Blasio FVD, Butt FA et al (2002) Submarine mass wasting on glacially-influenced continental slopes: processes and dynamics. In: Dowdeswell, JA, Cofaigh C (eds) Influenced sedimentation on high latitude continental margins. Geological Society of London, Special publication, vol 203, pp 73–87
3. Bolt BA (1975) Landslide hazard. Geological Hazard, Springer, New York, p 150
4. Varnes DJ (1984) Landslide hazard zonation: a review of principles and practice. Unesco, Paris
5. Brusden D (1984) Mudslides. In: Brusden D, Prior D (eds) Slope instability. Wiley, Chichester, pp 363–418
6. Crozier M (1986) Landslides-causes, consequences and environment. Croom Helm Ltd, London, pp 0.7097–0.7099
7. Hutchinson JN (1988) Mass movement. In: Fairbridge R (ed) The Encycl of Geomorp. Reinold, pp 688–695
8. Cruden D (1991) A simple definition of a landslide. Bull Int Assoc Eng Geol 43:27–29

9. Cruden DM, Varnes DJ (1996) Landslide types and processes. Special report, transportation research board. Nat Acad Sci 247:36–75
10. Shroder JF, Finkel RC, Kamp U (2011) The role of mass movements on landscape evolution in central Karakoram: discussion and speculation. Quat Intern 236(1–2):34–47
11. Schuster RL, Fleming RW (1986) Economic losses and fatalities due to landslides. Bul Am Assoc Eng Geol 23(1):11–28
12. Swanston DN, Schuster RL (1989) Long-term landslide hazard mitigation programs: structure and experience from other countries. Bul Am Assoc Eng Geol 26(1):109–113
13. Glade T (1998) Establishing the frequency and magnitude of landslide-triggering rainstorm events in New Zealand. Environ Geol 35:2–3
14. Rotaru A, Oajdea D, Raileanu P (2007) Analysis of the landslide movements. Int J Geol 1 (3):71–79
15. Scheidegger AE (1984) A review of recent work on mass movements on slopes and on rock falls. Ear Sci Rev 21(4):225–249
16. Keller EA (2000) Environmental geology, 8th edn. Prentice-Hall, Upper Saddle River
17. Aleotti P, Chowdhury R (1999) Landslide hazard assessment: summary review and new perspectives. Bull Eng Geol Environ 58:21–44
18. Zaki A, Chai HK, Razak HA, Shiotani T (2014) Monitoring and evaluating the stability of soil slopes: a review on various available methods and feasibility of acoustic emission technique. Comp Ren Geosci 346:223–232
19. Savvaidis PD (2003) Existing landslide monitoring system and techniques. School of rural and surveying engineering. The Aristotle University of Thessaloniki, pp 242–258
20. Pardeshi SD, Autade SE, Pardeshi SS (2013) Landslide hazard assessment: recent trends and techniques. Springer Publ 2:523
21. Jagtap KR, Aware SP (2015) Landslide pre-warning system based on wireless sensor network using zigbee-A review. Int conf on techn for sustain-Eng infor tech, manage and the environ, Faridabad, India. ISBN: 978-81-931039-7-5
22. Highland LM, Bobrowsky P (2008) The landslide handbook-A guide to understanding landslides, vol 1325. U.S. Geological Survey Circular, Reston. 129p
23. Baltzer A (1875) Uberbergstürze in den Alpen. Verlag der Schabelitz'schenbuchhandlung (C. Schmidt), Zurich, 50p
24. Stini J (1910) Die Muren. Verlag der Wagner'shen Universitätsbuchhandlung, Innsbruck (Debris flows, English translation by M. Jakob and N. Skermer, 1997, EBA Engineering Consultants, Vancouver, Canada, 106p
25. Sharpe CFS (1938) Landslides and related phenomena. Columbia University Press, New York. 1370p
26. Savage CN (1951) Mass-wasting, classification and damage in Ohio. Ohio J Sci 51(6):299–308
27. Varnes DJ (1954) Landslide types and processes. In: Eckel EB (ed) Landslides and engineering practice, special report 28. Highway Research Board. National Academy of Science, Washington, DC, pp 20–47
28. Varnes DJ (1978) Slope movement types and processes. In: Schuster RL, Krizek RJ (eds) Landslides, analysis and control, special report 176: transportation research board. National Academy of Science, Washington, DC, pp 11–33
29. Hutchinson JN (1968) Mass movement. In: Fairbridge RW (ed) Encyc of geomorph. Reinhold Publishers, New York, pp 688–695
30. Sassa K (1999) Introduction. In: Sassa K (ed) Landslides of the world. Kyoto University Press, Kyoto, pp 3–18
31. Hungr O, Leroueil S, Picarelli L (2014) The Varnes classification of landslide types, an update. Landslides 11:167–194
32. International Geotechnical Society's UNESCO Working Party on World Landslide Inventory (WP/WLI) (1991) A suggested method for a landslide summary. Bull Int Assoc Eng Geol 43:101–110

33. International Geotechnical Society's UNESCO Working Party on World Landslide Inventory (WP/WLI) (1993) A suggested method for describing the activity of a landslide. Bull Int Assoc Eng Geol 47:53–57
34. Carson MA, Kirkby MJ (1972) Hillslope forms and processes. Cambridge University Press, Cambridge
35. Hungr O, Evans SG, Bovis M et al (2001) Review of the classification of landslides of the flow type. Environ Eng Geosci VII:221–238
36. Goodman RE (1989) Introduction to rock mechanics. Wiley, New York
37. Eberhardt E, Preisig G, Gisсchig V (2016) Progressive failure in deep-seated rockslides due to seasonal fluctuations in pore pressures and rock mass fatigue. In: Aversa et al (eds) Landslides and engineered slopes. Experience, theory and practice. Asso Geot Ital, Rome
38. Chang KT, Ge L, Lin H (2015) Slope creep behavior: observations and simulations. Eniron Ear Sci 73(1):275–287
39. Rawat KT, Joshi V, Rawat BS et al (2011) Landslide movement monitoring using GPS technology: a case study of Bakthang landslide, Gangtok, East Sikkim, India. J Dev Agric Eco 3(5):194–200
40. Wangensteen B, Guðmundsson A, Eiken T et al (2006) Surface displacements and surface age estimates for creeping slope landforms in Northern and Eastern Iceland using digital photogrammetry. Geomorphology 80:59–79
41. Terzaghi K (1950) Mechanisms of landslides. Geological Society of America, Berkley, pp 83–123
42. Calcaterra D, Parise M (2010) Weathering as a predisposing factor to slope movements: an introduction. Geological Society of London, London, Engineering Geology Special Publications 23:1–4
43. Jaboyedoff M, Baillifard F, Bardou E et al (2004) The effect of weathering on Alpine rock instability. Q J Eng Geol Hydrol 37:95–103
44. Stoffel M, Tiranti D, Huggel C (2014) Climate change impacts on mass movements – case studies from the European Alps. Sci Tot Environ 493:1255–1266
45. Popescu ME (1994) A suggested method for reporting landslide causes. Bull Int Assoc Eng Geol 50:71–74
46. Fort M, Cossart E, Deline P et al (2009) Geomorphic impacts of large and rapid mass movements: a review. Geomorph Relief Proc Environ 1:47–63
47. Guzzetti F, Carrara A, Cardinali M et al (1999) Landslide hazard evaluation: an aid to a sustainable development. Geomorphology 31:181–216
48. Wieczorek GF, Snyder JB (2009) Monitoring slope movements. Geol Soc Am:245–271

Chapter 2
Dump Slope Stability

Tushar Gupta, T. N. Singh, and Dhananjay Verma

Abstract In order to mine out the valuable minerals from the depths of the earth's surface, huge amounts overlying material must be removed first. This overlying material, which in most cases is of no economic value to the mining operation, is called overburden (a mine waste), and has to be stored in the mine vicinity in order to keep on mining the underlying mineral effectively. Limited space available in a mining project renders it necessary for the overburden to be stored in form of dumps, which can reach huge dimensions as mining moves on to higher and higher stripping ratios. It hence becomes a necessity to ensure that these dumps are safe in all conditions and stages of mine working. Understanding of the mechanics and dynamics of the dumps and dump slopes therefore, becomes a crucial requirement for a mining engineers that will help them design safer dumps, simultaneously considering the economic aspects of mining. With this consideration, this chapter deals with a broad overview of the dumps and dump design, covering various details of the dumping methodologies, dump characterisations, and mechanics and dynamics of dump slope failure. Stress of this chapter is specifically focused on proper understanding of the various factors that affect the stability of the dump slopes.

2.1 Introduction

In today's mining industry, surface mining undoubtedly dwarfs the underground mining by a long margin in terms of production and productivity. For instance, almost 97% of United States' metal mining is open pit mining, producing around

T. Gupta
IITB-Monash Research Academy, IIT Bombay, Mumbai, India

T. N. Singh (✉)
Department of Earth Sciences, IIT Bombay, Mumbai, India
e-mail: tnsingh@iitb.ac.in

D. Verma
Geological Survey of India, Gandhinagar, India

S. P. Pradhan et al. (eds.), *Landslides: Theory, Practice and Modelling*, Advances in Natural and Technological Hazards Research 50,
https://doi.org/10.1007/978-3-319-77377-3_2

700 MT of ore as compared to 19MT of ore by underground mining methods [1]. In India, 92% of total coal production is from opencast mines [2]. Similar trends are observed around the world for major mineral exploration and mining operations. In conjugation to these huge amounts of production from open pit mines, there is a large requirement of waste material handling for the excavation to take place. In proportion to the stripping ratio of the mining operation, which is the amount of waste material handled per unit mineral excavated, huge amounts of overburden waste have to be removed and relocated in order to excavate the required mineral. The average stripping ratio for Indian open pit coal mining was 2.33 in the year 2015, implying that for the production of 612.44 MT of coal that year, 1427 MT of overburden was displaced from one location to another [2].

Historically, right from the time man started extracting valuable minerals from the rock, there has been a problem of waste management. In early days of mining, the waste management was simply done by moving the material out of the way, either down the hill slope or to any other available place. Many times, this waste material was thrown away in rivers, lakes etc., which, in due course of time, started getting blocked. Eventually laws were established to avoid such reckless practices and to regulate the disposal/storage of mine wastes [3].

Today, these huge amounts of overburden material are either casted back in to the pits are is dumped either inside the mine (in form of internal dumps) (Fig. 2.1) or outside the mine (external dumps). Since the mine economics favours internal dumping, and the land area available inside the mine is always limited, it is critical

Fig. 2.1 A typical internal dump in one of the coal mines in Northern coalfields limited

Fig. 2.2 Catastrophic damage after dump failure in Rajmahal Coal mines in 2016 [5]

to make steeper and higher dumps in order to put more and more material in the same amount of land area [4].

However, stability of these dumps is especially important for safe mine working, considering specifically that in most cases these dumps are very near to the haul roads or working face in opencast mines. Any failure in this case will result in catastrophic losses in terms of life, property and machinery as is seen in many of the past and recent examples such as in Rajmahal Coal Mine, Jharkhand in 2016 [5, 6] (Fig. 2.2), Sasti Mine in 2009, Jayant coal mine in 2008 [7], Kawadi Opencast mine in 2000, etc. [8–10]. Hence, it is imperative to know the intricacies of the dump slope design in order to plan the mining work ensuring safety as well as sound mine economics. This chapter will deal with basic understanding of the dumps, including the types of dump formation, the factors that affect the stability of the dump slope and the various modes in which the dump fails.

2.2 Dump Classification

Most mine dumps comprise of blasted out overburden that is highly heterogeneous in nature and the size of rocks can range from less clay size particles to big boulders (less than 0.1 mm to more than 1 m diameter). This is based on lots of factors including rock type, local geology, and blasting practices. Dumping operations that allow material to fall from some height undergo natural gravity segregation, however, obtaining an exact size distribution of material in a dump is very difficult [11]. Some especially designed dumps have layered positioning of set grade of material to allow formation of natural channels for water movement. In other cases the dump may have a layer of heavy rock covering the face of dump to avoid the fine grained particle flying off while dumping [12].

Shape of the dump is dependent on the topography of area on which the dump is made. Following are some of the shapes a mine dump can take:

2.2.1 Valley Fill

Valley Fill: As the name suggests it fills the valley. The filling starts from the higher level of the valley to the lower level of the valley (from upstream to downstream). In case the valley is not filled completely (partial valley fill), arrangement of culverts, rock drains etc. need to be provided for preventing water impoundment (Fig. 2.3).

2.2.2 Cross Valley Fill

Cross Valley fill: In this type of valley fill the layers cross the valley, in contrast to moving from upstream to downstream. These can be used as retention dams for fine coal or waste slurries, in which case proper drainage facilities are provided to ensure stability (Fig. 2.4).

2.2.3 Sidehill Fill

Sidehill fill: These are made on the existing natural or artificial slopes. These do not cross the valley bottom. They can be used to impound water in slopes or mine waste slurries (Fig. 2.5).

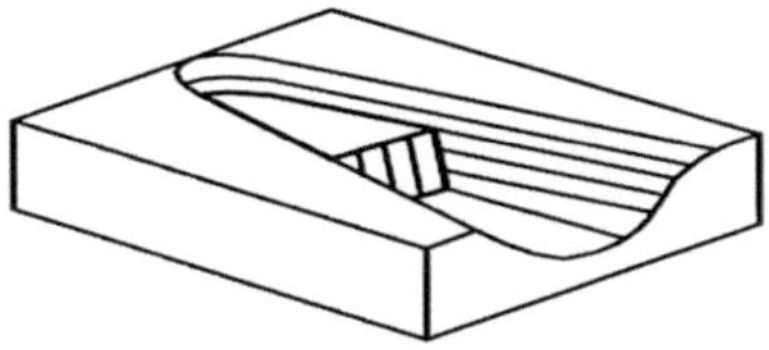

Fig. 2.3 Valley fill type of dumping [13]

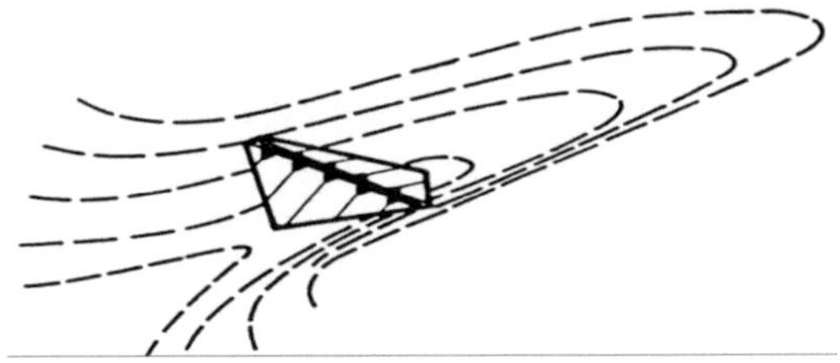

Fig. 2.4 Cross valley fill type of dumping [14]

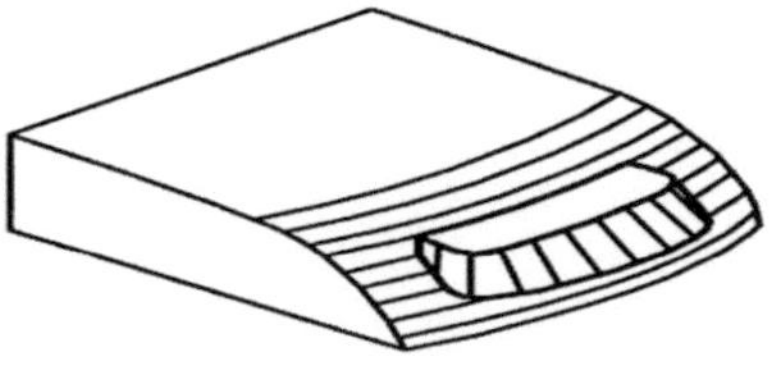

Fig. 2.5 Sidehill fill type of dumping [13]

2.2.4 Ridge Embankment

Ridge Embankment: These are made above an existing ridge, with both sides of the crest being touched by the waste material (Fig. 2.6).

2.2.5 Heaped

Heaped: These are generally constructed on level or gently dipping terrain. These can be stacked or piled up fills where material is filled in one lift after another. The dump has slopes on all sides (Fig. 2.7).

Apart from the shape a dump takes, dumps can also be classified based on the method of dumping.

2.2.6 End Dumping

End Dumping: This method simply involves dumping the waste rock on the dump face directly by the dumper or tippler. The layers of the material thus formed are parallel to the face (Fig. 2.8). Due to the movement of material from top to bottom of the dump face under gravity, some segregation of particle size takes place, with finer settling on the upper part of the dump while the coarser settling in the lower parts and near the toe of the dump.

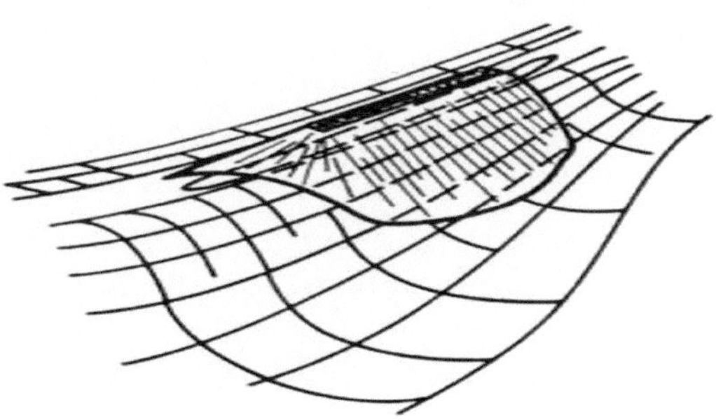

Fig. 2.6 Ridge embankment type of dumping [14]

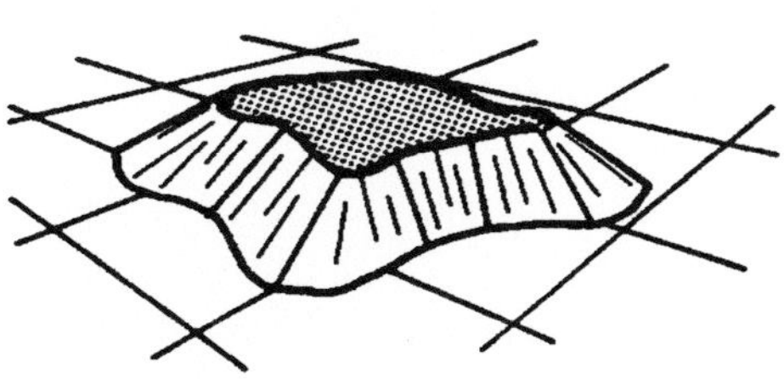

Fig. 2.7 Heaped type of dumping [13]

Fig. 2.8 Ending dumping in progress in one of the opencast coal mines [11]

Fig. 2.9 Dozer push dumping the overburden material in an opencast mine [15]

2.2.7 Push Dumping

Push Dumping: This method is like the end dumping, except that in this case the dumper places the waste material on the top surface of the dump, from which the material is levelled and pushed with help of dozers to fall on the face of the dump (Fig. 2.9). This is generally done in case of weak dump edges, where there is a

Fig. 2.10 Free dumping is seen in form of compressed lifts in an opencast mine [16]

danger of failure of the slope when the dumper is unloading on the edge. Particle size segregation is similar to that in case of end dumping.

2.2.8 *Free Dumping*

Free dumping: In this method, the material is dumped on the surface in form of small piles that are then levelled and compressed into layers to eventually form lifts. These lifts act as the base for the next lift when sufficiently large enough (Fig. 2.10). Since all material is compressed, there is no practical segregation of material size.

2.2.9 *Dragline/BWE Dumps*

Dragline/BWE dumps: Typically used in coal mines, these are small heighted dumps, dropped from some height by bucket, conveyor belt etc. (Fig. 2.11). There is no lift making, nor any compaction of material that takes place. Hence size segregation does not take place to that extent since the size of the dumps is also small.

Fig. 2.11 Dragline dumping the overburden material in an opencast mine [17]

2.3 Factors Affecting Dump Slope Stability

As discussed before it is very important to keep the dumps stable and safe for preventing any accidents and loss in life and property. A dump slope failure can cause unprecedented interruptions in mine production for e.g. by equipment burial in the debris, by blocking the haul road or the access way to the ore etc. Further to failure, the cost of clean-up and resumption of mine operation is substantial and will additionally be a burden to the mine economics. Hence it is important to understand the factors that affect the stability of dump slopes. Some of the major factors are:

2.3.1 Geometry of Dump

Geometry of dump: Perhaps one of the major factor that affects the stability of slope is the geometry of the dump. Specifically, parameters like the height of the dump, the overall slope of the dump, berm width/haul road width, etc. have a significant effect on the stability of slope [18]. Higher slopes and angles make the slope unsafe while the gentler slope gives a higher factor of safety as is shown in various studies [19] (Fig. 2.12).

Fig. 2.12 A comparative strain development model for (**a**) Dump with lesser height and lower overall slope angle. (**b**) Dump with more height and steeper overall slope

The slope of these dumps is dictated based on the method of dumping as well. The end dumping technique makes loosely compacted slopes at an angle that is very close to the natural angle of repose. At the angle of repose, the slope is critically safe (factor of safety is 1), and since the material is loose, there is high chance of failure of such slopes [3]. In contrast to this, the lift type or free dumping involves compaction of material and hence there is an increase in density and internal friction, resulting in more stable material that can support higher and steeper slopes. The slopes which can be made here are determined by simulation methods for allowable heights and angles, keeping the dump safe [4, 20].

2.3.2 Geotechnical Properties of Dump Material

Geotechnical properties of Dump material: The geotechnical properties of the dump material are the main factors that affect the stability of the dump slope. Properties like density, shear strength, grain size distribution, compression index, degradation behaviour, saturation etc. are some of the primary characteristics of the dump material that need to be known in order to design a safe dump. These properties are mostly calculated in labs (Fig. 2.13) by series of standard testing procedures laid out by local or global standardising organisations like ISRM, ASTM, ISO, BS etc., though some tests can also be performed on field.

Most of the geotechnical properties are based on the way dumps are made. An intact rock like sandstone, mudstone or shale will possess inherent rock properties such as compressive, tensile and shear strengths, density, porosity and permeability, inherent saturation etc. based on the location of the rock and its base geology. However, since the dump material is not intact and is in fact the fragmented form of the intact rock, most of these properties will be highly dependent on the grain size distribution, compaction, saturation etc. For instance, density of dump waste material will be highly dependent on the compaction of the material while the dump is formed. For the dumps made by end dumping method, the density will be less due to simple gravity compaction, while in case of lift dumping, the density will be much

a b

Fig. 2.13 Lab testing apparatus and setup for (**a**) Uniaxial compressive strength in a cubical sample, (**b**) Sieve analysis for grain size distribution study

higher as the dump is compacted in each lift stage. Density will also be based on the grain size distribution of dump material [3]. The finer grain size will give much denser dump materials than the coarser grain sizes. A gradual mix of finer and coarser grain size can give a much better packing efficiency, due to proper filling of pore spaces, as compared to the ones with only coarser grain sizes. This again is based on the type and purpose of dumping. The methods like end dumping and slide dumping involve gravity induced segregation of material, where the material is constantly added to the face of the dump, resulting in coarser particles to collect at the bottom and finer at the top, hence creating a density gradation from top to bottom. On the same lines, but in opposite trend, the free dumping, which uses lifts to create dump keep compacting the dump material at every step resulting in high densities, which keep on increasing as more number of lifts are added on the top of the dump. Typically, density of dump material may vary from 1.6 to 2.2 g per cubic centimetre based on the compaction.

Similar to density, shear strength of the dump material is also an important property affected by the variation of compaction and overlying load. The compaction of material changes its effective internal friction angle and cohesion. These factors are also affected by the degree to which the dump material is saturated. Based on the type of dump material, the cohesion and angle of friction will vary with moisture content of the dump. Increase in the surface moisture in some massive structured materials such as quarts increases the friction, while the same decreased in the minerals with layer like lattice such as mica [21]. It was also observed in many studies that the internal angle of friction increased with the increase in coarser grain sizes in a fine-grained soil sample, possibly due to better cross linking. However, for a given void ratio, the smallest size particles will have the highest internal angle of friction, which will decrease with further increase in grain size [12]. In most of the dump slope stability analysis, the simplest and most applicable model for the judging

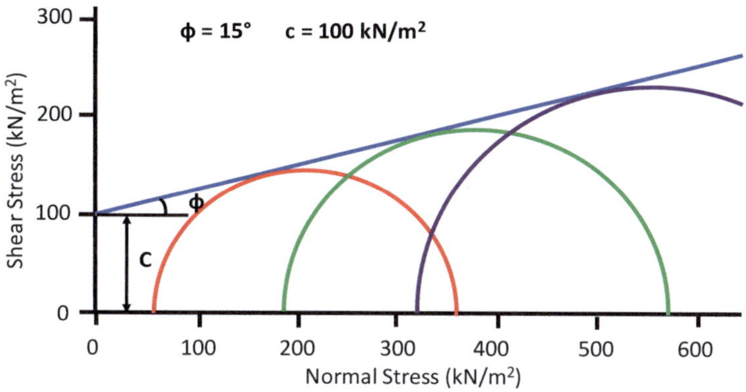

Fig. 2.14 Cohesion and angle of friction being calculated by the shear strength Mohr circles from the triaxial testing data. The blue line denotes the Mohr's failure envelope

the stability of dump slope is the Mohr-Coulomb failure criteria (Fig. 2.14), which effectively states the shear strength to be a function of the normal stress, internal angle of friction and the cohesion.

2.3.3 Geotechnical Properties of Foundation

Geotechnical properties of Foundation: Since the full dump must stand on it, the geotechnical properties of the foundation also play a key role in stability of the dump slope. In fact, one of the major causes of dump slope failure is the failure of the weak foundation base. The properties like saturation, porosity, permeability, shear strength of the foundation, are hence quite significant in slope stability. The foundation can deform in case it is made of saturated soil or similar zones resulting in a toe failure, or a complete circular failure involving both the dump as well as the foundation. Many times, it is difficult to predict the behaviour of foundation before the dumping starts as the base rock behaves differently under the load of tonnes of dump material [22]. Many types of foundation materials such as low plasticity silts and clays are expected to settle based on the consolidation of the in-situ conditions, as the dumping material is pilled over them with time. This settlement can cause serious complications on dump stability and the drainage structures that are made in the dump. Apart from that, the hydrological aspects change drastically as the dumping proceeds. Porosity and permeability are the factors that will decide the pore dynamics and the generation of pore water pressure. The water table will rise as the dump height increases resulting in saturation of bottom layers of foundation. This shall reduce the shear strength of foundation as compared to the no dumping conditions, and hence can result in unexpected failures by liquidation of the foundation. Hence detailed laboratory testing for the foundation material is also required including its classification, strength, permeability, consolidation etc.

2.3.4 Method of Dumping

Method of dumping: As we discussed in the section on dump classification, method of dumping plays a role in dictating the mechanical behaviour of dump. Based on the type of dumping, the geotechnical properties like the pore sizes, porosity, permeability, density, grain size distribution etc. vary from one dumping method to the next, hence affecting the stability of the dumps. Apart from geotechnical properties, the steepness of the slope is often dictated by the method of dumping as well. The layered dumping methods and the end dumping methods show contrasting characters when it comes to these properties. The design of the dump also dictates the hydrology of the dump and in extension the stability of dump slope. The presence of drainage channels provided in the dump design can act as relieve points for hydrological build up and can reduce the static water head that is often built in dumps that have high compaction scenarios. On the other hand, the loosely compacted end dumping techniques mostly provide enough pore spaces for the water to pass through to the base and in the foundation.

2.3.5 Hydrological Conditions

Hydrological Conditions: One of the important factors for dump slope stability that is often overlooked is the effect of water. Overburden is majorly blasted out rock and soil, which is generally highly porous in nature due to the end-dumping techniques. This results in easy percolation of water through the dump to its lower layers, which may result in accumulation of water if proper drainage routes are not made (Fig. 2.15). Accumulation of water in dumps can be disastrous and can lead to massive failures of dump slope as was seen in 2008 in western section of Jayant opencast mine, Northern coalfield Ltd. India [7].

Effect of accumulation of water in dumps can be threefolds. Firstly, water creates a hydrostatic pressure due to weight of the water itself, which increases the effective stress on the overall slope. Secondly, the presence of water creates pore pressure, which is the force of water present in voids of the overburden material. This pressure can be both positive (forcing the overburden particles away from each other) or negative (in form of suction/capillary pressure), and hence changes the dynamics of slope stability. Thirdly, oversaturation of water makes the particles loose due to dilution, which changes the internal properties of overburden material such as cohesion and internal angle of friction, hence causing instability in an originally stable slope [11]. Though proper drainage channels generally prevent the accumulation of water, these become ineffective with time due to build-up of fine soil at the base of dumps, transported from infiltration of rain water over time. Effective drainage hence becomes a key for keeping the hydrostatic pressure under check and keeping the dumps stable.

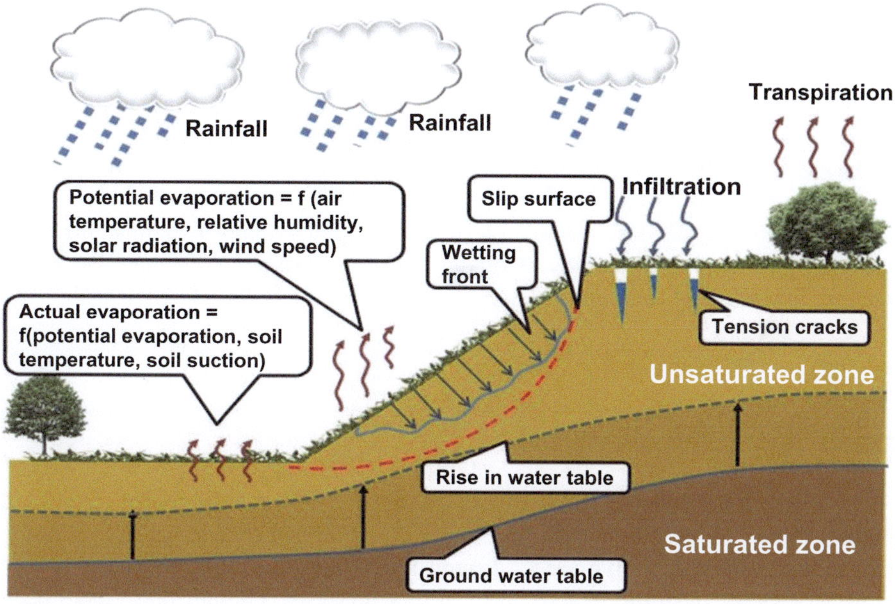

Fig. 2.15 Hydrodynamics depicting water movement and water cycle in a dump slope [23]

Fig. 2.16 Porosity and permeability are two different but important aspects in hydrology of a dump slope [24]

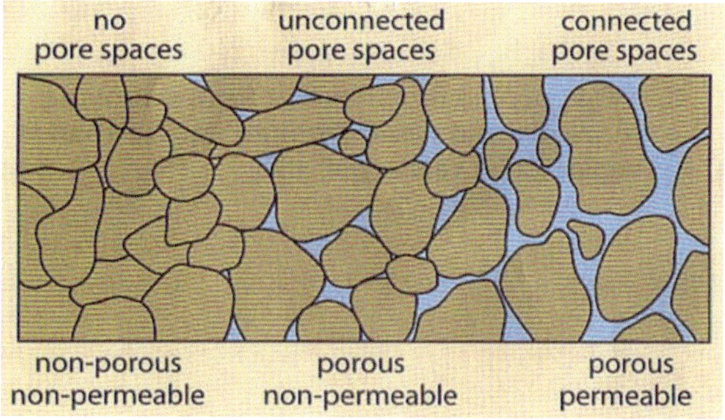

Apart from the hydrology of dump slopes, foundation hydrology also plays a vital role. The rise of the water table with increased dumping height makes the higher layers of foundation saturated. This results in decrease in shear strength and often a case of highly plastic behaviour of foundation resulting in massive failures due to dilation. Hence, factors like porosity, permeability, groundwater, and drainage are important considerations for designing a dump as well as for monitoring the stability of dump slopes. Porosity and permeability, both of which are responsible for the water statics and dynamics in the foundation (Fig. 2.16), will give idea of the void distribution and hence will be helpful in calculating water retention and static pressure heads.

Another important aspect of the hydrology of dump slopes is the rainfall and surficial erosion. Rain water will both infiltrate the dump material as well as will run off from the surface based on the porosity and permeability of dump material. Surficial runoffs result in surficial erosion and hence can be detrimental to the design of dump as a whole [25]. These erosions can choke the drain channels as well as make the dump structurally weak if the toe area of the dump is eroded. Hence erosion studies are also conducted for getting an idea of the effect of water movement on and around dumps. In this case, vegetation on the dump also plays a significant role as the roots of the vegetation acts as binder for the soil/overburden material, hence holding it against the erosion [26].

2.3.6 Static and Dynamic Forces

Static and Dynamic Forces: Apart from the local static properties of material and nature, there are other dynamic properties that also effect the stability of dump slope. In seismically active areas, ground movement due to earth quake is also kept as a factor in dump design and stability analysis. Though there have been many models for dynamic behavioural analysis of dump slopes, the exact effect on the stability of dump slopes is still debatable, mostly because of the unpredictable transfer of energy from ground to the dump and the effect it will have thereof. The greatest stability risk that the dump has in case of seismic activity is the liquefaction of foundation, resulting in progressive unavoidable failure.

Similar to the seismic activity in local area, one of the major form of dynamic forces active on the waste dumps in regular mining conditions is that by the blasting operations in mining [27]. Though modern-day blasting is well planned and studied for the effect of its vibrations on local structures, there is always a possibility of misfires and unregulated blasting resulting in detrimental effects on local mine structures, specifically the internal dumps which are at proximity to the blast zone (Fig. 2.17). Dumps these days are hence analysed for blast induced vibration study as well as other stability analysis for their safer design [28, 29]. Apart from blasting and seismic forces, some of the other loads such as moving of heavy earth moving machinery on or near the dump also play a role in stability of dump slopes.

Fig. 2.17 Blast vibrations can cause failures in the dump slopes and hence are designed and analysed accordingly (NCL, India)

2.4 Modes of Failure

Any negative factor in the above discussed aspects affecting the stability of the dump slopes can lead to slope failure. The way, or mode, in which a slope fails is dictated by which of the factors are not up to the mark to make the slope stable. The prediction for the mode of failure is also important in case the failure is unavoidable and remedial measures have to be planned in advance. Based on the kinetics of failure there are three basic kinds of failure movements as seen in Fig. 2.18. In regressive failure, there are short term displacement cycles at a decelerating rate on the removal of the failure initiation event such as external loading, blasting vibration etc. On the other hand, in the progressive cycle, the displacement cycles are accelerating and won't stop by themselves unless some stabilisation measures are introduced [30]. Various combinations of factors result in various modes of failures in dump slopes. Some of the main failure modes discussed in various literatures are mentioned here.

2.4.1 Surface or Edge Slide

Surface or edge slide: These are some of the most common type of failure in mine dumps, where a layer of material, parallel to the surface of the slope, fails and translates down the slope (Fig. 2.19). These are generally seen in end dumping and

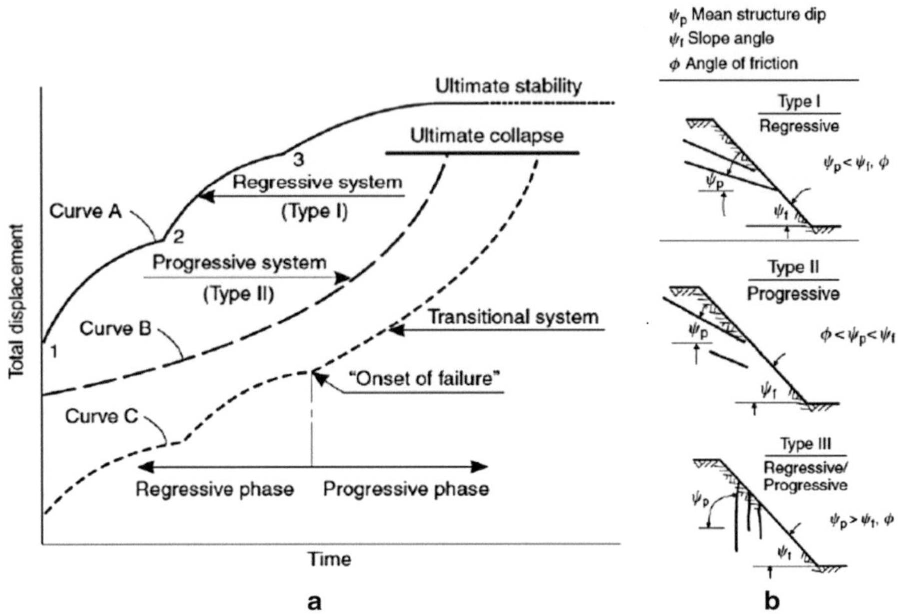

Fig. 2.18 (**a**) Displacement vs time plot for repressive and progressive displacement cycles during the slope failure. (**b**) Corresponding structural conditions for slope kinetics [30]

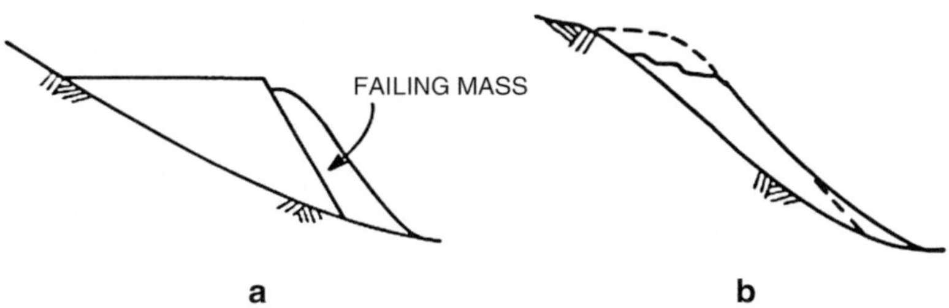

Fig. 2.19 (**a**) Surface slump failure, (**b**) Edge slide failure [31]

push dumping methods and happen mostly due to over steepening of the crest regions of the dump. The steepening may happen due to higher cohesion of material, especially post rains, when the steepening is also coupled with weakening of the toe region of the dumps. A variation of this is the plane failure of the dumps where sliding occurs along a single plane either at an angle to the slope or parallel to it.

2.4.2 *Shallow Flow Slides*

Shallow flow slides: These are shallower as compared to the surface slumping. The failure occurs here due to complete or partial saturation of dump material due to rains

Fig. 2.20 Shallow flow slide failure

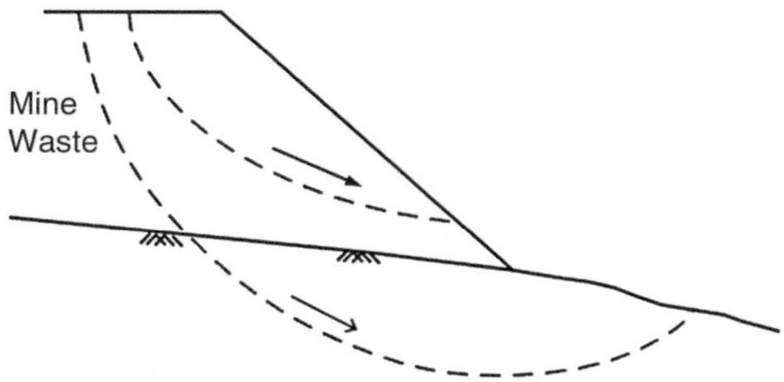

Fig. 2.21 Rotational circular failure [31]

or snow melt in very cold regions, resulting in reduction in shear strength of the material [14]. Mostly restricted to the crest of the dump, the material translates from the top to bottom of the dump increasing the toe region (Fig. 2.20).

2.4.3 *Rotational Circular*

Rotational circular: Rotational failure again is one of the major failure modes seen in dumps, where the mass fails along a curved surface (Fig. 2.21). This is especially seen in dumps with excessive dump heights or loading materials. This can also happen during rains, especially due to loss of toe support and during earthquakes. Rotational failure may be limited to failure of the dump material or can extend to the foundation of the dump if the foundation material is weak soil type or due to high pore-water pressure development [14]. Creep is also characterised as a wider form of circular failure.

2.4.4 *Block Translation*

Block Translation: This happens in all the conditions similar to the rotational circular failure, however the translation generally occurs along a plane which is either in the foundation, or at the interface of foundation and dump material (Fig. 2.22). This type

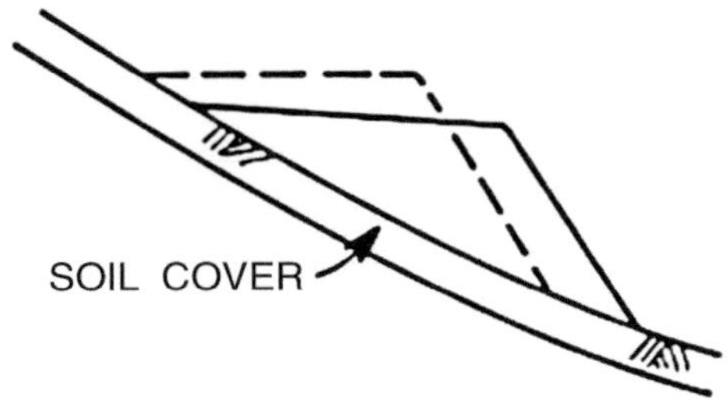

Fig. 2.22 Block translation failure [11]

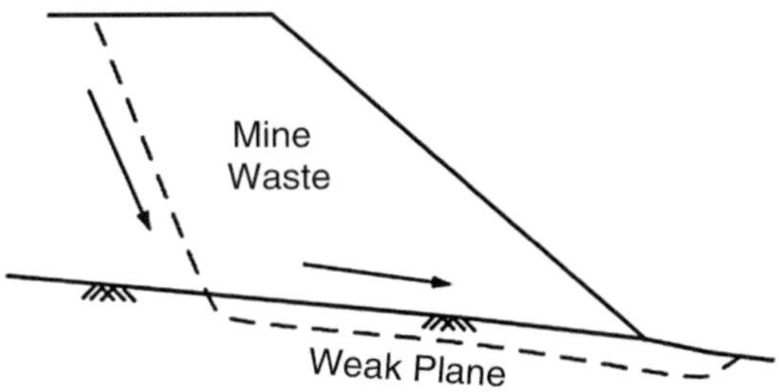

Fig. 2.23 Base failure (spreading) [31]

of failure is characterised by movement of the whole dump along the weak plane, which is further favoured by increasing steepness of the foundation.

2.4.5 Base Failure (Spreading)

Base Failure (spreading): This is a progressive type of failure where there a weak failure plane in the foundation which combines with the shear failure plane of the dump. This is specially seen in steeper foundations where the foundation base translates in advance of the dump failure and keeps failing at an incremental rate [31] (Fig. 2.23).

2.4.6 Liquefaction

Liquefaction: Liquefaction happens when the effective stresses on the liquefiable material are effectively reduced to zero, which results in complete loss of shear strength. This quick sand like condition may happen due to seismic activity, cyclic loading process or due to excess pore water pressure as a result of heavy rains and water accumulation [14]. If the foundation faces liquefaction, whole dump may end up in progressive failure (Fig. 2.24).

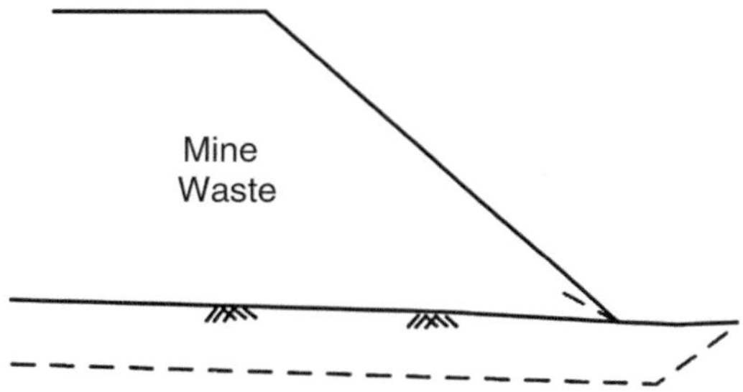

Fig. 2.24 Liquefaction of base [31]

2.5 Conclusion

A brief overview of the dumps and the dump slope stability was presented in this chapter. Starting from the understanding of the importance of the dump slope stability in today's mining scenario, various models of dumping were analysed, both based on the shape of the dump, as well as based on the method of dumping. It was seen that some of the geotechnical and hydrological properties of the dump are directly dependent on the method of dumping itself. Consequently, the dumping method was discussed as one of the factors that affect the stability of dump slopes.

Factors affecting the stability of the dump slopes were discussed and compared further. Important factors such as the geometry of the dumps, the geotechnical properties of the dump material as well as the dump's foundation were identified to be crucial parameters affecting the dump slope stability. However, hydrology of the dump slope system was not neglected and its effect on the stability of the dump slope was specifically discussed, especially recognizing the uncontrolled watery conditions to be one of the primary causes of dump failure. The effect of dynamic forces such as seismic activity, blasting vibrations and heavy earth moving machinery movement, on the stability of dumps was also discussed. Correlating these factors to the failure conditions, various modes of dump failure were finally analysed and understood in context of the dump slope stability.

References

1. Chadwick J (2008) Going underground, or not. Int Mining:48–54
2. GOI (2015) In: C.C.s. Organisation (ed) Provisional coal statistics. Ministry of Coal, Kolkata
3. Darling P (2011) SME mining engineering handbook, vol 1. Society of Mining, Metallurgy and Exploration Inc. (SME), Englewood
4. Gupta T, Yellishetty M, Singh T (2015) Optimization of ash content in overburden dumps: a numerical approach. In: Proceedings of MPES 2015-Smart innovation in mining. The Southern African Institute of Mining and Metallurgy Johannesburg, South Africa. pp 997–1004
5. Saran B, Kumar P (2016) Jharkhand coal mine collapse: eleven workers killed, over 50 trapped, in Hindustan Times, Hindustan Times, New Delhi

6. GOI (2016) Update on ECL Rajmahal coal mine accident. In: M.o.C. Government of India (ed). http://pib.nic.in
7. Sharma S, Roy I (2015) Slope failure of waste rock dump at jayant opencast mine, India: a case study. Int J Appl Eng Res 10(13):33006–33012
8. Rai R et al (2012) Sensitivity analysis of internal dragline dump stability: finite element analysis. Geotech Geol Eng 30(6):1397–1404
9. Kainthola A et al (2011) A coal mine dump stability analysis—a case study. Geomaterials 1 (01):1
10. Gupta T et al (2014) Sensitivity analysis of coal rib stability for internal mine dump in opencast mine by finite element modelling. Geotech Geol Eng 32(3):705–712
11. NMEICT (2014) Mine waste dump. In: Rajesh R (ed) Slope engineering. The National Mission on Education through Information and Communication Technology, Varanasi
12. Orman M, Peevers R, Sample K (2011) Waste piles and dumps. In: SME mining engineers handbook, pp 667–680
13. Zahl EG et al (1992) Waste disposal and contaminant control. In: Hartman HL (ed) SME, mining engineering handbook. SME, Littleton
14. Cadwell J, Allan M, Bruce V (1981) The simplified analysis of mine waste embankments, US Forest Services
15. Latimer C (2014) Caterpillar launches new dozer. In: Australia mining. Prime creative media: Online
16. Asianet (2012) Mining dumps may cause pollution hazards
17. Bizj (2016) Titan America business grows with construction resurgence. S Fla Bus J. http://media.bizj.us/view/img/529551/titan-dragline*180.jpg
18. Singh P et al (2013) Estimation of critical parameters for slope instability in an in-pit mine dump. SGAT Bull 14(1):34–44
19. Verma D et al (2011) Evaluation of open pit mine slope stability analysis. Int J Earth Sci Eng 4 (4):590–600
20. Vandre B (1980) Tentative engineering guide: stability of non water impounding mine waste embankments. U.S. Forest Service, Ogden
21. Lewis JG (1956) Shear strength of rockfill. In: 2nd Australia-New Zealand conference on soil mechanics and foundation engineering
22. Singh A, Singh T (2006) Assessing instability of coal mine waste dump. Ind Miner Ind J 113–118
23. Rahardjo H, Satyanaga A, Leong EC (2011) Unsaturated soil mechanics for slope stabilization
24. PGCE (2014) Rock and geology time. Weebly: PGCE Geography Classroom
25. Cho SE, Lee SR (2002) Evaluation of surficial stability for homogeneous slopes considering rainfall characteristics. J Geotech Geoenviron 128(9):756–763
26. Waldron L (1977) The shear resistance of root-permeated homogeneous and stratified soil. Soil Sci Soc Am J 41(5):843–849
27. Singh T, Singh V (2005) An intelligent approach to prediction and control ground vibration in mines. Geotech Geol Eng 23(3):249–262
28. Khandelwal M, Singh T (2007) Evaluation of blast-induced ground vibration predictors. Soil Dyn Earthq Eng 27(2):116–125
29. Kumar M, Jain AK, Singh TN (2009) Blast vulnerability analysis of waste dump during production blasting. J Mines Metals Fuels 57(10):313–316
30. Broadbent CD, Zavodni ZM (1982) Influence of rock structure on stability. In: Stability in surface mining, vol 2. Society of Mining Engineers, Denver
31. BCMDC, B.C.M.D.C.P.E.L. (1991) Investigation and design of mine dumps: interim guidelines. The Committee

Chapter 3
Shear Strength Behaviour of Jointed Rock Masses

Mahendra Singh

Abstract Rocks encountered in civil and mining engineering structures are generally jointed in nature. The presence of joints renders anisotropy in rock and makes them weaker in their engineering response. Assessment of shear strength response of such jointed rocks, subject to given stress state, is a challenging task. Large size field tests are very expensive and time consuming and hence not feasible for majority projects. The best alternative available is to use indirect methods to describe the shear strength behaviour of jointed rocks.

The present articles presents some of the most widely used techniques developed during last few decades, using which the shear strength response of jointed rock can be assessed with reasonable accuracy. Relatively simple tests and observations are required for applying these techniques and hence input data can be procured without much difficulty. The shear strength response is divided into two broad categories i.e. strength behaviour of joints and strength behaviour of jointed rock mass. Shear strength models described in this article cover linear as well as non-linear strength response. Classification systems are widely used to characterize the rock masses in the field. It has been explained, how, these classification systems could be used to assess the shear strength response of the rock masses.

Keywords Jointed rock · Shear strength · Strength criteria · Classification systems

3.1 General

While analysing structures like tunnels, underground caverns, landslides, road cuts and foundations of heavily loaded structures like dams, bridges situated in or on rocks, the engineers and geologists are often required to assess the shear strength of

M. Singh (✉)
Department of Civil Engineering, IIT Roorkee, Roorkee, India

S. P. Pradhan et al. (eds.), *Landslides: Theory, Practice and Modelling*,
Advances in Natural and Technological Hazards Research 50,
https://doi.org/10.1007/978-3-319-77377-3_3

jointed rock masses. The discontinuities e.g. joints, foliations and bedding planes are invariably present in rock masses and induce planes of weakness in the mass. While shearing, failure may occur due to a complex combination of sliding on pre-existing discontinuities, shearing of rock substance, translation and/or rotation of intact rock blocks. Consequently, the jointed rock is quite incompetent and anisotropic in strength and deformational behaviour. In addition, the strength behaviour of jointed rock is non-linear with increase in confining pressure. The assessment of shear strength response is therefore, an extremely difficult task. The present article discusses in brief, how the shear strength of the rocks can be assessed in the field.

From application point of view, two broad categories may occur: In the first category, a single or a set of planar persistent discontinuities exists in the rock. Sliding may occur along the discontinuities depending on the kinematics of the problem. These types of the conditions are commonly encountered in case of slopes. The second category pertains to the failure of the rock mass as a whole. The potential failure surface lies partly along discontinuity surfaces, and partly through the intact rock. Sliding, rotation, translation, splitting or shearing of intact rock blocks occur at the time of failure. The rock mass may behave isotropically or anisotropically depending upon the number, orientation and spacing of discontinuities and level of confining stress. These types of failure are common in tunnels and other underground structures. These two broad categories are discussed in the following sections.

3.2 Discontinuity Shear Strength

In case of shear strength along the surface of discontinuity, the shear strength is represented as a function of normal stress across the failure surface and the shear strength parameters. The most widely used shear strength model is linear Coulomb's model. To account for non-linearity in shear strength response, Patton, Ladanyi-Archambault and Barton's models are commonly used.

3.2.1 Coulomb's Model

Coulomb's shear strength parameters cohesion, c_j and friction angle, ϕ_j are used to estimate the shear strength for given normal stress. The shear strength parameters may be obtained by performing direct shear tests on the discontinuity surfaces. Direct shear tests are conducted for various normal loads and shear stress vs. shear displacement plots are obtained. From these plots, the peak and residual shear strength of the joints are obtained. The failure envelopes of peak and residual

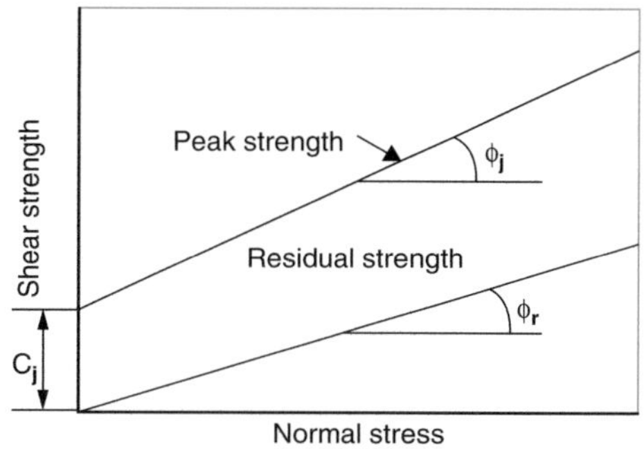

Fig. 3.1 Failure envelopes of shear strength for rough rock joints

shear strength are plotted (Fig. 3.1). The shear strength of the discontinuity is defined as:

$$\tau_f = c_j + \sigma_n \tan \phi_d \quad \text{(Peak strength)} \tag{3.1}$$

$$\tau_f = \sigma_n \tan \phi_r \quad \text{(Residual strength)} \tag{3.2}$$

Where τ_f is the shear strength along the discontinuity; σ_n is the effective normal stress over the discontinuity; ϕ_j is the peak friction angle of the discontinuity surface; c_j is the peak cohesion of the discontinuity surface, and ϕ_r is residual friction angle for discontinuity surface.

3.2.2 Patton's Bi-Linear Shear Strength Model

Patton [1] recognised the importance of failure modes and suggested a bi-linear shear strength model. It was postulated that at low normal stress level, sliding occurs along the asperities of the joint surfaces, and at high normal stress, the shearing of the asperities takes place. The model is expressed as:

$$\tau_f = \sigma_n \tan\left(\phi_\mu + \mathrm{i}\right) \quad \text{for low } \sigma_n \quad (\textit{Sliding}) \tag{3.3}$$

$$\tau_f = c_j + \sigma_n \tan\left(\phi_r\right) \quad \text{for high } \sigma_n \quad (\textit{Shearing}) \tag{3.4}$$

Where i defines the roughness angle, ϕ_μ is basic friction angle and ϕ_r is the residual friction angle.

In the field, it is very difficult to assess the normal stress level at which transition from sliding to shearing takes place. In reality, there is gradual transition from sliding to shearing and as such, there is no distinct and clear-cut normal stress level, which defines the boundary between the two failure modes.

3.2.3 Ladanyi and Archambault Criterion

Based on the principle of strain energy, Ladanyi and Archambault [2] equated the external work done in shearing a jointed rock to the internal energy stored in the rock and expressed the shear strength of a joint or rock mass as:

$$\tau_f = \frac{\sigma_n(1 - a_s)\left(\dot{v} + \tan\phi_\mu\right) + a_s(c_i + \sigma_n \tan\phi_i)}{1 - (1 - a_s)\,\dot{v}\,\tan\phi_\mu} \tag{3.5}$$

where τ_f is the shear strength; a_s is sheared area ratio equal to A_s/A; A is the total area of joint surface; A_s is the sheared area of joint surface; σ_n is the mean applied normal stress; $\dot{v}$ is the rate of dilation at failure; ϕ_μ is the basic friction angle of joint surface, and c_i, ϕ_i are Mohr-Coulomb parameters for intact rock.

The sheared area ratio, a_s and dilation rate, $\dot{v}$ are estimated as:

$$a_s = 1 - \left(1 - \frac{\sigma_n}{\sigma_{trn}}\right)^{K_1} \tag{3.6}$$

$$\dot{v} \approx \left(1 - \frac{\sigma_n}{\sigma_{trn}}\right)^{K_2} \tan i \tag{3.7}$$

Where K_1 and K_2 are equal to 1.5 and 4 respectively; σ_{trn} is the brittle – ductile transition stress, which may be taken equal to the UCS of intact rock; and i is the initial roughness of the joints.

3.2.4 Barton's JRC-JCS Model

Barton's model, also known as JRC-JCS model is very simple and is the most widely used strength criterion for assessing shear strength along discontinuity surfaces. The shear strength of a joint is expressed as:

$$\tau_f = \sigma_n \tan\left[\phi_r + \mathrm{JRC}\log_{10}\frac{\mathrm{JCS}}{\sigma_n}\right] \tag{3.8}$$

Where JRC is the joint roughness coefficient, which is a measure of the initial roughness (in degrees) of the discontinuity surface. JRC is assigned a value in the range of 0–20, by matching the field joint surface profile with the standard surface profiles on a laboratory scale of 10 cm [3] as shown in Fig. 3.2. JCS is the joint wall compressive strength of the discontinuity surface, and σ_n is the effective normal stress acting across the discontinuity surface.

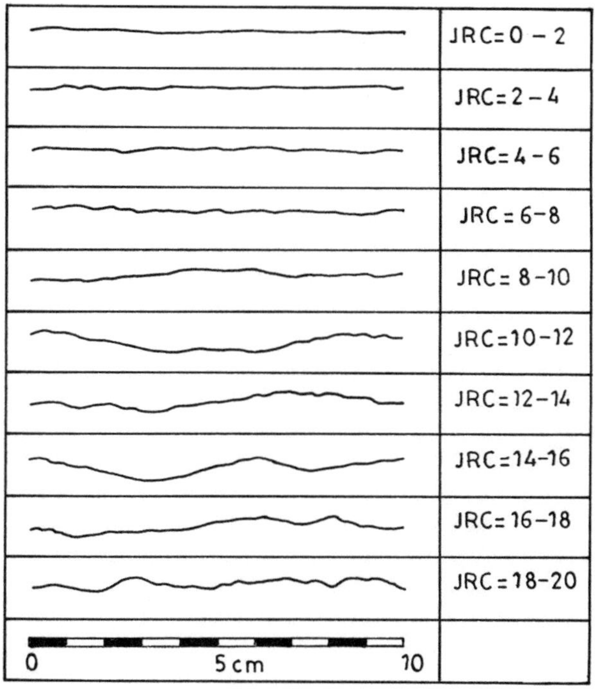

Fig. 3.2 Roughness profiles to estimate JRC [3]

3.3 Shear Strength of Rock Mass

Depending on the scale of structure, geometry of discontinuities and interlocking conditions, the failure may occur due to complex interaction of sliding, rotation, splitting, shearing and translation of rock blocks. In such cases, the jointed rock mass may be replaced with an equivalent continuum for mechanical analysis. Several strength criteria are used to express the strength behaviour of such rock masses. Some of the strength criteria are discussed in the followings in two sub-headings i.e. linear and non-linear strength criteria.

3.3.1 Linear Mohr-Coulomb Criterion

The rock mass is treated as an isotropic continuum and the shear strength along the failure surface is expressed as follows:

$$\tau_f = c_m + \sigma_n \tan \phi_m \tag{3.9}$$

Where c_m and ϕ_m are Mohr-Coulomb shear strength parameters for jointed rock or rock mass. The values of c_m and ϕ_m may be obtained from field shear tests on rock mass. Alternatively, classification approaches provide a rough estimate of the shear strength and some of these approaches are given below.

3.3.1.1 Rock Mass Rating (RMR)

The RMR classification system for rock masses was suggested by Bieniawski [4–6] to characterise the quality of the rock mass. The following parameters of the rock mass are used to classify the mass:

(a) UCS of intact rock material,
(b) Rock Quality Designation (RQD),
(c) Spacing of discontinuities,
(d) Condition of discontinuities, and
(e) Groundwater condition.

Basic RMR is obtained based on above five parameters and then rating is adjusted based and orientation of discontinuities with respect to the structure. The values of shear strength parameters c_m, ϕ_m are presented in Table 3.1 [5].

Mehrotra [7], based on experience from Indian project sites, observed that the shear strength is under-predicted by expressions suggested by Bieniawski [5]. Figure 3.3 may be used for assessing the shear strength parameters of rock masses especially for slopes.

3.3.1.2 Q System

The classification system Q [8, 9] is very popular for characterisation of rock mass. The rock mass quality index Q, is defined as:

$$Q = \left(\frac{RQD}{J_n}\right)\left(\frac{J_r}{J_a}\right)\left(\frac{J_w}{SRF}\right) \tag{3.10}$$

Where, RQD is the rock quality designation [10]; J_n is the joint set number; J_r is the joint roughness number; J_a is the joint alteration number; J_w is the joint water reduction factor, and SRF is stress reduction factor.

The shear strength parameters are obtained as:

$$c_m = \left(\frac{RQD}{J_n}\right)\left(\frac{1}{SRF}\right)\left(\frac{\sigma_{ci}}{100}\right) \text{ MPa} \tag{3.11}$$

$$\phi_m = \tan^{-1}\left(\frac{J_r}{J_a}J_w\right) \tag{3.12}$$

where c_m is the cohesion of the undisturbed rock mass; ϕ_m, the friction angle of the mass; and σ_{ci} is the uniaxial compressive strength of intact rock material.

Table 3.1 Mohr-Coulomb parameters from RMR [5]

Class number	I	II	III	IV	V
Cohesion of rock mass (kPa)	>400	300–400	200–300	100–200	<100
Friction angle of rock mass (deg)	>45	35–45	25–35	15–25	<15

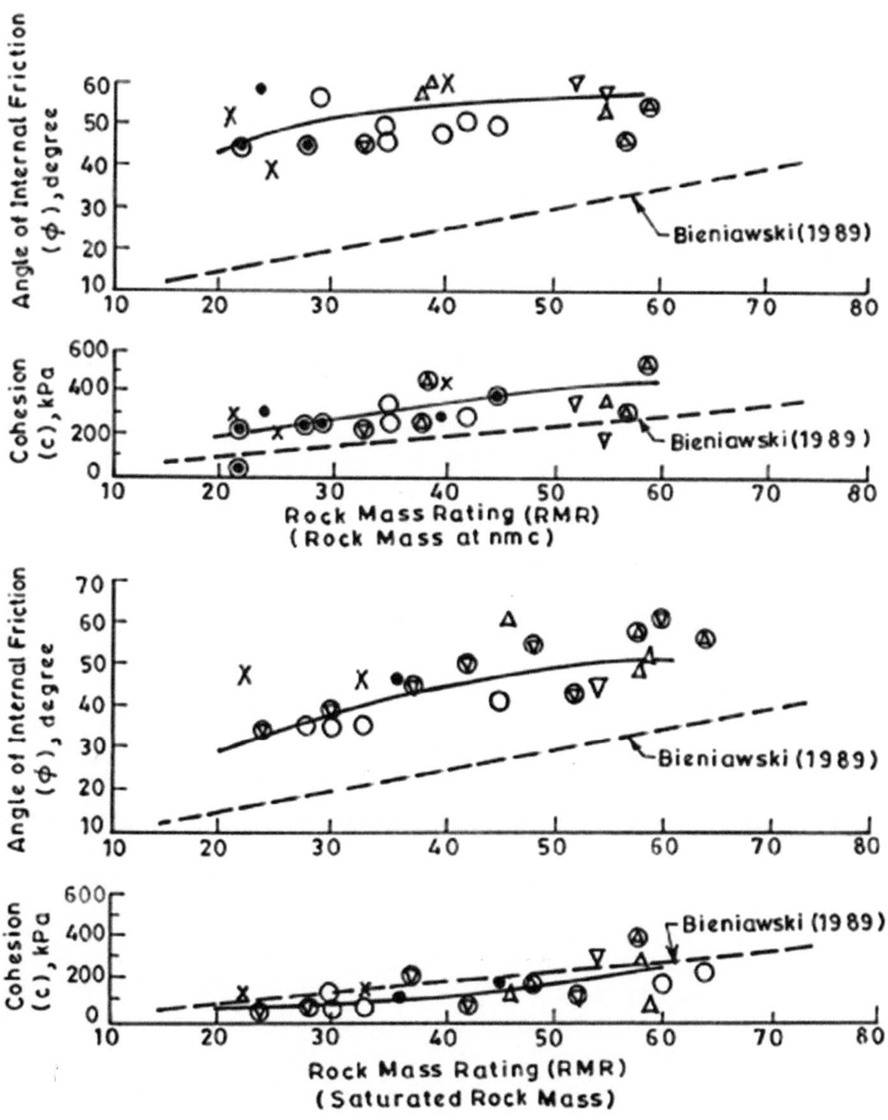

Fig. 3.3 Estimation of friction angle of rock mass from RMR [7]

If used for slopes, an overestimation in the strength may be expected. For slopes, this author personally feels that the relationship between shear strength parameters and RMR as suggested by Mehrotra [7] will be more appropriate for the Himalayan rock masses. The relationships were developed based on extensive in-situ direct shear tests on saturated rock masses in Himalayas.

3.3.2 Non-linear Strength Criteria

The Mohr-Coulomb strength criterion considers the rock mass shear strength as a linear function of normal stress σ_n. In reality, the shear strength response is highly non-linear and the parameters c_m, ϕ_m change with the range of confining pressure used in estimating these parameters. To resolve this issue, several non-linear strength criteria have been proposed for jointed rocks and rock masses. Some of them are presented in the following section.

3.3.2.1 Empirical Criteria Based on RMR and Q

Mehrotra [7] utilised results of large number of in-situ direct shear tests on rock masses in the Himalayas, and suggested the non-linear variation of shear strength as:

$$\frac{\tau_f}{\sigma_{ci}} = A\left(\frac{\sigma_n}{\sigma_{ci}} + B\right)^C \tag{3.13}$$

where A, B and C are empirical constants and depend on RMR or Q. Their values for different moisture contents, RMR and Q index are presented in Table 3.2.

3.3.2.2 Hoek-Brown Strength Criterion

Hoek-Brown [11] proposed a non-linear strength criterion for intact rocks as follows:

$$\sigma_1 = \sigma_3 + \sqrt{m_i \sigma_{ci} \sigma_3 + \sigma_{ci}^2} \tag{3.14}$$

where σ_1 is the effective major principal stress at failure; σ_3 is the effective minor principal stress at failure; m_i is a criterion parameter; and σ_{ci} is the UCS of the intact rock, which is also treated as a criterion parameter. The parameters m and σ_{ci} are obtained by fitting the criterion into the laboratory triaxial test data. If triaxial test data is not available the approximate values of parameters m_i can also be obtained from Table 3.3 [12].

The criterion was extended to heavily jointed isotropic rock masses [11]. The latest form of the criterion [13] is expressed as:

$$\sigma_1 = \sigma_3 + \sigma_{ci}\left(m_j \frac{\sigma_3}{\sigma_{ci}} + s_j\right)^a \tag{3.15}$$

Where m_j is an empirical constant, which depends upon the rock type; and s_j is an empirical constant, which varies between zero (for crushed rock) to one (for intact

Table 3.2 Shear strength parameters for jointed rock masses [7]

Rock type quality	Limestone	Slate, xenolith, phyllite	Sandstone, quartzite	Trap, metabasic
Good rock mass	NMC	NMC	NMC	NMC (S_{av} = 0.30)
RMR = 61–80	A = 0.38, B = 0.005, C = 0.669	A = 0.42, B = 0.004, C = 0.683	A = 0.44, B = 0.003, C = 0.695	A = 0.50, B = 0.003, C = 0.698
Q > 10	Saturated (S = 1)	Saturated (S = 1)	Saturated (S = 1)	Saturated (S = 1)
	A = 0.35, B = 0.004, C = 0.669	A = 0.38, B = 0.003, C = 0.683	A = 0.43, B = 0.002, C = 0.695	A = 0.49, B = 0.002, C = 0.698
Fair rock mass	NMC	NMC	NMC (S_{av} = 0.15)	NMC (S_{av} = 0.35)
RMR = 41–60	A = 2.60, B = 1.25 , C = 0.662	A = 2.75, B = 1.15 , C = 0.675	A = 2.85, B = 1.10 , C = 0.685	A = 3.05, B = 1.00 , C = 0.691
Q = 2–10	Saturated (S = 1)	Saturated (S = 1)	Saturated (S = 1)	Saturated (S=1)
	A = 1.95, B = 1.20, C = 0.662	A = 2.15, B = 1.10, C = 0.675	A = 2.25, B = 1.05, C = 0.688	A = 2.45, B = 0.95, C = 0.691
Poor rock mass	NMC (S_{av} = 0.25)	NMC (S_{av} = 0.40)	NMC (S_{av} = 0.25)	NMC (S_{av} = 0.15)
RMR = 21–40	A = 2.50, B = 0.80, C = 0.646	A = 2.65, B = 0.75, C = 0.655	A = 2.85, B = 0.70, C = 0.672	A = 3.00, B = 0.65, C = 0.676
Q = 0.5 – 2	Saturated (S = 1)	Saturated (S = 1)	Saturated (S=1)	Saturated (S=1)
	A = 1.50, B = 0.75, C = 0.646	A = 1.75, B = 0.70, C = 0.655	A = 2.00, B = 0.65, C = 0.672	A = 2.25, B = 0.50, C = 0.676
Very poor rock mass	NMC	NMC	NMC	NMC
RMR < 21	A = 2.25; B = 0.65, C = 0.534	A = 2.45; B = 0.60, C = 0.539	A = 2.65; B = 0.55, C = 0.546	A = 2.90; B = 0.50, C = 0.548
Q < 0.5	Saturated (S = 1)	Saturated (S = 1)	Saturated (S = 1)	Saturated (S = 1)
	A = 0.80, B = 0.0, C = 0.534	A = 0.95, B = 0.0, C = 0.539	A = 1.05, B = 0.0, C = 0.546	A = 1.25, B = 0.0, C = 0.548

S Degree of saturation, *NMC* Natural moisture content, S_{av} Average value of degree of saturation

rock) depending upon the degree of fracturing. The expressions for m_j and s_j are given as:

$$m_j = m_i \exp\left(\frac{GSI - 100}{28 - 14D}\right) \tag{3.16}$$

$$s_j = \exp\left(\frac{GSI - 100}{9 - 3D}\right) \tag{3.17}$$

Table 3.3 Approximate estimation of parameter m_i [12]

Rock type	Class	Group	Texture Coarse	Medium	Fine	Very fine
Sedimentary	Clastic		Conglomerate	Sandstone	Siltstone	Claystone
			(22)	19	9	4
				Greywacke		
				(18)		
	Non-clastic	Organic		Chalk		
				7		
				Coal		
				(8–21)		
		Carbonate	Breccia	Sparitic Limestone	Micritic Limestone	
			(20)	(10)	8	
		Chemical		Gypstone	Anhydrite	
				16	13	
Metamorphic	Non Foliated		Marble	Hornfels	Quartzite	
			9	(19)	(24)	
	Slightly Foliated		Migmatite	Amphibolite	Mylonites	
			(30)	(25–31)	(6)	
	Folaited[a]		Gneiss	Schists	Phyllites	Slate
			33	4–8	(10)	9
Igneous	Light		Granite		Rhyolyte	Obsidian
			33		(16)	(19)
			Granodiorite		Dacite	
			(30)		(17)	
	Dark		Diorite		Andesite	
			(28)		19	
			Gabbro	Dolerite	Basalt	
			27	(19)	(17)	
			Norite			
			22			
	Extrusive Pyroclastic type		Agglomerate	Breccia	Tuff	
			(20)	(18)	(15)	

Note: Values in parenthesis are estimates

[a]These values are for intact rock specimens tested normal to bedding or foliation. The value will be significantly different if failure occurs along a weakness plane

where *D* is a factor which depends upon the degree of disturbance to which the rock mass has been subjected by blast damage and stress relaxation. It varies from zero for undisturbed in situ rock masses to one for very disturbed rock masses. For blasted rock slopes, D is taken in the range 0.7–1.0.

GSI is the Geological Strength Index [13, 14] which depends on the structure of mass and surface characteristics of the discontinuities (Fig. 3.4).

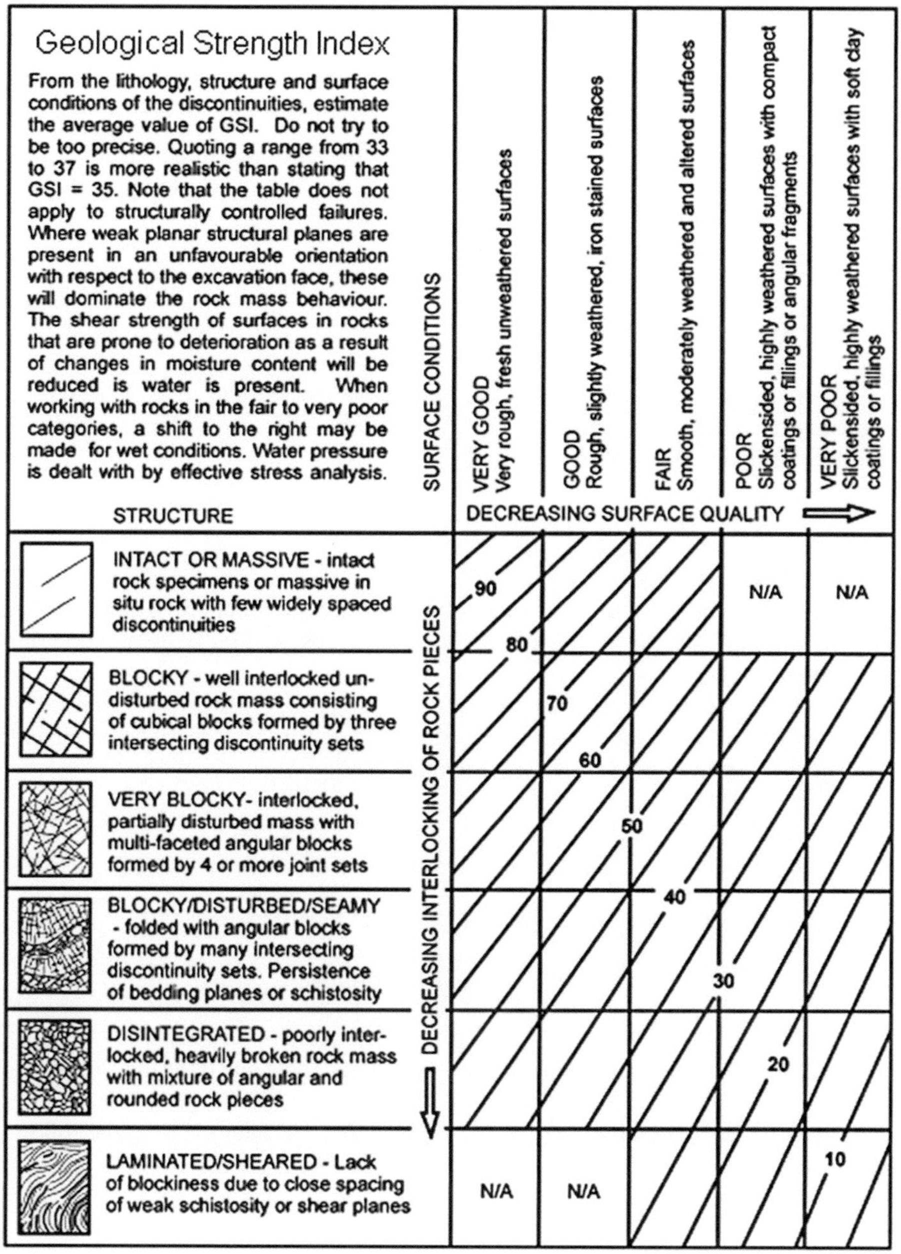

Fig. 3.4 Estimation of geological strength index [15]

The index a is obtained as:

$$a = \frac{1}{2} + \frac{1}{6}\left(e^{-GSI/15} - e^{-20/3}\right) \tag{3.18}$$

The limitation of the GSI approach is that the GSI is estimated only from geological features and disturbance to the mass, and no measurements e.g. joint mapping are done in the field.

3.3.2.3 Ramamurthy Criterion

Based on extensive triaxial tests conducted on rocks and model materials, Ramamurthy and co-workers [16–19] suggested the following non-linear strength criterion for intact isotropic rocks:

$$\left(\frac{\sigma_1 - \sigma_3}{\sigma_3 + \sigma_t}\right) = B_i\left(\frac{\sigma_{ci}}{\sigma_3 + \sigma_t}\right)^{\alpha_i} \tag{3.19}$$

where σ_3 and σ_1 are the minor and major principal stresses at failure; σ_t is the tensile strength of intact rock; σ_{ci} is the UCS of the intact rock; and α_i, B_i are the criterion parameters.

Parameters α_i and B_i are criterion parameters and are suggested to be obtained by fitting the criterion into the laboratory triaxial test data for intact rock. Alternatively, the following approximate correlations may be used:

$$\alpha_i = 2/3; \text{ and } B_i = 1.1\left(\frac{\sigma_{ci}}{\sigma_t}\right)^{1/3} \text{ to } 1.3\left(\frac{\sigma_{ci}}{\sigma_t}\right)^{1/3} \tag{3.20}$$

The criterion was extended to jointed rocks and rock masses as:

$$\left(\frac{\sigma_1 - \sigma_3}{\sigma_3}\right) = B_j\left(\frac{\sigma_{cj}}{\sigma_3}\right)^{\alpha_j} \tag{3.21}$$

where α_j and B_j are the criterion parameters for jointed rock; and σ_{cj} is the UCS of the jointed rock.

The criterion parameters α_j and B_j are suggested to be obtained from the following correlations:

$$\frac{\alpha_j}{\alpha_i} = \left(\frac{\sigma_{cj}}{\sigma_{ci}}\right)^{0.5} \tag{3.22}$$

$$\frac{B_i}{B_j} = 0.13\exp\left[2.037\left(\frac{\sigma_{cj}}{\sigma_{ci}}\right)^{0.5}\right] \tag{3.23}$$

Where σ_{cj} is the UCS of the rock mass.

3.3.2.4 Modified Mohr Coulomb Criterion

The Mohr-Coulomb criterion, though most widely used criterion, has the limitation in that the non-linear strength response of rocks is not captured by this criterion. Singh and Singh [20] used critical state concept for rocks [21] and suggested the Modified Mohr Coulomb (MMC) criterion to incorporate non-linearity in strength behaviour. The advantage of MMC is that the parameters c_m and ϕ_m are retained as such. The criterion is expressed as:

$$(\sigma_1 - \sigma_3) = \sigma_{cj} + \frac{2 \sin \phi_{m0}}{1\text{-} \sin \phi_{m0}} \sigma_3 - \frac{1}{\sigma_{ci}} \frac{\sin \phi_{m0}}{(1\text{-} \sin \phi_{m0})} \sigma_3^2 \text{ for } 0 \le \sigma_3 \le \sigma_{ci} \quad (3.24)$$

Where σ_{ci} and σ_{cj} are the UCS of the intact rock and rock mass respectively; ϕ_{m0} is the friction angle of the rock mass corresponding to very low confining pressure range ($\sigma_3 \rightarrow 0$) and can be obtained as:

$$\sin \phi_{m0} = \frac{(1 - \text{SRF}) + \frac{\sin \phi_{i0}}{1\text{-} \sin \phi_{i0}}}{(2\text{-SRF}) + \frac{\sin \phi_{i0}}{1\text{-} \sin \phi_{i0}}} \quad (3.25)$$

Where SRF = Strength Reduction Factor = σ_{cj}/σ_{ci}; ϕ_{i0} is friction angle for the intact rock and is obtained by conducting triaxial strength tests on intact rock specimens at low confining pressures ($\sigma_3 \rightarrow 0$).

If triaxial test data on intact rock is not available, the following non-linear form of the criterion may be used [22, 23]:

$$\sigma_1 = \text{A}(\sigma_3)^2 + (1 - 2\text{A}\sigma_{ci})\sigma_3 + \sigma_{cj}; \qquad \sigma_3 \le \sigma_{ci} \quad (3.26)$$

Where A is criterion parameter and may be estimated from the experimental value of σ_{ci}, using the following expressions:

$$\text{For average } \sigma_1 \qquad \text{A} = -1.23(\sigma_{ci})^{-0.77} \quad (3.27)$$

$$\text{For lower bound } \sigma_1 \quad \text{A} = -0.43(\sigma_{ci})^{-0.72} \quad (3.28)$$

For design purposes, the lower bound values of σ_1 are recommended to be used.

3.3.3 *Rock Mass Strength (σ_{cj})*

An important input to the strength criteria for rock masses is the UCS of rock mass σ_{cj}. The following approaches may be used to determine the UCS of the rock mass:

(i) Joint Factor concept, J_f
(ii) Rock quality designations, RQD
(iii) Rock mass quality, Q
(iv) Rock mass rating, RMR
(v) Modulus ratio concept (Strength reduction factor)

3.3.3.1 Joint Factor Concept

Ramamurthy and co-workers have defined a weakness coefficient that characterises the effect of fracturing on rocks and termed it Joint Factor [16, 18, 24–26]. The Joint Factor considers the combined effect of frequency, orientation, shear strength of joints, and is defined as:

$$J_f = \frac{J_n}{n\,r} \tag{3.29}$$

where, J_n = joint frequency, i.e., number of joints/metre; n is inclination parameter, depends on the inclination of sliding plane with respect to the major principal stress direction (Table 3.4); r is a parameter for joint strength; it is obtained from direct shear tests conducted along the joint surface at low normal stress levels and is given by:

$$r = \frac{\tau_j}{\sigma_{nj}} = \tan \phi_j \tag{3.30}$$

Where τ_j is the shear strength along the joint; σ_{nj} is the normal stress across the joint surface; and ϕ_j is the equivalent value of friction angle incorporating the effect of asperities [27]. The tests should be conducted at very low normal stress levels so that the initial roughness is reflected through this parameter. For cemented joints, the value of ϕ_j includes the effect of cohesion intercept also. In case the direct shear tests are not possible and the joint is tight, a rough estimate of ϕ_j may be obtained from Table 3.5 [27].

Table 3.4 Joint inclination parameter n [16]

Orientation of joint θ°	Inclination parameter n	Orientation of joint θ°	Inclination parameter n
0	1.00	50	0.071
10	0.814	60	0.046
20	0.634	70	0.105
30	0.465	80	0.460
40	0.306	90	0.810

θ = Angle between the normal to the joint plane and major principal stress direction

Table 3.5 Values of joint strength parameter, r for different values of σ_{ci} (After Ramamurthy [16, 27])

Uniaxial compressive strength of intact rock, σ_{ci} (MPa)	Joint strength parameter, r	Remarks
2.5	0.30	Fine grained micaceous to coarse grained
5.0	0.45	
15.0	0.60	
25.0	0.70	
45.0	0.80	
65.0	0.90	
100.0	1.0	

Table 3.6 Joint strength parameter, r for filled-up joints at residual stage (After Ramamurthy [16, 27])

Gouge material	Friction angle (ϕ_j)	$r = \tan \phi_j$
Gravelly sand	45°	1.00
Coarse sand	40°	0.84
Fine sand	35°	0.70
Silty sand	32°	0.62
Clayey sand	30°	0.58
Clay silt		
Clay – 25%	25°	0.47
Clay – 50%	15°	0.27
Clay – 75%	10°	0.18

Table 3.7 Empirical constant 'a' for estimating σ_{cj}

Failure mode	Coefficient a
Splitting/shearing	−0.0123
Sliding	−0.0180
Rotation	−0.0250

If the joints are filled with gouge material and have reached the residual shear strength, the value of r may be assigned from Table 3.6 [27].

The UCS of the rock mass is obtained as:

$$\sigma_{cj} = \sigma_{ci} \exp(a\ J_f) \tag{3.31}$$

where a is an empirical coefficient equal to −0.008.

Singh [25] and Singh et al. [26] suggested that the failure of the rock mass under uniaxial stress condition may occur due to various failure modes i.e. splitting, shearing, sliding and rotation. The values for different modes of failure are presented in Table 3.7. The failure mode may be decided as per guideline given below [25, 26]. If it is not possible to assess the failure mode, an average value of the empirical constant, 'a' may be taken equal to −0.017.

Let θ be the angle between the normal to the joint plane and the major principal stress direction:

(i) For $\theta = 0$–$10°$, the failure is likely to occur due to *splitting* of the intact material of blocks.
(ii) For $\theta = 10°$ to $\approx 0.8\ \phi_j$, the mode of failure shifts from splitting (at $\theta = 10°$) to sliding (at $\theta \approx 0.8\ \phi_j$).
(iii) For $\theta = 0.8\phi_j$ to 65°, the mode of failure is expected to be sliding only.
(iv) For $\theta = 65$–$75°$, the mode of failure shifts from sliding (at $\theta = 65°$) to rotation of blocks (at $\theta = 75°$).
(v) For $\theta = 75$–$85°$, the mass fails due to rotation of blocks only.
(vi) For $\theta = 85$–$90°$, the failure mode shifts from *rotation* at $\theta = 85°$ to *shearing* at $\theta = 90°$.

3.3.3.2 Rock Quality Designation, RQD

Zhang [28] has proposed the following correlation for obtaining UCS of rock mass as a function of RQD. It may be noted that joint attributes like frequency and surface roughness have not been given any consideration in this approach.

$$\frac{\sigma_{cj}}{\sigma_{ci}} = 10^{(0.013\,RQD - 1.34)} \tag{3.32}$$

3.3.3.3 Rock Mass Quality, Q

Based on extensive database, Singh et al. [29] have proposed correlations of rock mass strength, σ_{cj} with Q by analysing block shear tests in the field.

$$\sigma_{cj} = 0.38\gamma Q^{1/3}\text{MPa} \quad \text{for slopes} \tag{3.33}$$

$$\sigma_{cj} = 7\gamma Q^{1/3}\,\text{MPa} \quad \text{for tunnels} \tag{3.34}$$

Barton [9] modified the above equation for tunnels and suggested the expression:

$$\sigma_{cj} = 5\gamma\left(\frac{Q\sigma_{ci}}{100}\right)^{1/3}\text{MPa} \quad \text{for tunnels} \tag{3.35}$$

Where γ is the unit weight of rock mass in gm/cm^3; and Q is the Barton's rock mass quality index.

3.3.3.4 Rock Mass Rating, RMR

Rock Mass Rating (RMR) may be used to get the shear strength parameters c_m and ϕ_m from RMR [4–6] and the rock mass strength σ_{cj} may be obtained as:

$$\sigma_{cj} = \frac{2c_m \cos\phi_m}{1\text{-}\sin\phi_m} \tag{3.36}$$

Ramamurthy [19] has suggested that the shear strength parameters recommended by Bieniawski [4–6] appear to be on lower side resulting in very low values of σ_{cj}.

The other commonly used correlations with RMR are as follows:

(i) Kalamaras and Bieniawski [30]

$$\frac{\sigma_{cj}}{\sigma_{ci}} = \exp\left(\frac{\text{RMR} - 100}{24}\right) \tag{3.37}$$

(ii) Sheorey [31]

$$\frac{\sigma_{cj}}{\sigma_{ci}} = \exp\left(\frac{RMR - 100}{20}\right) \tag{3.38}$$

3.3.3.5 Strength Reduction Factor

In the opinion of this author the best estimates of rock mass strength, σ_{cj} can only be made in the field through large size field-testing in which the mass may be loaded upto failure to determine rock mass strength. It is, however, rarely feasible. An alternative will be to get the deformability properties of rock mass by stressing a limited area of the mass upto a certain stress level, and then relate the ultimate strength of the mass to the laboratory UCS of the rock material through a strength reduction factor, SRF. Singh and Rao [32] have shown that the Modulus Reduction Factor, MRF and Strength Reduction Factor, SRF are correlated with each other by the following expression approximately:

$$SRF = (MRF)^{0.63} \tag{3.39}$$

$$\Rightarrow \frac{\sigma_{cj}}{\sigma_{ci}} = \left(\frac{E_j}{E_i}\right)^{0.63} \tag{3.40}$$

Where SRF is the ratio of rock mass strength to the intact rock strength; MRF is the ratio of rock mass modulus to the intact rock modulus; σ_{cj} is the rock mass strength; σ_{ci} is the intact rock strength; E_j is the elastic modulus of rock mass; and E_i is the intact rock modulus available from laboratory tests and taken equal to the tangent modulus at stress level equal to 50% of the intact rock strength.

It is recommended that field deformability tests should invariably be conducted on project sites. The elastic modulus of rock mass, E_j may be obtained in the field by conducting uniaxial jacking tests [33] in drift excavated for the purpose. The test consists of stressing two parallel flat rock faces (usually the roof and invert) of a drift by means of a hydraulic jack [7]. The stress is generally applied in two or more cycles as shown in Fig. 3.5. The second cycle of the stress deformation curve is used for computing the field modulus as:

$$E_j = \frac{m(1 - \upsilon^2)P}{\sqrt{A}\ \delta_e} \tag{3.41}$$

where E_j is the elastic modulus of the rock mass in kg/cm^2; υ is the Poisson's ratio of the rock mass (= 0.3); P is the load in kg; δ_e is the elastic settlement in cm; A is the area of plate in cm^2; and m is an empirical constant (=0.96 for circular plate of 25 mm thickness).

The size of the drift should be sufficiently large as compared to the plate size so that there is little effect of confinement. The confinement may result in over prediction of the modulus values.

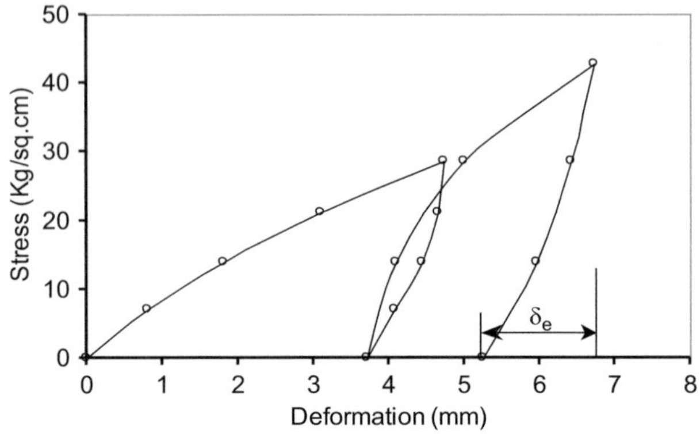

Fig. 3.5 Field modulus of elasticity [33]

A number of methods for assessing the rock mass strength, σ_{cj} have been discussed above. It is desirable that more than one method be used for assessing the rock mass strength and generating the failure envelopes. A range of values will thus be obtained and design values may be taken according to experience and confidence of the designer.

3.4 Concluding Remarks

Assessment of shear strength behaviour of jointed rocks and rock masses is a difficult task. At the time of failure, the strength may be mobilised along a dominating persistent discontinuity or through blocks of the rock mass. Accordingly, discontinuity shear strength or rock mass strength will govern the design. Some of the approaches available for obtaining the shear strength of an individual discontinuity or of a mass as a whole have been discussed in the present article. The strength behaviour is known to vary non-linearly with confinement and hence special emphasis has been given to non-linear strength criteria. In real life problems, heterogeneity and uncertainty are very common. It may be expected that shear strength obtained from different approaches will vary over a range. It is advisable parametric analysis be done to examine the behaviour of the structure for the range of values to gain more confidence in the design.

Acknowledgement The author gratefully acknowledges the contributions made by Dr. T. Ramamurthy, Professor (Retd.) and Dr. K.S. Rao Professor from IIT Delhi; Dr. Bhawani Singh, Professor (Retd.), Prof. M. N. Viladkar and Prof. N. K. Samadhiya from IIT Roorkee, Roorkee for their valuable technical inputs during the research presented in this paper. Some part of the research presented in this paper was conducted under the research projects (Project No. DST-209-CED, IIT Roorkee, 2005-08; and DST-697-CED, IIT Roorkee, 2013-14) sponsored by Department of Science and Technology (DST), New Delhi. The author sincerely puts on record the appreciation for the financial support from DST New Delhi, and the co-operation and encouragement from Dr. Bhoop Singh, Director NRDMS, DST, New Delhi, in carrying out research related to slope stability problems.

References

1. Patton FD (1966) Multiple modes of shear failure in rock. Ist Cong ISRM Lisbon 1:509–513
2. Ladanyi B, Archambault G (1972) Evaluation of shear strength of a jointed rock mass. 24th Int Geol Cong Sect 13D:249–270
3. Barton NR, Choubey V (1977) The shear strength of rock joints in theory and practice. Rock Mech 10(1–2):1–54
4. Bieniawski ZT (1973) Engineering classification of jointed rock masses. S Afr Inst Civ Eng 15 (12):335–344
5. Bieniawski ZT (1989) Engineering rock mass classifications. Wiley, New York
6. Bieniawski ZT (1993) Classification of rock masses for engineering: the RMR system and future trends. In: Rock testing and site characterization. Pergamon Press, Oxford, pp 553–573
7. Mehrotra VK (1992) Estimation of engineering parameters of rock mass. PhD thesis. University of Roorkee, Roorkee
8. Barton NR, Lien R, Lunde J (1974) Engineering classification of rock masses for the design of tunnel support. Rock Mech 6(4):189–239
9. Barton NR (2002) Some new Q–value correlations to assist in site characteristics and tunnel design. Int J Rock Mech Min Sci 39:185–216
10. Deere DU (1963) Technical description of rock cores for engineering purpose. Rock Mech Eng Geol 1:18–22
11. Hoek E, Brown ET (1980) Empirical strength criterion for rock masses. J Geotech Eng Div ASCE 106(GT9):1013–1035
12. Hoek E (2000) Practical rock engineering. http://www.rocscience.com/roc/Hoek/Hoeknotes2000.htm
13. Hoek E, Carranza-Torres C, Corkum B (2002) Hoek-Brown failure criterion – 2002 edition. NARMS-TAC Conf Tor 1:267–273
14. Hoek E, Brown ET (1997) Practical estimates or rock mass strength. Int J Rock Mech Min Sci 34(8):1165–1186
15. Marinos V, Marinos P, Hoek E (2005) The geological strength index: applications and limitations. Bull Eng Geol Environ 64:55–65
16. Ramamurthy T (1993) Strength and modulus response of anisotropic rocks. In: Compressive rock engineering principle, practice and projects, vol 11. Pergamon Press, Oxford, pp 313–329
17. Ramamurthy T (1994) Strength criterion for rocks with tensile strength. Ind Geotech Conf, Warangal, pp 411–414
18. Ramamurthy T, Arora VK (1994) Strength prediction for jointed rocks in confined and unconfined states. Int J Rock Mech Min Sci 13(1):9–22
19. Ramamurthy T (2014) Strength, modulus and stress-strain responses of rocks, Engineering in Rocks for Slopes, Foundations and Tunnels, vol 3. Prentice-Hall of India Pvt. Ltd, New Delhi, pp 93–137
20. Singh M, Singh B (2012) Modified Mohr–Coulomb criterion for non-linear triaxial and polyaxial strength of jointed rocks. Int J Rock Mech Min Sci 51:43–52
21. Barton N (1976) The shear strength of rock and rock joints. Int J Rock Mech Min Sci Geomech Abstr 13:255–279
22. Singh M, Singh B (2004) Critical state concept and a strength criterion for rocks. Asian rock mechanics symposium: contribution of rock mechanics to the new century, Kyoto, Japan, 3:877–880
23. Singh M, Rao KS (2005a) Bearing capacity of shallow foundations in anisotropic non Hoek-Brown rock masses. ASCE J Geotech Geo-environ Eng 131(8):1014–1023
24. Arora VK (1987) Strength and deformational behaviour of jointed rocks. PhD thesis, IIT Delhi
25. Singh M (1997) Engineering behaviour of jointed model materials. Ph.D. Thesis, IIT, New Delhi
26. Singh M, Rao KS, Ramamurthy T (2002) Strength and deformational behaviour of jointed rock mass. Rock Mech Rock Eng 35(1):45–64

27. Ramamurthy T (2001) Shear strength responses of some geological materials in triaxial compression. Int J Rock Mech Min Sci 38:683–697
28. Zhang L (2009) Estimating the strength of jointed rock masses. Int J Rock Mech Min Sci 43 ((4)):391–402
29. Singh B, Viladkar MN, Samadhiya NK et al (1997) Rock mass strength parameters mobilised in tunnels. Tunn Undergr Space Technol 12(1):47–54
30. Kalamaras GS, Bieniawski ZT (1993) A rock mass strength concept for coal seams. 12th Ground control in mining conference, Morgantown, pp 274–283
31. Sheorey PR (1997) Empirical rock failure criteria. Balkema, Rotterdam
32. Singh M, Rao KS (2005b) Empirical methods to estimate the strength of jointed rock masses. Eng Geol 77:127–137
33. IS:7317 (1974) Code of practice for uniaxial jacking test for modulus of deformation of rocks

Chapter 4
Rockfall: A Specific Case of Landslide

Tariq Siddique, S. P. Pradhan, and V. Vishal

Abstract Rockfall is a specific case of mass wasting that occurs frequently in mountainous regions and when it occurs along transportation corridor or near populated areas it can pose significant hazards. Rockfall is a freefall type of movement generally from steep cliffs or slopes. After initiation of rockfall, type of movement or trajectories of falling blocks largely depends upon certain factors like potential falling blocks, prevailing geometry and geomechanical properties of interacting surfaces. Various other natural and anthropogenic causative and triggering factors have been briefly described here. During preliminary investigation stage, vulnerable zones can be screened out by using Rockfall Hazard Rating schemes. A comprehensive review of major existing rockfall hazard rating system is presented. The delineated hazardous zones should be evaluated in detail using comprehensive site specific studies by modeling techniques which provide much better insight of the problem. Earlier rockfall studies were conducted using in-situ tests and physical modeling which includes lot of time, money and man-power. Later, the development of software for rockfall simulation reduced ambiguity, cost, time and expenditure. Such techniques are used worldwide extensively and have achieved immense popularity among researchers working on rockfall studies. Few such software have been discussed in this chapter. A summary on remedial and protection measures has been presented.

Keywords Rockfall · Rockfall hazard rating · Rockfall simulation · Landslides

T. Siddique
Department of Earth Sciences, IIT Roorkee, Roorkee, India

Department of Geology, Aligarh Muslim University, Aligarh, India

S. P. Pradhan (✉)
Department of Earth Sciences, IIT Roorkee, Roorkee, India
e-mail: sppradhanfes@iitr.ac.in

V. Vishal
Department of Earth Sciences, IIT Bombay, Mumbai, India
e-mail: v.vishal@iitb.ac.in

S. P. Pradhan et al. (eds.), *Landslides: Theory, Practice and Modelling*,
Advances in Natural and Technological Hazards Research 50,
https://doi.org/10.1007/978-3-319-77377-3_4

4.1 Introduction

Slope failures depend upon geometry of slope, material and triggering factors involved. Most of the popular and widely used classification schemes of mass movements include fall, topple, slide, spread and flow. According to Cruden and Varnes [1] the movement of rocks on a surface can occur in three ways: rockslide, rock toppling and rock fall. In rockslides, motion is resisted by shear forces while maintaining contact of the rock mass with the slope surface. Rock toppling is triggered by the rotation of a large block about its base. Rockfall occurs when a rock or boulder becomes dislodged from an exposed face and moves downwards by a combination of rolling, sliding, free fall or bouncing under the influence of gravity [1]. Detached blocks of varying sizes fall freely under the influence of gravity from sub-vertical or steep cliffs. Rockfall occur frequently in mountainous region having steep slopes, quarries and mines forming flight trajectories that are hazardous and cause large accidents and fatalities [2–4]. Transportation corridors such as highways and railways are vulnerable to rockfall whenever they cut across mountains [5–8]. Therefore, it is very significant to investigate the areas experiencing frequent rockfall especially in precarious slopes which are continuously experiencing dynamic loading.

4.2 Rockfall

Rockfalls is a natural process often associated with slope erosion. Rock fragment detached from a cliff or bedrock by sliding, toppling or falling that follows a vertical or nearly vertical trajectory and proceeds down slope by bouncing and flying [9]. Rockfall may be defined as fast gravitational movement of rock boulders by rolling, tumbling or sliding down a hill slope [10]. Depending upon mean slope gradient, detached blocks can have different modes of motion viz. freefall, bouncing, rolling, and sliding (Fig. 4.1) depending upon mean slope gradient [11, 12]. Piteau and Clayton [13] and Giani [14] had discussed transition between bouncing and rolling. However, Bozzolo et al. [15] illustrated transition between rolling and

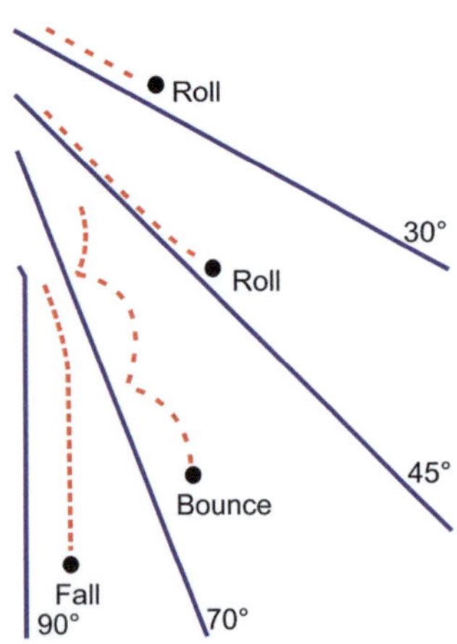

Fig. 4.1 General modes of motion of rocks during their descent on slopes related to the mean slope gradients [11]

sliding. The occurrence of these different modes of block movement is strongly influenced by the slope angle [11].

In steep slopes, free fall is mostly observed whereas for intermediate slopes, rockfall propagation is a succession of free falls and rebounds while for gentle slopes, the prevalent motion types are rolling or sliding [16]. However, in a practical scenario falling, bouncing, sliding and rolling motions are closely linked with each other and sometimes it is very difficult to separate them or to place a boundary in between. Any free falling rock block would involve bouncing to bridge the intervals. Depending upon the shape of falling blocks, topography, catchment features etc. the type of motion may change from falling motion to sliding and rolling. In rockfall the block size may range from small pebbles to massive blocks weighing possibly hundreds of tonnes [17]. However, there is no accepted size of rock characteristic of a rockfall [18]. Rapp [19] and Whalley [15] classified rockfall on the basis of block size or volume. Extremely large rocks can cause catastrophic damage to property, severe accidents, inconvenience along transportation routes and even fatalities. Even a rockfall comprising of merely small pebbles could act as a significant hazard for traffic and life [20].

4.3 Causes and Triggering Factors of Rockfall

Generally rockfalls are generated by factors like climatic, geological and biological events that cause disequilibrium in the forces acting on the rock mass. These events may include increase in pore pressure by infiltration of water; erosion due to heavy precipitation; freeze and thaw processes; chemical disintegration of rocks; generation of discontinuities in the rock mass due to any tectonic activity; widening of pre-existing cracks or generation of discontinuities or fractures due to growing roots of plants, burrowing of animals or blasting. Rockfalls may be initiated by meteorological factors such as rainfall, biological factors like trees and animals, and by vibrations induced by blasting and earthquakes. Causes of rockfall can be classified as structural and environmental factors. In case of the structural factors, a highly fractured discontinuous loose rock mass must exist on the slope and that should be sufficiently fissured in order to produce potentially unstable blocks. Secondly, the slope on which the blocks are present must be steep enough in order to promote instability and falling of blocks [22]. Environmental factors include triggering forces that are responsible for influencing instability of the slopes. Both, chemical and physical weathering are among the major triggering factors causing rock fall. Percolation of water along discontinuities causes erosion, it also reduces the shear strength of rock mass to a great extent posing instability to the slope. Differential weathering of rock mass can cause large unsupported overhangs by removal of only weak portion of rock mass. Vibrations induced due to blasting and seismicity can act as a potential triggering factor for slope instability and rockfall. Once a rockfall has been initiated, its behavior is affected majorly by slope geometry, slope roughness and lateral variation. According to Pfeiffer and Bowen [23] surface irregularities can change the angle at which a rock hits the surface and are therefore significant in determining the bounce height and trajectories of the falling blocks.

4.4 Rockfall Assessment

Assessment of rockmass by rating system is the first step towards screening of critical slopes or zones for detailed site specific evaluation. Such approaches are very significant and cost effective particularly during initial stages of the investigation. Over many decades, abundant qualitative methods have been introduced. Rockfall hazard is the probability of occurrence of rockfall event of given magnitude or intensity over predefined time period and within a given area [24–26]. To reduce inevitable slope failures and to gain safe design, proper investigation and slope characterization is an efficient method [27]. The concept of rockfall hazard rating was first introduced for Canadian Pacific railways by Brawner and Wyllie [7]. This qualitative scheme categorised rockfall from class A to F, in which slopes falling in class A are the most hazardous. This hazard rating system served as the foundation for rockfall hazard assessment. Since then several rockfall hazard rating system have been proposed (Table 4.1). Selection of different methods available for hazard evaluation depend upon the depth of understanding of the individuals by considering the merits and demerits of each method.

Table 4.1 Major existing rockfall hazard rating systems

Rating system	Authors with their year of publication	Place
Highway Rockfall Hazard Rating System (HRHRS)	Wyllie (1987) [28]	Oregon (USA)
Rockfall Hazard Rating System (RHRS)	Pierson et al. (1990) [29]	Oregon (USA)
Ontario Rockfall Hazard Rating System (RHRON)	Franklin and Senior (1997) [30]	Ontario (Canada)
Modified Rockfall Hazard Rating System (mRHRS)	Budetta (2004) [31]	Italy
Falling Rock Hazard Index (FRHI)	Singh (2004) [32]	–
Missouri Rockfall Hazard Rating System (MORFH-RS)	Maerz et al. (2005) [33]	Missouri (USA)
Rockfall Hazard Rating Matrix for Ohio (RHRM)	Woodard et al. (2005) [34]	Ohio (USA)
Rockfall Hazard Classification and Mitigation System (RHCMS)	Pierson et al. (2005) [35]	Montana (USA)
UDOT Rockfall Hazard Rating System (UDOT-RHRS)	Pack et al. (2006) [36]	Utah (USA)
Tennessee Rockfall Hazard Rating System (TRHRS)	Mauldon et al. (2007) [37]	Tennessee (USA)
Colorado Rockfall Hazard Rating System (CRHRS)	Russell et al. (2008) [38]	Colorado (USA)
Modified Colorado Rockfall Hazard Rating System (modified CRHRS)	Santi et al. (2009) [39]	Colorado (USA)
New Rockfall Rating System (NRSS)	Saroglou et al. (2012) [40]	Peloponnese (Greece)
Rockfall Hazard Rating System of India (RHRSI)	Ansari et al. (2013) [21]	India

4.5 Parameters in Rockfall Rating System

In mountainous regions several geological, geometrical, geotechnical and climatic factors contribute for the initiation of major rockfall events. Worldwide acceptable and majorly used rockfall hazard rating systems have been illustrated in Table 4.1. These rating schemes have some minor differences in consideration of influencing, triggering and controlling parameters and their scores. CRHRS by Russell et al. [38] is considered to be the most comprehensive among all. Santi et al. [39] published extension of CRHRS as modified CRHRS. The parameters included in modified CRHRS have been discussed here. The parameters considered are broadly categorized as: slope conditions, climatic conditions, geologic and traffic conditions.

4.5.1 Slope Conditions

This category includes parameters related to dimension and characteristics of slope.

4.5.1.1 Slope Height

It represents the vertical height of the slope which is measured from the road to the highest point of potential rockfall source. Blocks falling from higher elevation have more potential energy than the lower, thus higher the falling potential elevation much hazardous the slope. In field, the vertical slope height can be obtained by Eq. 4.1:

$$\text{Total slope height} = \frac{(X)\sin\alpha\ \sin\beta}{\sin(\alpha-\beta)} + \text{H.I} \tag{4.1}$$

Where,

X: Horizontal distance between angle measurements
H.I: Height of the instrument from the ground
α and β: Angle of inclination from horizontal at two different locations
All the components have been illustrated in Fig. 4.2.

4.5.1.2 Rockfall Frequency

This includes the frequency of rockfall occurring at a particular site within a specific time period. Such information are provided by Department of Transportation in the US context.

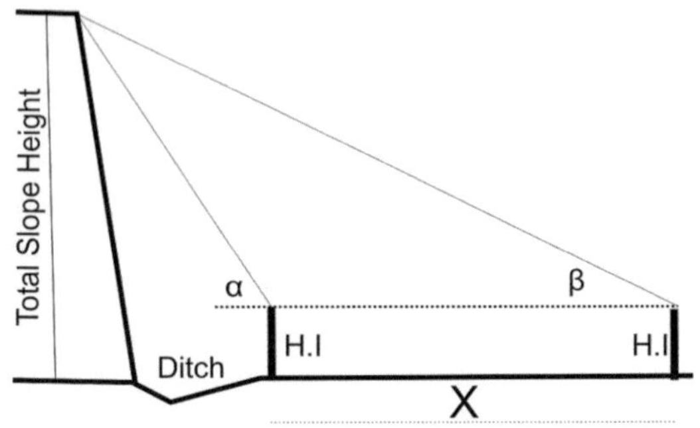

Fig. 4.2 Slope height measurement

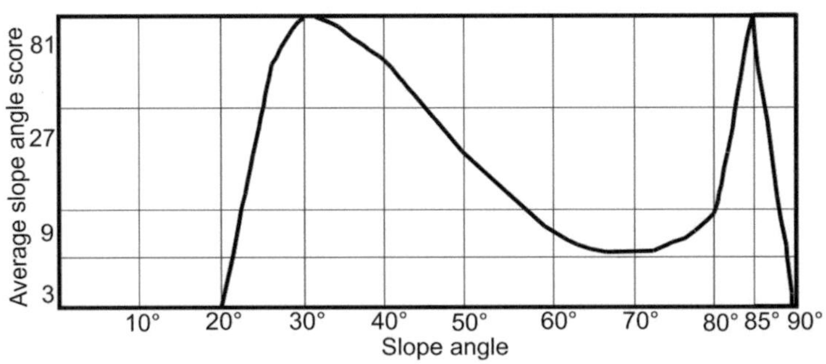

Fig. 4.3 Slope angle score [33]

4.5.1.3 Average Slope Angle

Slopes having higher angle have higher degree of slope instability. But a recent research conducted by Maerz et al. [33] revealed that slopes having angle near to 30° and 85° tend to give higher runout distance during rockfall (Fig. 4.3). Therefore, scores for average slope angles will not increase linearly with increasing angle. Such modifications in consideration of average slope angle have been incorporated in RHRS for Indian rockmass also [21].

4.5.1.4 Launching Features

This is a qualitative parameter which depends upon the smoothness of slope profile and provides significant clue for rockfall occurrence. Santi et al. [39] described different categories of launching features:

None: Relatively smooth slope with little or no topographic variation along slope profile.

Minor: Small topographic variations that can cause launching of boulders, such as presence of small ridges and benches extending <0.6 m from the slope surface.

Many: Several topographic variations that can cause launching of boulders, such as presence of ridges or benches from 0.6 to 1.8 m from the slope surface.

Major: Highly irregular slope profile with large variations in topography, benches and ridges >1.8 m.

4.5.1.5 Ditch Catchment

It can be defined as the effectiveness of ditch to restrict prevailing rockfall to reach the roadway, railway or public places. Ditch catchment depends upon several factors:

(a) Height and angle of slope
(b) Height, width and depth of ditch
(c) Probable size and quantity of potential falling blocks
(d) Launching features

This is also subjective parameter and largely depends upon the perception and experience of investigator. However, Ritchie [11] had given a formula (Eq. 4.2) to measure the effectiveness of ditch and proposed a ditch design chart (Fig. 4.4).

$$\text{Ditch dimension effectiveness} = 100 \times \frac{(\text{Da} + \text{Wa})}{(\text{Dr} + \text{Wr})} \tag{4.2}$$

Where,

Da is actual depth of the ditch
Wa is actual width of the ditch
Dr. is Ritchie design depth based on slope height and angle (from Fig. 4.4)
Wr is Ritchie design width based on slope height and angle (from Fig. 4.4)

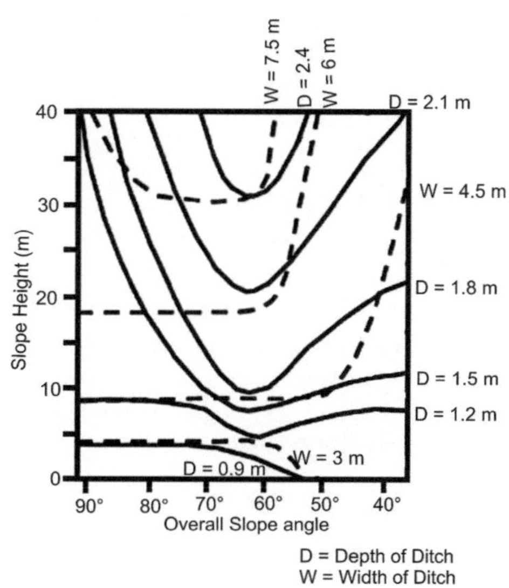

Fig. 4.4 Ditch design criteria [11]

4.5.2 *Climatic Conditions*

Climatic conditions include precipitation, seepage, exposure to sunlight, freeze and thaw cycles.

4.5.2.1 Annual Precipitation

Annual precipitation specifies the actual amount of rainfall that occurred round the year. It can be defined in terms of low medium and high but to remove the subjectivity of the parameter Pierson et al. [35] proposed an equation to calculate the exact score for annual precipitation (Eq. 4.3).

$$\text{Exact score for annual precipitation} = 3^{\left(\frac{\text{Annual Precipitation in mm}}{254}\right)} \tag{4.3}$$

4.5.2.2 Annual Freeze/Thaw Cycles

Freeze and thaw cycles widen pre existing discontinuities and also generate secondary discontinuities within the rock mass. Freezing cycles can be defined as the annual average number of days when the daily temperature fluctuates above and below freezing point [41]. Accordingly the site can be scored.

4.5.2.3 Seepage/Water

Presence of water on slope adversely affects the inherent condition of rock mass by reducing its shear strength. It also increase pore water pressure and internal erosion. Seepage or presence of water on the slope is rated qualitatively in which the minimum score is for dry condition and maximum for flowing.

4.5.2.4 Slope Aspect

It has been evidenced that slope facing towards sunlight experiences more freeze and thaw cycles [42–44]. In contrast, slopes that are faced in such a way that they remain under shade for significant time period during a day experience least temperature variations. So, the rating of slope aspect varies as per the location of the country. In the Indian context, the slope aspect scores given in RHRS for Indian rock mass given by Ansari et al. [21] can be used. The slope aspect may be affected by other climatic factors, presence of vegetation, rate of evaporation, distribution of soil and bedrock [39].

4.5.3 Geologic Conditions

Geologic conditions include type of rocks, some parameters related to condition and orientation of discontinuities and block size. According to Santi et al. [39] three types of geological materials have been considered:

(a) Sedimentary rocks, where undercutting and differential erosion tend to control rockfall. It includes degree of undercutting, jar slake and degree of interbedding.
(b) Crystalline rocks, where the rock mass in homogeneity and fractures tend to control rockfall. This includes a variety of igneous and metamorphic rocks which have been scored on the basis of homogeneity, small faults, veins, schistosity, fabric and shear zones. Degree of overhang has been considered in this category. Overhang due to differential erosion leads to undercutting in crystalline rock as weaker schist or shear zones erode into the slope. Many rockfall hazard systems, self explanatory definitions have been used for different grades of weathering. Note that here weathering of intact rock is considered, and not the weathering grade of discontinuities which is rated in another category.
(c) Block-in-matrix materials (Colluvium, glacial till, debris flow deposits etc.), where erosion of matrix and subsequent raveling of larger blocks tends to control rockfall.

Apart from rock types, there are some general characteristics like block size or volume, number of discontinuity sets, persistence, aperture, weathering conditions, block shape and vegetation are considered in distinct rockfall hazard rating schemes. Exact scores for block size and block volume can be calculated by following Eqs. 4.4 and 4.5 respectively [21]. The block size/volume is an important parameter to assess the degree of damage, to estimate kinetic energy and the maximum runout distance of the potential falling blocks. Larger blocks have much higher kinetic energy and runout distance as compared to the smaller ones. During their journey they also act as a triggering agents for rockfall initiation of smaller blocks.

$$\text{Exact score for block size} = 3^{\left(\frac{\text{block size in meters}}{0.3}\right)} \tag{4.4}$$

$$\text{Excat score for block volume} = 3^{\left(\frac{\text{block volume in meters}}{0.3}\right)} \tag{4.5}$$

Block shape gives an idea about the mechanism of movement of potential falling blocks. According to Vandewater et al. [45] hazard associated with rounded blocks are greater as compared to tabular blocks. The number of discontinuity set present within the rock mass greatly influence the infiltration, seepage, frost wedging, chemical alteration, physical weathering, slope instability etc.. Persistence means the length of discontinuity present within the slope. Higher persistence will tend to impart instability to the rock mass. Orientation of discontinuities is one of the most crucial parameters for quick assessment for slope instability. Daylight condition of discontinuities gives rise to potential rockfall where dip direction of discontinuity is parallel or near parallel to the slope inclination, provided the dip amount of discontinuity is less than the slope angle. Apart from the daylight condition, a comprehensive idea about the orientation factors can be judged by assessing the probability of planar, toppling and wedge failures to occur by means of kinematic tool. In debris/soil slopes, presence of vegetation over slope enhances stability. However, it has been noticed that in case of hard rock slopes growing roots of big trees may increase the aperture of discontinuities and consequently slope instability may become much pronounced in such conditions.

4.5.4 *Traffic Conditions*

The slope, climatic and geologic factors are related to some prime factors related to initiation and controlling of rockfalls. Scores obtained by above are to be adjusted with the risk that may be produced due to likelihood of accidents or injuries. Traffic factors include the following parameters that are related with the traffic conditions of associated transportation route:

4.5.4.1 Sight Distance

Sight distance can be defined by the following formula (Eq. 4.6).

$$\text{Sight distance} = 100 \times \left(\frac{\text{Actual sight distance}}{\text{Required decision sight distance}}\right)\% \tag{4.6}$$

According to Pierson and Van Vickle [46], the actual sight distance can be defined as the minimum distance on a roadway when an object of 15 cm is placed on the edge of the road is visible to a driver. Sight distance depends upon the speed of the vehicle (Table 4.2). Generally, speed is estimated by the posted speed limit on the associated road. Exact score for decision sight distance can be calculated by Eq. 4.7.

$$\text{Exact score for decision sight distance} = 3^{\left(\frac{120-\%\text{decision sight distance}}{20}\right)} \tag{4.7}$$

Table 4.2 Required decision sight distance based on posted speed limits [46]

Speed limit (Kmph)	Required decision sight distance (m)
40	114
48	137
56	160
64	183
72	206
80	229
89	267
97	305
105	1320

4.5.4.2 Average Vehicle Risk

Average vehicle risk gives an idea about the time for which a vehicle will remain in rockfall prone zone. It is calculated using the (Eq. 4.8) and the exact score for average vehicle risk can be obtained by Eq. 4.9.

$$\mathrm{AVR} = 100 \times \mathrm{ADT}\left(\frac{\mathrm{cars}}{\mathrm{day}}\right) \times \frac{\mathrm{SL\ (Kms)}}{24\left(\frac{\mathrm{hrs}}{\mathrm{day}}\right)} \times \mathrm{PSL}\left(\frac{\mathrm{Km}}{\mathrm{hr}}\right)\% \tag{4.8}$$

where,

AVR: Average vehicle risk
ADT: Average daily traffic
SL: Slope length
PSL: Posted speed limit

$$Exact\ score\ for\ AVR = 3^{(\%AVR)} \tag{4.9}$$

4.5.4.3 Number of Accidents

The number of accidents caused due to rockfall and landslides along highways is recorded by the transportation officials of the state or country and accordingly the scores can be given to the cut slopes.

4.6 Rockfall Simulation

Earlier research on rockfall was carried out by in-situ tests and physical modeling techniques which involve high costs. With passage of time and better understanding of rockfall behavior through physical and in-situ tests, computer simulation programs were developed and tested to compute trajectories of rockfall. Such advancements in rockfall studies are widely acceptable and had gained immense popularity all over the globe and have replaced the expensive and statistically ambiguous rockfall tests. Rockfall modeling includes simulation of falling rock trajectories to estimate velocity, bounce height, maximum runout distance etc. Numerical simulation of rockfall trajectories is based on Newtonian mechanics and provides reasonable insight of rockfall trajectories, velocities, and runout distances [47]. Over many decades, a varietiy of computer programs have been developed by incorporating different assumptions. Some major contributions have been illustrated in Table 4.3.

Azzoni et al. [63] proposed a mathematical model based on rigid body mechanics, codified for computer use called as CADMA that predicts fall trajectories and the relevant outputs like kinetic energy, potential energy, bounce height and runout

Table 4.3 Major existing rockfall models

Model/program	Author(s) with year of publication	Spatial dimensions	Approach
N.N.	Ritchie (1963) [11]	2-D	LM
Discrete Element Method	Cundall (1971) [49]	2-D	RB
Computer Rockfall Model	Piteau and Clayton (1976) [13]	2-D	LM
N.N.	Azimi et al. (1982) [50]	2-D	LM
N.N.	Falcetta (1985) [51]	2-D	RB
Rotolamento Salto Massi	Bassato et al. (1985) [52]	2-D	LM
ROCKSIM	Wu (1985) [53]	2-D	LM
SASS	Bozzolo and Pamini (1986) [54]	2-D	H
EBOUL-LMR	Descoeudres and Zimmemann (1987) [55], Labiouse et al. (2001) [56]	3-D	RB
PROPAG/CETE Lyon	Rochet (1987) [57]	2-D	LM
N.N.	Hungr and Evans (1988) [58]	2-D	LM
CRSP (4.0)	Pfeiffer and Bowen (1989) [23], Jones et al. [59]	2-D	H
N.N.	Van Dijke and Van Westen [60]	2-D	LM
N.N.	Kobayashi et al. (1990) [61]	2-D	RB
Rotomap	Scioldo (1991) [62]	3-D	LM
CADMA	Azzoni et al. (1995) [63]	2-D	H
Rockfall (Dr. Spang)	Spang and Sonser (1995) [64]	2-D	RB
ROFMOD 4.1	Zinggeler et al. (1990) [65], Krummenacher and Keusen (1996) [66]	2-D	H
3-D-GEOTEST-Zinggler	Krummenacher et al. (2008) [67]	3-D	H
RocFall	Stevens (1998) [68]	2-D	LM
Sturzgeschwindigkeit	Meissl (1998) [69]	2-D	LM
Mobyrock	Paronuzzi and Artini (1999) [70]	2-D	LM
STONE	Guzzetti et al. (2002) [48]	3-D	LM
STAR3-D	Dimnet (2002) [71], Le Hir et al. (2006) [72]	3-D	RB
Rocky3	Dorren and Seijimonsbergen (2003) [73]	2.5-D	H
HY-STONE	Crosta et al. (2004) [74], Frattini et al. (2008) [75], Agliardi et al. (2009) [76]	3-D	H
RockyFor	Dorren et al. (2004) [77], Dorren et al. (2006) [78], Bourrier et al. (2009) [79]	3-D	H
RAMMS::Rockfall	Christen et al. (2007) [80]	3-D	RB
RockFall Analyst	Lan et al. (2007) [81]	3-D	LM
PICUS-ROCKnROLL	Woltjer et al. (2008) [82], Rammer et al. (2007) [83]	3-D	LM
CONEFALL	Jaboyedoff and Labiouse (2011) [84]	3-D	–
Trajec 3D	Basson (2012) [85]	3-D	RB

Modified from Guzzetti et al. [48] and Volkwein et al. [16]
LM Lumped mass, *RB* Rigid body, *H* Hybrid

distance of falling blocks. Many rock fall simulators have been proposed for trajectory analyses. These different models can be classified into rigid body, lumped-mass and hybrid methods [14, 58]. Rigid body methods consider the falling block as body with its own shape and volume and solve fundamental equations of dynamics and also consider all possible modes of block movement [49, 51, 55, 63]. Lumped-mass methods consider that the block have either no mass or mass concentrated at one point and they also do not consider shape and movement of blocks [48, 50, 58, 68]. The hybrid methods are advantageous as they are faster and can easily simulate free flight by considering the geometry and geomechanical characteristics of slope and falling block to model the impact [23, 54, 59, 61, 77, 86]. According to Dattola et al. [87], in lumped mass, boulders are described by material point and impact by means of translational restitution coefficients. The rigorous methods consider actual shape and dimensions of falling blocks and the impact with the basal layer by means of moment balance equation and contact laws. It makes this approach much accurate however it is time consuming. In the hybrid approach, the geometry of boulder is taken into account by means of constitutive rules. This method describes the contact by non-linear, coupled and irreversible relationships.

The Mechanical properties of falling blocks and interacting slope surface also plays significant role in guiding trajectories of falling blocks. The degree of rockfall depends upon many factors such as geometry of the slope, type of bedrock or slope surface, mechanical properties of rock, physical and chemical weathering [88–90]. Freeze and thaw cycles of water along discontinuities are very common process that facilitates rockfall [91–94]. Coupling of topographical, geological, climatological, time and anthropogenic factors largely controls the occurrence and intensity of rockfall [95]. Azzoni and Freitas [96] highlighted some aspects of rockfall phenomena and also discussed methods for determination of parameters relevant to the prediction of rockfall trajectories. Some important parameters that guide rockfall trajectories are:

4.6.1 Coefficient of Restitution

Coefficient of restitution (COR) is the ratio of final to initial velocity of two objects after and before collision (Eq. 4.10). Usually it ranges from 0 to 1 and would be 1 for perfectly elastic and 0 for perfectly inelastic collision.

$$\text{COR} = \frac{\text{Relative velocity after collision}}{\text{Relative velocity before collision}} \tag{4.10}$$

Usually the coefficient of restitution is always less than one because the initial kinetic energy is being lost during rotational kinetic energy, plastic deformation and heat. It can be 1, if the object gains energy during collision.

4.6.2 Coefficient of Rolling Friction

It can be defined as the tangent of the angle of the slope at which the block can be considered to move with steady velocity [22].

4.6.3 Slope Roughness

Irregularities on the slope face plays an important role in significant variation in slope angle. Differential weathering and small failures along slope cause such variations which became very important for future rockfalls. Slope roughness can be defined as the deviation from the average slope angle.

Majority of researchers have conducted 2D, 3D rockfall simulation and rockfall hazard rating systems for rockfall investigation and assessment [31, 33, 95–101]. Rockfall studies have also been conducted by stereoscopic oblique aerial photography [102], geomatics and airborne laser scanning [103], gigapixel photography [104], terrestial laser scanning technique [105], LIDAR technique [106], multisensor remote sensing [107]. Volkwein and Klette [108] presented an in-situ device Local Positioning System (LPS) to validate and calibrate rockfall simulation software. In this method, an assembly of sensors is installed in falling blocks which record 3D accelerations and rotational velocities. However, many researchers, practitioners, professionals have widely used computer programs for rockfall studies. There are certain pitfalls and limitations raised due to site conditions and dynamic behavior of falling blocks. Site conditions may include variation in potential falling block at a particular slope, correct perception and identification of most hazardous zone etc. Behavior of falling blocks may change due to unpredictable falling trajectories or path posed due to local variations in topography and geology of the slope. Uncertainties may arise if fallen blocks got broken during impact before bouncing. Despite such problems rockfall modeling technique is the most efficient, economic and reliable for rockfall risk assessment.

4.7 Remedial Measures or Protection from Rockfall

Remedial measures for rockfall can be taken by stabilizing the slope and/or by protection from falling blocks. Both are technically different but in both the cases, the prime aim is to reduce damage to transportation routes causing inconvenience during transportation, loss of property and life. Stabilization involves preventing the movement of rock material from its initial place while protective measures are used to deal with rocks that are already in motion [109]. Stabilization of slopes can be done by rock mass reinforcement techniques, by trim blasting of overhangs and complete removal of unstable material. It is important that the appropriate method is used for the particular conditions at each site. Another effective technique to minimize the hazard of rockfall is to allow the rockfall to occur and then control the distance and direction of travel [110]. Dorren and Berger [111] suggested that protective effect of forests against rockfall cannot be neglected in risk management as considerable slope length and diameter of tree can provide sufficient resistance against falling blocks. In mountainous region, trees on the slope act as natural barrier that can prevent snow avalanches and rockfall [112, 113]. Extensive studies have been conducted to evaluate the capacity of trees to dissipate energy during rock fall [73, 114]. Dorren and Berger [115] determined a relationship of the amount of energy dissipated by tree, horizontal distance between the impact centre and vertical central axis of the tree during rockfall. Cost-effective means of controlling rockfall, is to catch falling blocks within the ditch. Sometimes, ditch catchment measure becomes ineffective as it requires enough space for its construction. Efficient and economic control over rockfall can be done by installing barriers made of concrete, gabion or any earth materials. Such barriers can be constructed either to enhance the performance of ditches or to form desired catchment areas. When barriers are used in combination with ditch, they become very efficient in controlling the falling blocks. Ghoussoub et al. [116] conducted experimental and numerical studies and developed a model i.e. curtain to determine technical parameters and mechanical properties of barriers to be used for rockfall protection. Martin [117] summarized various rockfall stabilization, protection and warning methods and rockfall monitoring systems as follows:

4.7.1 Stabilization Methods

1. Excavation
2. Scaling
3. Trimming
4. Groundwater control and drainage
5. Rock mass reinforcement

    5.1 Shotcrete and mortar
    5.2 Dental treatment
    5.3 Rock bolt, dowels and anchors
    5.4 Buttresses and bulk heads
    5.5 Retaining wall and tie back wall
    5.6 Anchored beam and strapping
    5.7 Beam and cable walls
    5.8 Cable nets, lashing and chains

4.7.2 Protection Methods

1. Relocation, tunnels and sheds
2. Interception and shaped ditches
3. Interception and shaped berms
4. Catch walls
5. Draped and pinned mesh
6. Catch fences and catch nets

4.7.3 Warning Methods

1. Patrols and signs
2. Electric fences and wires
3. Warning lights and sirens

4.7.4 Monitoring Systems

1. Precise surveys extensometers, inclinometers, tilt meters, load cells system in combination with protection

References

1. Cruden DM, Varnes DJ (1996) Landslide types and processes, Special Report, Transportation Research Board, Nat Acad Sci 247:36–75
2. Heim A (1932) Bergsturz und Menschenleben. Fretz und Wasmuth, Zurich
3. Hoek E, Bray JW (1981) Rock slope engineering, Revised, 3rd edn. The Institution of Mining and Metallurgy, London
4. Crosta GB, Agliardi F (2003) A methodology for physically based rockfall hazard assessment. Nat Hazards Earth Syst Sci 3(5):407–422
5. Whalley WB (1984) Rockfalls. In: Brunsden D, Prior DB (eds) Slope instability. Wiley, Chichester, pp 217–256
6. Bunce CM, Cruden DM, Morgenstern NR (1997) Assessment of the hazard from rock fall on a highway. Can Geotech J 34:344–356
7. Brawner CO, Wyllie D (1976) Rock slope stability on railway projects. Am Railway Eng Assoc 656:449–474
8. Hungr O, Evans S, Hazzard J (1999) Magnitude and frequency of rockfalls and rock slides along the main transportation corridors of south-western British Columbia. Can Geotech J 36:224–238
9. Varnes DJ (1978) Slope movement types and processes. In: Schuster RL, Krizek RJ (eds) Landslides: analysis and control. Special Report 176. Transportation and Road Research Board, National Academy of Science, Washington, DC, pp 11–33
10. Selby MJ (1995) Hillslope material and processes. Oxford University Press, Oxford, 359 p
11. Ritchie AM (1963) Evaluation of rockfall and its control. Highw Res Rec 17:13–28

12. Descoeudres F (1997) Aspects géomécaniques des instabilités de falaisesrocheuseset des chutes de blocs. Soc Suisse Mécan Sols Roch 135:3–11
13. Piteau DR, Clayton R (1977) Discussion of paper "Computerized design of rock slopes using interactive graphics for the input and output of geometrical data". In: Design Methods in Rock Mechanics, Minneapolis. In Proceedings of the 16th symposium on rock mechanics, American Society of Civil Engineers, New York, 6263
14. Giani GP (1992) Rock slope stability analysis. AA Balkema, Rotterdam
15. Bozzolo D, Pamini R, Hutter K (1988) Rockfall analysis – a mathematical model and its test with field data, In Proc. 5th Int. symposium on landslides, Lausanne 1:555–563
16. Volkwein A, Schellenberg K, Labiouse V, Agliardi F, Berger F, Bourrier F, Dorren LKA, Gerber W, Jaboyedoff M (2011) Rockfall characterisation and structural protection – a review. Nat Hazards Earth Syst Sci 11:2617–2651
17. Spang RM (1987) Protection against rockfalls-stepchild in the design of rock slopes. In proceedings of the 6th International Congress on Rock Mechanics, Montreal: 551–557
18. Douglas GR (1980) Magnitude Frequency study of rockfall in Co. Antrim, N Ireland. Earth Surf Process 5:123–129
19. Rapp A (1960) Recent development in the mountain slopes in Karkevagge and surroundings, Northern Scandinavia. Geogr Ann 42:1–158
20. Bjerrum L, Jorstad F (1968) Stability of rock slopes in Norway, publication 79. Norwegian Geotechnical Institute, Oslo, pp 1–11
21. Ansari MK, Ahmad M, Singh R, Singh TN (2013) Rockfall hazard rating system for Indian rockmass. Int J Earth Sci Eng 6(1):18–27
22. Statham I (1979) A simple dynamic model of rockfall: some theoretical principles and model and field experiments, In: ISMES: international colloquium on physical and geomechanical models, Bergamo, pp 237–258
23. Pfeiffer T, Bowen T (1989) Computer simulation of rockfalls. Bull Assoc Eng Geol 26:135–146
24. Fell R (1994) Landslide risk assessment and acceptable risk. Can Geotech J 31(2):261–272
25. Fell R, Hartford D (1997) Landslide risk management. In: Cruden DM, Fell R (eds) Landslide risk assessment, Balkema, Rotterdam, pp 51–109.
26. Jaboyedoff M, Baillifard F, Hantz D, Heidenreich B, Mazzoccola D (2001) Terminologie. In: Carere, Ratto, Zanolini (eds) Prévention des mouvements de versants et des instabilites de falaises, pp 48–57.
27. Siddique T, Alam MM, Mondal MEA, Vishal V (2015) Slope mass rating and kinematic analysis of slopes along the national highway-58 near Jonk, Rishikesh, India. J Rock Mech Geotech Eng 7(5):600–606
28. Wyllie DC (1987) Rock slope inventory system, In: Proceedings of the Federal Highway Administration Rockfall Mitigation Seminar FHWA Region 10
29. Pierson LA, Davis SA, Van Vickle R (1990) Rockfall hazard rating system –implementation manual, Federal Highway Administration (FHWA), Report FHWA-OR-EG 90–01, FHWA, US Department of Transportation
30. Franklin JA, Senior SA (1997) The Ontario rockfall hazard rating system. In: Proceedings of the conference on engineering geology and environment, Athens, pp 647–658
31. Budetta P (2004) Assessment of rockfall risk along roads. Nat Hazards Earth Syst Sci 4:71–81
32. Singh A (2004) FRHI: a system to evaluate and mitigate rockfall hazard in stable rock excavations. J Inst Eng 85:62–75
33. Maerz NH, Youssef A, Fennessey TW (2005) New risk- consequence rockfall hazard rating system for Missouri highways using digital image analysis. Environ Eng Geosci 11:229–249
34. Woodard MJ, Baker MJ, Shakoor A (2005) Development of a rock fall hazard rating matrix for Ohio, USA, In: Proceedings of Geoline, Lyon
35. Pierson LA, Beckstrand DL, Black BA (2005) Rockfall hazard classification and mitigation system, Montana department of transportation, final report, FHWA/MT-05-011/8176

36. Pack R, Boie K, Mather S, Farrell J (2006) UDOT rockfall hazard rating system: final report and user's manual, UT-06.07. Utah State University, Logan
37. Mauldon M, Drumm EC, Dunne WM, Bateman N, Rose B, Kim M (2007) Rockfall management system for Tennessee. Tennessee Department of Transportation Division of Material and Tests, Nashville. 301p
38. Russell CP, Santi PM, Humphrey JD (2008) Modification and statistical analysis of the Colorado rockfall hazard rating system. Colorado Department of Transportation, DTD Applied Research and Innovation Branch (No. CDOT-2008-7)
39. Santi PM, Russell CP, Higgins JD, Spriet JI (2009) Modification and statistical analysis of the Colorado rockfall hazard rating system. Eng Geol 104(1):55–65
40. Saroglou H, Marinos V, Marinos P, Tsiambaos G (2012) Rockfall hazard and risk assessment: an example from a high promontory at the historical site of Monemvasia Greece. Nat Hazards Earth Syst Sci 12:1823–1836
41. Lienhart DA (1988) The geographic distribution of intensity and frequency of freeze/ thaw cycles. Bull Assoc Eng Geol 25:465–469
42. Flatland R (1993) Application of the rockfall hazard rating system to the rock slopes adjacent to U.S. 50 and state route 28 on the east side of lake Tahoe, M.S. thesis, University of Nevada, Reno, 318 p
43. Mazzoccola DF, Hudson JA (1996) A comprehensive method of rock mass characterization for indicating natural slope instability. Q J Eng Geol 29:37–56
44. Watters RJ (1998) Modification to the rockfall hazard rating system for successful mitigation in mountainous terrain as a result of climate and slope aspect considerations. Association of Engineering Geologists 41st Annual Meeting, Programs with Abstracts. 41: 134
45. Vandewater CJ, Dunne WM, Mauldon M, Drumm EC, Batemann V (2005) Classifying and assessing the geologic contribution to rockfall hazard. Environ Eng Geosci 11:141–154
46. Pierson LA, Vickle V (1993) Rockfall hazard rating system: participant's manual: Federal Highway Administration Publication SA-93-057, 104 p
47. Turner AK, Duffy JD (2012) Modeling and prediction of rockfall. In: Turner AK, Schuster RL (eds) Rockfall: characterization and control. Transportation Research Board, National Research Council, Washington, DC, pp 334–406
48. Guzzetti F, Crosta G, Detti R, Agliardi F (2002) STONE: a computer program for the three dimensional simulation of rock-falls. Comput Geosci 28:1079–1093
49. Cundall P (1971) A computer model for simulating progressive, large scale movements in blocky rock systems, In: International society of rock mechanics, 1, Paper No. II-8, Nancy
50. Azimi C, Desvarreux P, Giraud A, Martin-Cocher J (1982) Méthode de calcul de la dynamique des chutes de blocs – Application à l'étude du versant de la montagne de La Pale (Vercors). Bull de liaison des lab des ponts et chauss'ees 122:93–102
51. Falcetta J (1985) Un nouveau mod'ele de calcul de trajectoires de blocs rocheux. Revue Franc¸aise de G'eotechnique 30:11–17
52. Bassato G, Cocco S, Silvano S (1985) Programma di simulazione per lo scoscendimento di blocchirocciosi. Dendronatura 6(2):34–36
53. Wu S (1985) Rockfall evaluation by computer simulation. Transp Res Rec 1031:1–5
54. Bozzolo D, Pamini R (1986) Simulation of rock falls down a valley side. Acta Mech 63:113–130
55. Descoeudres F and Zimmermann T (1987) Three-dimensional dynamic calculation of rockfalls. In: Sixth International Congress on Rock Mechanics, International Society for Rock Mechanics, Montreal, pp 337–342
56. Labiouse V, Heidenreich B, Desvarreux P, Viktorovitch M, Guillemin P (2001) Etudes trajectographiques. In: Carere K, Ratto S, Zanolini F, (eds) Prevention des mouvements de versants et des instabilit'es de falaises, Aosta, pp 155–211
57. Rochet L (1987) Development of numerical models for the analysis of propagation of rockfalls. In: 6th International Congress on Rock Mechanics 1:479–484

58. Hungr O, Evans S (1988) Engineering evaluation of fragmental rockfall hazards. In: 5th International Symposium on Landslides, Lausanne, Switzerland. Balkema, Rotterdamm, 1:685–690
59. Jones CL, Higgins J, Andrew R (2000) Colorado Rockfall Simulation Program Version 4.0, Technical report, Colorado Department of Transportation, Denver
60. Van Dijke J, Van Westen C (1990) Rockfall hazard: a geomorphological application of neighbourhood analysis with ILWIS. ITC J 1:40–44
61. Kobayashi Y, Harp E, Kagawa T (1990) Simulation of rockfalls triggered by earthquakes. Rock Mech Rock Eng 23:1–20
62. Scioldo G (1991) Slope instability recognition, analysis, and zonation. In: Rotomap: analisistatisticadelrotolamentodeimassi, Milano, pp 81–84
63. Azzoni A, de Freitas MH (1995) Experimentally gained parameters, decisive for rock fall analysis. Rock Mech Rock Eng 28(2):111–124
64. Spang R, Sonser T (1995) Optimized rockfall protection by "Rockfall". In 8th International Congress on Rock Mechanics, Tokyo, 3:1233–1242
65. Zinggeler A, Krummenacher B, Kienholz H (1990) Steinschlagsimulation in Gebirgsw¨aldern. Berichte und Forschungen der GeographischesInstitut der Universit¨at Freiburg. 3:61–70
66. Krummenacher B, Keusen H (1996) Rockfall simulation and hazard mapping based on Digital Terrain Model (DTM). European Geol 12:33–35
67. Krummenacher B, Schwab S, Dolf F (2008) Assessment of natural hazards by three calculations of rockfallbehaviour. In: Volkwein A, Labiouse V, Schellenberg K (eds) Interdisciplinary workshop on rockfall protection. Swiss Federation Research Institute WSL, Morschach
68. Stevens WD (1998) Rockfall: a tool for probablistic analysis, design of remedial measures and prediction of rockfalls, Master's thesis, University of Toronto
69. Meissl G (1998) Modellierung der Reichweite von Felsst¨urzen: FallbeispielezurGIS-gest¨utztenGefahrenbeurteilung, Ph.D. thesis, Institutf¨urGeographie.Univ. Innsbruck
70. Paronuzzi P, Artini E (1999) Unnuovoprogramma in ambiente Windows per la modellazionedellacadutamassi. Geol Tec e Ambient 1(99):13–24
71. Dimnet E (2002) Mouvementet collisions de solidesrigidesoud'eformables, Ph.D. thesis, EcoleNationale des Ponts et Chauss'ees, Paris
72. Le Hir C, Dimnet E, Berger F (2006) 'Etude de la trajectographie des chutes de blocs en forêts de montagne. Bull Lab PontsChauss'ees 263(264):85–101
73. Dorren LKA (2003) A review of rockfall mechanics and modelling approaches. Prog Phys Geogr 27(1):69–87
74. Crosta G, Agliardi F, Frattini P, Imposato S (2004) A three dimensional hybrid numerical model for rockfall simulation. In: Geophysical research abstracts, vol 6. General Assembly of European Geosciences Union
75. Frattini P, Crosta G, Carrara A, Agliardi F (2008) Assessment of rockfall susceptibility by integrating statistical and physically based approaches. Geomorphology 94:419–437
76. Agliardi F, Crosta GB, Frattini P (2009) Integrating rockfall risk assessment and countermeasure design by 3D modelling techniques. Nat Hazards Earth Syst Sci 9:1059–1073
77. Dorren LKA, Maier B, Putters US, Seijmonsbergen AC (2004) Combining field and modelling techniques to assess rockfall dynamics on a protection forest hillslope in the European Alps. Geomorphology 57:151–167
78. Dorren LKA, Berger F, Putters US (2006) Real-size experiments and 3-D simulation of rockfall on forested and non forested slopes. Nat Hazards Earth Syst Sci 6:145–153
79. Bourrier F, Dorren L, Nicot F, Berger F, Darve F (2009) Towards objective rockfall trajectory simulation using a stochastic impact model. Geomorphology 110:68–79
80. Christen M, Bartelt P, Gruber U (2007) RAMMS – a modelling system for snow avalanches, debris flows and rockfalls based on IDL. PFG Photogrammetrie – Fernerkundung – Geoinformation 4:289–292
81. Lan H, Martin D, Lim C (2007) RockFall analyst: a GIS extension for three-dimensional and spatially distributed rockfall hazard modeling. Comput Geosci 33:262–279

82. Woltjer M, Rammer W, Brauner M, Seidl R, Mohren G, Lexer M (2008) Coupling a 3D patch model and a rockfall module to assess rockfall protection in mountain forests. J Environ Manag 87:373–388
83. Rammer W, Brauner M, Dorren L, Berger F, Lexer M (2007) Validation of an integrated 3D forest – rockfall model. Geophysical Research Abstract, Vienna
84. Jaboyedoff M, Labiouse V (2011) Technical note: preliminary estimation of rockfall runout zones. Nat Hazards Earth Syst Sci 11:819–828
85. Basson FRP (2012) Rigid body dynamics for rock fall trajectory simulation. American Rock Mechanics Association, Geomechanics Symposium held in Chicago, USA, pp 12–267
86. Azimi C, Desvarreux P (1977) Calcul de chutes de blocs etv'erificationsurmod'eler'eduit. Association pour le d'eveloppement des recherchessur les glissements de terrain, Grenoble
87. Dattola G, Crosta GB, Prisco CD (2016) Impact of blocks on deformable layers: influence of block rotation and size. In: Aversa et al (eds) Landslides and engineered slopes, experience, theory and practice. Associazione Geotecnia Italiana, Rome. ISBN:978-1-138-029880
88. Day RW (1997) Case studies of rockfall in soft versus hard rock. Environ Eng Geosci 3 (1):133–140
89. Schumm SA, Chorley RJ (1964) The fall of endangering rock. Am J Sci 262:1041–1054
90. Dorren LKA, Seijmonsbergen A (2003) Comparison of three GIS based models for predicting rockfall runout zones at a regional scale. Geomorphology 56:49–64
91. Coutard JP, Francou B (1989) Rock temperature measurements in two alpine environments: implications for frost shattering. Arct Alp Res 21:399–416
92. Grove JM (1972) The incidence of landslides avalanches and floods in western Norway during the little ice age. Arct Alp Res 4:131–138
93. Matsuoka N, Sakai H (1999) Rockfall activity from an alpine cliff during thawing periods. Geomorphology 28(3):309–328
94. McCarroll D, Shakesby RA, Matthews JA (1998) Spatial and temporal patterns of late Holocene rockfall activity on a Norwegian talus slope: a lichenometric and simulation modeling approach. Arct Alp Res 30(1):51–60
95. Vishal V, Siddique T, Purohit R, Phophliya MK, Pradhan SP (2017) Hazard assessment in rockfall-prone Himalayan slopes along National Highway-58, India: rating and simulation. Nat Hazards 85(1):487–503
96. Azzoni A, Freitas DMH (1995) Experimentally gained parameters, decisive for rock fall analysis. Rock Mech Rock Eng 28(2):111–124
97. Sun S, Li P, Li S, Zhang Q, Hu C (2011) Rockfall hazard assessment on Wangxia rock mass in Wushan (Chongqing, China). Geotech Geol Eng. https://doi.org/10.1007/s10706-017-0203-2
98. Andrianopoulos A, Saroglou H, Tsiambaos G (2013) Rockfall hazard and risk assessment of road slopes. Bull Geol Soc Greece 47(3):1664–1673
99. Ansari MK, Ahmad M, Singh TN (2014) Rockfall risk assessment for pilgrims along the circumambulatory pathway, Saptashrungi Gad Temple, Vani, Nashik Maharashtra, India. Geomatics Nat Hazard Risk 5(1):81–92
100. Budetta P, Nappi M (2013) Comparison between qualitative rockfall risk rating systems for a road affected by high traffic intensity. Nat Hazards Earth Syst Sci 13:1643–1653
101. Regmi AD, Cui P, Dhital MR, Zou Q (2016) Rock fall hazard and risk assessment along Araniko Highway, Central Nepal Himalaya. Environ Earth Sci 75:1112
102. Nichol D (2006) Rockfall geohazard assessment and protection measures on the highway network North whales, The Geological Society of London, IAEG2006: 320
103. Monnet JM, Clouet N, Bourrier F, Berger F (2010) Using geomatics and airborne laser scanning for rockfall risk zoning: a case study in the French Alps, The 2010 Canadian Geomatics Conference and Symposium of Commission I (ISPRS), Jun 2010, Calgary Alberta
104. Stock GM, Bawden GW, Green JK, Hanson E, Downing G, Collins BD, Bond S, Leslar M (2011) High-resolution three-dimensional imaging and analysis of rock falls in Yosemite Valley, California. Geosphere 7(2):573–581

105. Abellán A, Oppikofer T, Jaboyedoff M, Rosser NJ, Lim M, Lato MJ (2014) Terrestrial laser scanning of rock slope instabilities. Earth Surf Process Landf 39(1):80–97
106. Collins BD, Stock GM (2012) Lidar-based rock-fall hazard characterization of cliffs, GeoCongress, ASCE, Oakland, pp 3021–3030
107. Kromer R, Hutchinson J, Lato M, Gauthier D, Edwards T (2015) A multi-sensor remote sensing approach for railway corridor ground hazard management. Geophysical Research Abstracts. EGU2015-8200
108. Volkwein A, Klette J (2014) Semi-automatic determination of rockfall trajectories. Sensors 14 (10):18187–18210
109. Fookes PG, Sweeny M (1976) Stabilization and control of local rock falls and degrading rock slopes. Quat J Eng Geol 9:37–55
110. Badger TC and Lowell S(1992) Rockfall control Washington State. In: Rockfall Prediction and control and landslide case histories, transportation research record. National Research Council, Washington, DC, 1342:14–19
111. Dorren LKA, Berger F (2012) Integrating forests in the analysis and management of rockfall risks: experiences from research and practice in the Alps. In: Eberhardt et al (eds) Landslides and engineered slopes: protecting society through improved understanding. Taylor and Francis Group, London, pp 117–127
112. Brang P (2001) Resistance and elasticity: promising concepts for the management of protection forests in the European Alps. For Ecol Manag 145:107–119
113. Berger F, Quetel C, Dorren LKA (2002) Forest: a natural protection mean against rockfalls, but with which efficiency? In: Proceedings of the International Congress. Interpravent 2002 in the Pacific Rim – Matsumoto, Japan. Congress Publication, 2:815–826
114. Perret S (2005) Rockfall-forest interaction: inventory, analysis and simulation of rockfall activity in mountain forests, Ph.D. Thesis, University of Berne, Berne
115. Dorren LKA, Berger F (2005) Stem breakage of trees and energy dissipation at rockfall impacts. Tree Physiol 26:63–71
116. Ghoussoub L, Douthe C, Sab K (2014) Analysis of the mechanical behaviour of soft rockfall barriers. RocExs 2014 – 5th interdisciplinary workshop on rockfall protection, Italy
117. Martin DC (1988) Rockfall control: an update. Technical note. Bull Assoc Engg Geol 25 (1):137–144

Part II
Landslide Monitoring and Prediction

Chapter 5
Study of Jointed and Weathered Rock Slopes Through the Innovative Approach of InfraRed Thermography

Giovanna Pappalardo and Simone Mineo

Abstract InfraRed Thermography (IRT) is presented herein as a support methodology during the rock mass survey. Although this technique is widely used in several scientific fields, its direct application for such purposes is still pioneering. In this review paper, the outcomes of the most recent researches on the application of IRT to the rock mechanics are reported and commented, with particular reference to the study of thermograms and to the development of a Cooling Rate Index (CRI), useful for the evaluation of the jointing condition of the rock.

Three application cases are commented to demonstrate the reliability of such methodology in the geomechanics, taking into account both bare-jointed rock masses and highly weathered rock slopes. Achieved results lay the foundation for future researches aiming at a refined and improved survey methodology, which would be a useful support in the geomechanical analysis of heavily fractured rock masses.

5.1 Introduction

The stability of rock slopes is mainly controlled by the jointing of rock mass along with the condition of its discontinuities. Geostructural survey is a necessary procedure for the slope modelling and the stability analysis, because it allows the direct evaluation of both the degree of fracturing of the rock and its peculiar characteristics. Nevertheless, according to the growth of technology, satisfactory outcomes are nowadays achieved also through remote surveys, which are often considered important support methodologies to the field campaign. In this light, InfraRed Thermography (IRT), whose utility is known in several fields of science, has recently been employed in the rock mechanics for various purposes. In

G. Pappalardo · S. Mineo (✉)
Department of Biological, Geological and Environmental Sciences, Università degli Studi di Catania, Catania, Italy
e-mail: smineo@unict.it

S. P. Pradhan et al. (eds.), *Landslides: Theory, Practice and Modelling*, Advances in Natural and Technological Hazards Research 50,
https://doi.org/10.1007/978-3-319-77377-3_5

laboratory, it is a useful tool for the indirect estimation of the porosity of artificially heated rock specimens based on their cooling rate and for the identification of persistent voids [1–3]. In the field, it supported the detection of specific items along the unstable slopes, such as crevasses, caves and open fractures [4–6] and it was also used as a complementary methodology during the assessment of rockslide susceptibility scenarios [7, 8].

In this paper, a review on the most recent outcomes about the application of IRT for the survey of unstable rock slopes is reported, with the aim of presenting a general overview on some applicative cases of such pioneering practice. In particular, three cases will be reported and commented, such as the survey of heavily jointed rock masses, presented by Mineo et al. [9] and Pappalardo et al. [10]. They proposed IRT for the study of the discontinuity systems, with particular reference to those characterized by highly persistence, in order to understand what kind of satisfactory information such methodology could bring in a rock slope modeling. Moreover, they analysed the cooling behaviour of rock masses introducing the Cooling Rate Index, which well correlates to the main geomechanical parameters of the rock.

A further application of IRT discussed in this paper concerns slopes affected by landslides at a highly loosened and weathered rock masses, where Mineo et al. [11] and Mineo and Pappalardo [12] highlighted potentially unstable spots characterized by a different thermal behaviour.

The last case is focused on the analysis peculiar unstable areas of rock slopes, where an IRT survey allowed highlighting portions of the rock that can be involved in future rockfalls.

5.2 The InfraRed Thermography Technique

InfraRed Thermography (IRT) is a technique allowing the estimation of the surface temperature of an object by exploiting its emitted infrared radiation. This is linked to a property that all forms of matter with temperature above the absolute zero have, i.e., the thermal emissivity, whose intensity is function of the temperature of the material itself. As all the electromagnetic radiations, thermal emissivity falls within a precise portion of the electromagnetic spectrum, mostly comprised between the Ultra Violet and Infrared bands [e.g. 13, 14] (Fig. 5.1); therefore it is mostly invisible for the human eye.

According to the known optical principles, the radiation can be deflected, focused with a lens, or reflected from surfaces [15]. Based on the amount of energy reflected or absorbed by a body, emissivity ranges from 0 (when all the energy is reflected) to 1 (when all the energy is absorbed). The last condition is related the "black bodies", which are capable to re-emit 100% of the absorbed energy. However, these cases are purely theoretical, since bodies capable to reflect or absorb all the radiation do not exist in nature. All the materials have intermediate emissivity values and are known as "grey bodies" [9].

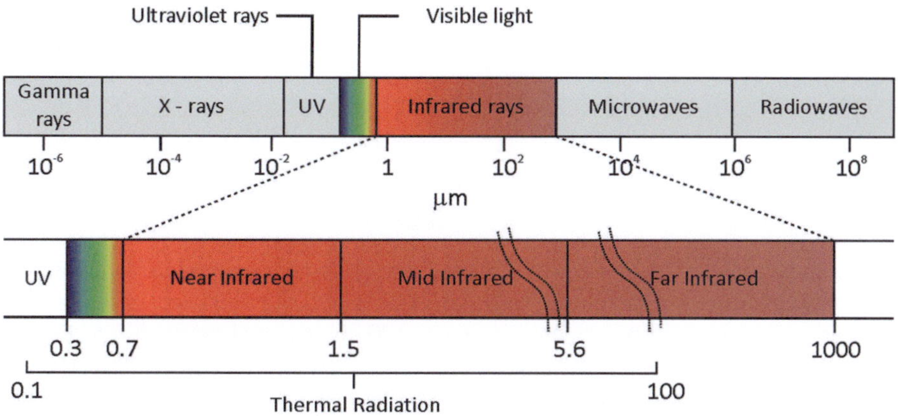

Fig. 5.1 Representation of the electromagnetic spectrum with emphasis on the infrared band [10]

From the mathematical point of view, emissivity is the ratio between the infrared radiation emitted by a body and the radiation emitted by a black body at the same temperature. According to the Stefan-Boltzman law (Eq. 5.1), there is a proportionality between the whole energy emitted by an object (J) and the Stefan-Boltzman constant (σ), to the surface temperature of the body (T) to the fourth power and to its emissivity (ε) [15].

$$J = \varepsilon\sigma T^4 \tag{5.1}$$

A thermal camera can build images using the infrared radiation, as it is a device operating in the range of wavelengths as long as 13–14 µm. It provides colour-scaled pictures showing the variation of temperature within the shot area. Such images are called "thermograms" and consist of a matrix of pixels labelled with a surface temperature value.

The radiation captured by the camera does not depend only on the surface temperature of the shot body, but it is also influenced by the surrounding ambient (i.e. parasite radiation). Moreover, it can be both partially absorbed/dispersed by the atmosphere and reflected by the object [e.g. 16, 17].

5.3 Application of IRT in the Geomechanical Study of Heavily Jointed Rock Slopes

This section provides a concise description of IRT shooting campaigns carried out by Mineo et al. [9] and Pappalardo et al. [10] with the aim of collecting the most recent outcomes on the applicability of such methodology to the survey of heavily fractured rock masses. Due to the pioneering aspect of this research, most of the attention is focused on what kind of information can be retrieved by the analysis of thermograms acquired both at different stages of a day and under different seasonal conditions.

5.3.1 Method

The campaign described herein was carried out by using a thermal camera with the following features: temperature accuracy ±2 °C or ±2%; range of assessable temperature comprised between −20° and +120 °C; spectral range of 7.5–13 μm; Field of view 25 × 19°; focal length of 18 mm; Noise Equivalent Temperature Difference (NETD) of 50 mK. Sensing device was an uncooled Focal Plane Array with IR resolution of 320 × 240 pixels.

Thermal images of rock masses (Fig. 5.2), belonging to different lithologies (dolostones, limestones and porphyroids), were taken in the dry season with the help of a tripod and an incorporated laser pointer to preserve the constant framing (Fig. 5.3a). The different lithologies have in common the tectonic history and, therefore, the presence of numerous discontinuity systems, which strongly affect the geomechanical quality of rock masses (e.g. [18]).

The geostructural setting of all the outcrops was previously studied, according to ISRM [19], to provide a reliable reference dataset in terms of orientation of discontinuities and main geostructural parameters [20–22]. To ensure the lowest number of variables, all the rock slopes chosen for this test face NE, therefore they are usually hit by the sun radiation during the morning, while they are in shadow from midday and during afternoon.

Based on the daylight condition, the set up methodology consisted in four different measurement stages during the same day:

1. *t1*: morning shooting, when the slope faces were directly hit by sun radiation.
2. *t2*: images shot at noon, when the slope had just gone under shadow.
3. *t3*: images taken in the afternoon, when the cooling phase of the slope had already started for some hours.
4. *t4*: images shot at night, when the impact of parasite radiation is minimum, since no artificial illumination was lighting the slope face.

   The same campaign was repeated in the cold season, to compare thermal images and to study differences occurring between outcomes (Fig. 5.3b).

5.3.2 Analysis of Thermograms

The analysis of thermograms shows that *t1* images are strongly influenced by the sun radiation, which, due to the orientation of the slope face, was right hitting the indented rock masses. Under this condition the thermal output is “disturbed” by the uneven warming of the slope face, as jutting sector of the outcrop are radiated by

Fig. 5.2 Representative rock masses of selected lithologies with corresponding geostructural stereonet

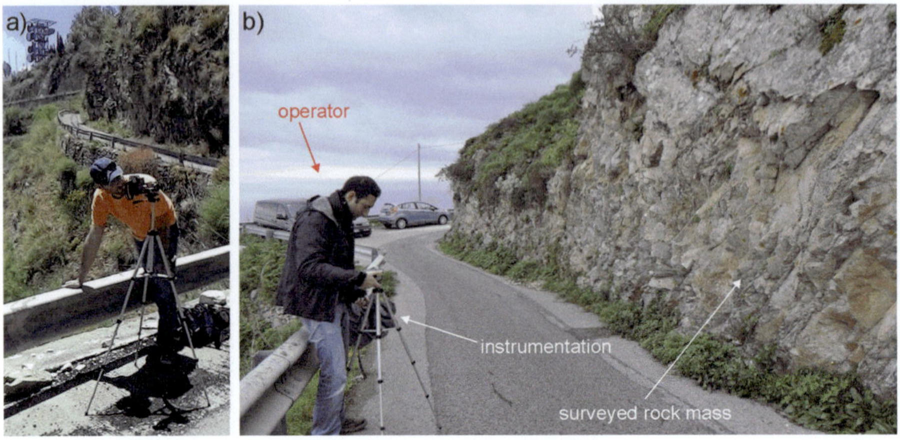

Fig. 5.3 Field operation during (**a**) summer and (**b**) winter campaigns [10]

the sunrays (resulting in positive anomalies/warm areas), thus gaining heat and shadowing the hollow portions (negative anomalies/cold areas).

During the subsequent shooting sessions (*t2, t3, t4*), shadowed rock masses slowly release heat through their discontinuity systems and the IRT images provide interesting hints on the degree of fracturing of the rock, with an increasing quality as time passes.

At these stages, warm areas in thermograms retrace fractures and hollow portions of the rock, while colder sectors are proper of low fractured planes, weathered portions and jutting rocks. It is underlined that *t2* and *t3* were carried out in daylight condition, therefore the influence of parasite radiations cannot be excluded although rock masses were already shadowed. At nighttime (darkness condition), the best definition among the IRT images is achieved (Fig. 5.4).

Figure 5.4 shows that, with reference to *t4* images, positive linear anomalies mark the main discontinuity system, emphasizing their geometrical relationship; for example, the "X" shapes at dolostones, defined by two intersecting persistent and open sets, is clearly visible in the thermogram. Warm sectors are also related to crushed and/or hollow portions, proving that IRT allows a good definition of the crushed rock, thus representing a useful help during a traditional geostructural survey, especially at slopes with closely spaced discontinuities.

Porphyroids *t4* thermograms show how the highest temperatures retrace the main discontinuities along with an intensely jointed zone in the upper portion of the image. The dip-slope plane occurring in the center of the thermogram is a cold area due to its low fracturing and smooth surface. This is one of the main unstable planes, giving rise to potential planar sliding failures, and its identification is a key point during a stability analysis.

Moreover, it has been noticed that the aperture of fractures is another element affecting the thermal output, as maximum surface temperatures were found at open discontinuities (even mm). Such temperature values decrease at filled or closed

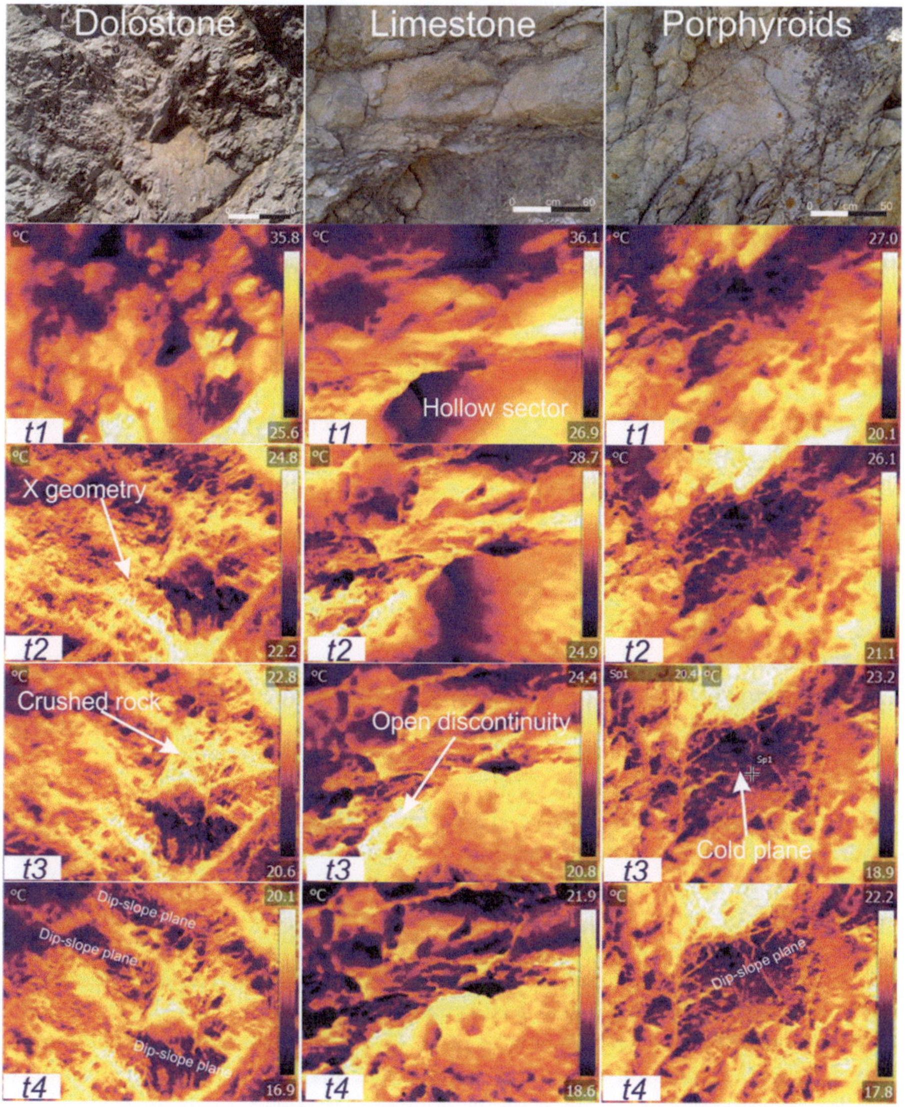

Fig. 5.4 Thermal outcomes for representative rock masses of the studied lithologies at the 4 measurement stages

fractures (Fig. 5.4). Similarly, the intensity of thermal anomalies can be related to the persistence of fractures, starting from the assumption that a persistent discontinuity concurs to the heat exchange between the rock mass and the external ambient.

By reiterating the daily campaign on the same outcrops during winter, an interesting difference in the quality of thermograms can be outlined with respect to the summer images. In particular, winter thermograms show a poor definition and sometimes are characterized by a blurred effect (Fig. 5.5). It happens because, during

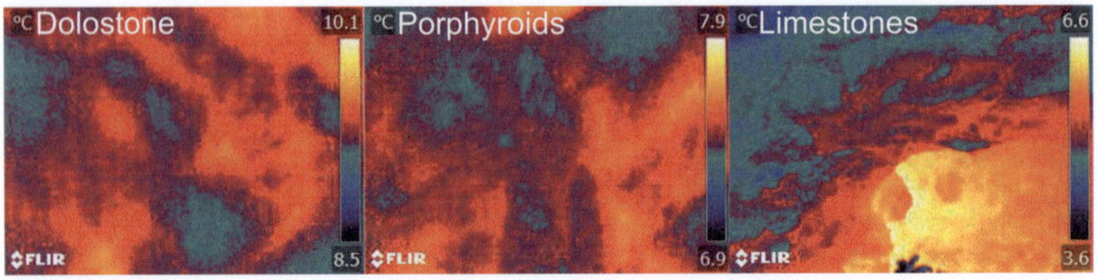

Fig. 5.5 Winter representative thermograms, showing the blurred effect due to the low warming of the rock masses (Modified after [10])

winter, rock masses are less heated and the ambient temperature is sensibly lower during the day, therefore winter images need a deeper processing before being compared to summer images.

5.3.3 The Cooling Rate Index

The digital processing of an IRT image permits the assessment of the maximum, minimum and average surface temperatures of the surveyed object(s). While maximum and minimum are temperature values belonging to the hottest and coldest pixels respectively, the average surface temperature is representative of the entire shooting area. Therefore, in this case, we can consider it as the "average" surface temperature of the surveyed portion of rock mass. Based on such values, recorded at different times, the cooling attitude of the rock mass can be monitored. In this light, according to Newton's Law of Cooling, the cooling of an object in a colder ambient is described by a curve; the slope of the tangent to the curve at any point gives the rate of the temperature loss. On a two-variable diagram, the plot of average surface temperature estimated at each outcrop against time describes the cooling of the rock through a curve with a decreasing trend (Fig. 5.6).

Based on this outcome, Pappalardo et al. [10] have calculated a specific index representing the variation of surface temperature per unit of time and named it Cooling Rate Index (CRI), which can be computed by Eq. 5.2.

$$\mathrm{CRI} = \Delta T/\Delta t \tag{5.2}$$

Where ΔT is the variation of temperature and Δt is the considered time window. Therefore, the higher CRI, the faster the rock mass cooling.

With reference to surveyed rock masses, cooling curve of dolostones and limestones show decreasing trends, while in the porphyroids curve a slight heating phase between *t1* and *t2* is outlined. This is caused by the local setting of the outcrops, since porphyroid rock masses are partly shadowed by a protruding rock portion on

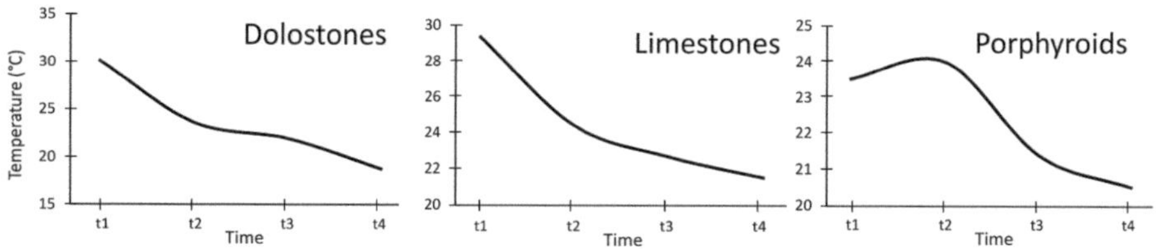

Fig. 5.6 Representative cooling curves presented by [10] for a the studied lithologies

the top of the outcrop at *t1*. In this case, the projected shadow leads to a slower warming of the rock, which lasts until *t2*, when the cooling phase begins (Fig. 5.6).

According to acquired data, the cooling phase of dolostones is characterized by CRI ranging between 0.65 and 0.84 °C per hour.

Limestone outcrop is affected by a CRI value around 0.41° per hour, while porphyroids have an average CRI of 0.58 °C per hour.

Achieved values are indicative of a faster cooling at dolostones, which are the most fractured rock masses showing also open discontinuities and crushed portions. This consideration leads to hypothesize that there is a link between CRI and the degree of fracturing of the rock.

Therefore, a further challenge of the research reviewed herein is to ascertain the validity of this relationship. In this light, statistical correlations between CRI and two important geomechanical indexes, i.e. the Rock Quality Designation (RQD) [23] and the Volumetric Joint count (Jv) [24], previously estimated for each geostructural station, were carried out.

Resulting CRI vs Jv scatterplot shows a positive linear trend indicating that the rock mass cools down faster as Jv increases (i.e. an increasing rock fracturing). This outcome is in accordance with the field condition of the outcrops, as dolostones (highest Jv) are characterized by the highest CRI and the best fitting is achieved by calculating the Cooling Rate only within the nighttime (*t3-t4*), when the temperature difference between rock outcrop and external environment is maximum (Fig. 5.7a). According to this result, the cooling of rock masses accelerate proportionally to their degree of fracturing, resulting in higher CRI at the most fractured outcrops.

Correspondingly, the correlation between the daily CRI and RQD shows a negative trend, with a satisfactory best fit for *t3-t4* data (Fig. 5.7b). A low CRI corresponds to a high RQD (i.e. good geomechanical quality of the rock); this is in accordance with the principle that intact rock (i.e. rock with no visible fractures) is not a good conductor of heat.

The above reported correlations prove the reliability of CRI as a potential index for the remote assessment of the state of fracturing of a rock mass. Therefore, further survey at outcrops with different properties is currently being performed to enrich the statistics of such new survey methodology.

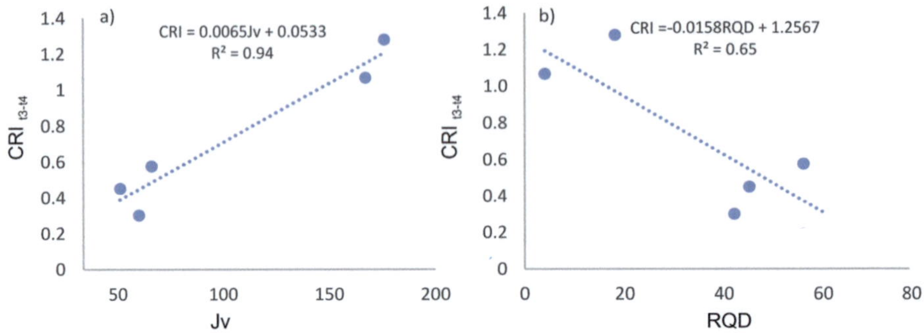

Fig. 5.7 Statistical correlations proposed by [10] between (**a**) CRI_{t3-t4} and Jv; (**b**) CRIt3–t4 and RQD

5.4 Application of IRT to Weathered Rock Slopes

This section reports on an Infrared Thermography shooting campaign aimed at individuating potentially unstable areas along a crystalline weathered rock slope characterized by a high predisposition to fail. Several landslides, classifiable as rockfalls and debris-falls according to [25, 26], repeatedly affected the studied slope, even after the execution of remedial works. Mineo et al. [11] proposed an integrated approach to study the instability mechanisms affecting this slope and highlighted that the poor geomechanical condition of the rock is not only related to the rock face, but persists inside the outcrop.

In this case, the analysis of thermograms was aimed at mapping the potentially unstable portions of the slope according to the difference of temperature related to the elements of the outcrop. This kind of survey proved a helpful tool during the stability study of rock slopes, especially where it is logistically hard to accomplish direct surveys.

5.4.1 *IRT Shooting Campaign*

The studied outcrop is a weathered paragneiss rock mass, characterized by scars of previous rockfalls, well visible among partly vegetated sectors (Fig. 5.8a). The surface temperature of the slope, along with its variations, can be employed as an indicator to recognize the different elements occurring at the slope face according to Wu et al. [5] and Mineo and Pappalardo [12].

In this case, the shooting campaign was carried out during the dry season, in night time, when the possible influence of parasite radiations could be neglected.

The minimum and maximum surface temperatures were about 14 °C and 29 °C, respectively.

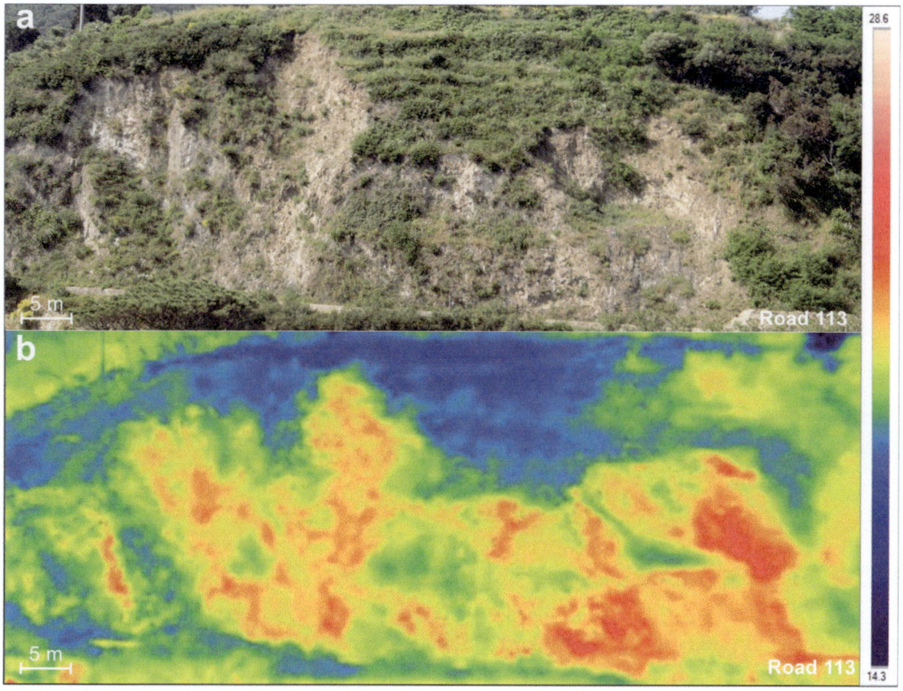

Fig. 5.8 Comparison between a digital photo and an IRT image of the rock slope taken from the same shooting point: (**a**) photo taken in daylight; (**b**) IRT image taken in nighttime [11]

The lowest values are found along the vegetated upper part of the outcrop, while maximum values are found at the source areas of past rockfalls, that is the place where the rock is bared (Fig. 5.8b). In particular, by highlighting different ranges of surface temperatures, six different frames were considered pointing out the elements of the slope. Such approach is aimed at linking the different temperature values to the physical elements (rock, vegetation, and debris) along the slope (Fig. 5.9).

According to [11, 12], a first differentiation based on selected ranges of temperature can be carried out as follows:

1. 18–19 °C: vegetated areas;
2. 19–20 °C: weathered and fractured, partly vegetated rock;
3. 20–21 °C: talus and debris;
4. 21–22 °C: bare, fractured rock.

In order to enhance the thermal contrast between the mentioned elements, the considered temperature range was constricted between 19 and 24 °C (Fig. 5.10). In this way, the contribution of shrubs and trees on the top of the slope (lowest surface temperature) was avoided and the bare rock (warmest regions) was emphasized. Assuming that the bare rock crops out at the source areas of already occurred events, such sectors may be regarded as the currently stable portions where the unstable

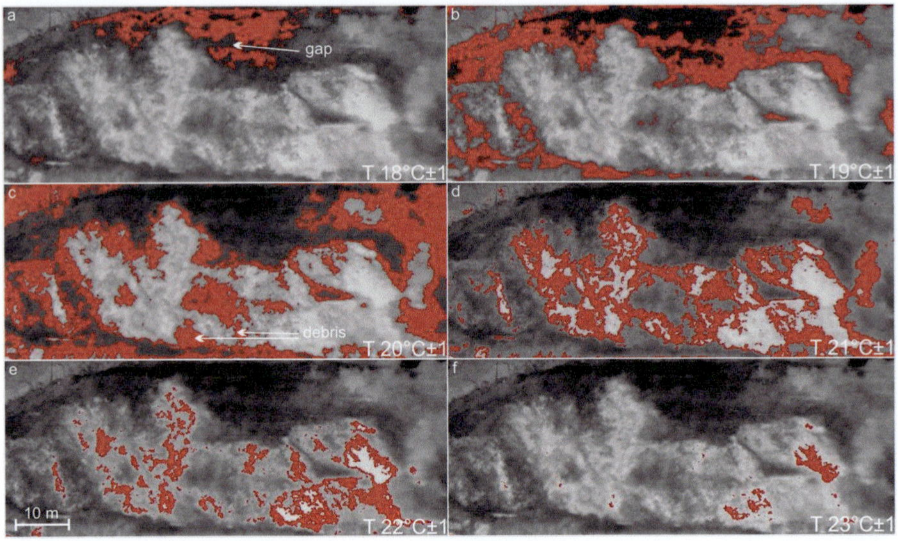

Fig. 5.9 Slope portions with homogeneous temperature range [11]

material has already failed. Therefore, particular attention is focused on all the other elements lying between these two extremes.

In fact, lowest surface temperatures label partly vegetated and weathered sectors of the slope due to the effect of the local vegetation and of a natural moisture rate. Indeed, weathered rocks, especially when broken down to sand and silt, tend to absorb humidity from the atmosphere favoring the growth of plants and shrubs. By the radical apparatus, these concur in the disaggregation of the rock enhancing its instability.

According to these considerations, Mineo and Pappalardo [12] found 11 potentially unstable sectors along the surveyed slope face, with average surface temperatures ranging between 19.3 and 20.7 °C (Fig. 5.10c).

Identified sectors mostly border the source areas of latest events suggest that, there are still some rock materials which may fail, which was un-stabilized by the occurred landslides. In particular, a brief description of the 11 sectors is reported as follows.

1 Located above the crown of a landslide occurred in 2008, this area holds an intensely weathered rock material populated by vegetation. The main criticism is related to the basal sapping process caused by the loss of the already failed material, especially in the steepest portions.

2/3 These two similar sectors lie between two denudation surfaces. In particular, sector 3 seems to hold a great volume of material, whose failure could occur along the same sliding surface of the adjacent events.

4 Partly vegetated rock material, not involved in previous landslide events.

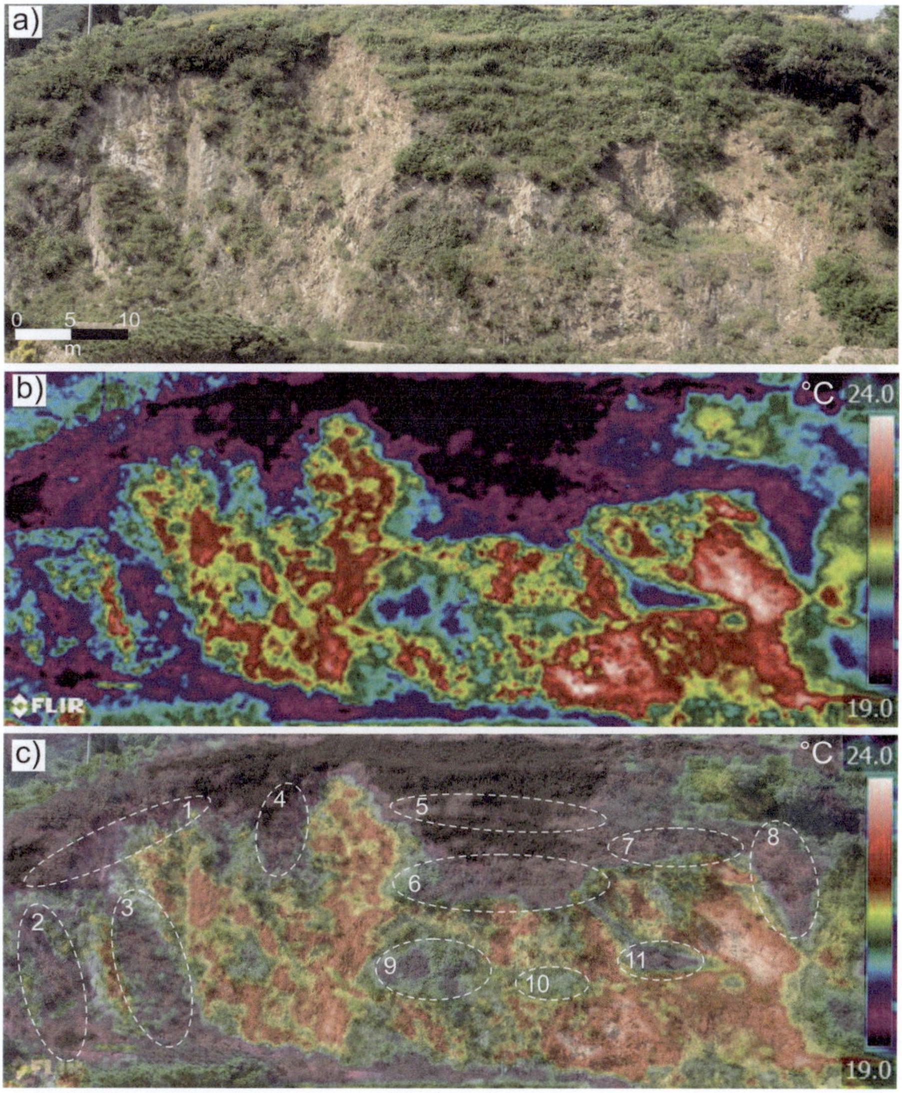

Fig. 5.10 (**a**) Photo of the studied rock slope taken in daylight; (**b**) 19–24 °C thermogram of the slope; (**c**) Thermogram overlapped to the digital photo of the slope. *Dashed ellipses* indicate the most critical sectors. Numbers are the ID of each sector [12]

5 This portion of rock mass is identified by a cold region, caused by a morphological step at the top of the slope. Since this sector has not been involved in previous landslides, this step could also be considered as a potential detachment point for rock material.

6/7/8 Similar in features to sector 1.

Fig. 5.11 Map of the unstable sectors of the slope showing the potential evolution of the instability [12]

9/10 Remnants of weathered and partly vegetated rock material in the middle of the slope. Due to their peculiar morphological features, these sectors could also hold a rate of accumulated talus, which is likely to be pushed down the slope by heavy rainfalls.

11 This is a particular sector since the negative thermal anomaly highlights a sub-flat morphology. Apparently, it does not seem to have a relevant feature, but it lays on a protruding portion of the rock mass composed of weathered and fractured rock. Rainwater, flowing on the slope face, may stagnate on such flat morphology and infiltrate within the rock mass, thus favoring weathering and disaggregation of the already poor rock.

Mineo and Pappalardo [12] graphically summarized achieved outcomes in a map, obtained by overlapping the most critical areas to a digital image of the slope and showing the identified most potentially unstable sectors (Fig. 5.11).

5.5 Application of IRT for the Study of Unstable Areas for Potential Reactivation

According to the results reported in Sect. 5.3, a further utility of the application of IRT to the analysis of rock masses was found in the study of a slope affected by sliding of wedges. The case was taken into account and was initially presented by Pappalardo et al. [27] and it is about a dolostone outcrop already affected by a rockfall, which left an evident scar on the slope face. The source area of the wedge is bordered by two intersecting discontinuities planes (S1-S2) and by a third plane (TC) acting as a tension crack since it strikes almost parallel to the face (Fig. 5.12a). It is likely that the initial rock volume slides along the intersection line and its movement was enhanced by TC.

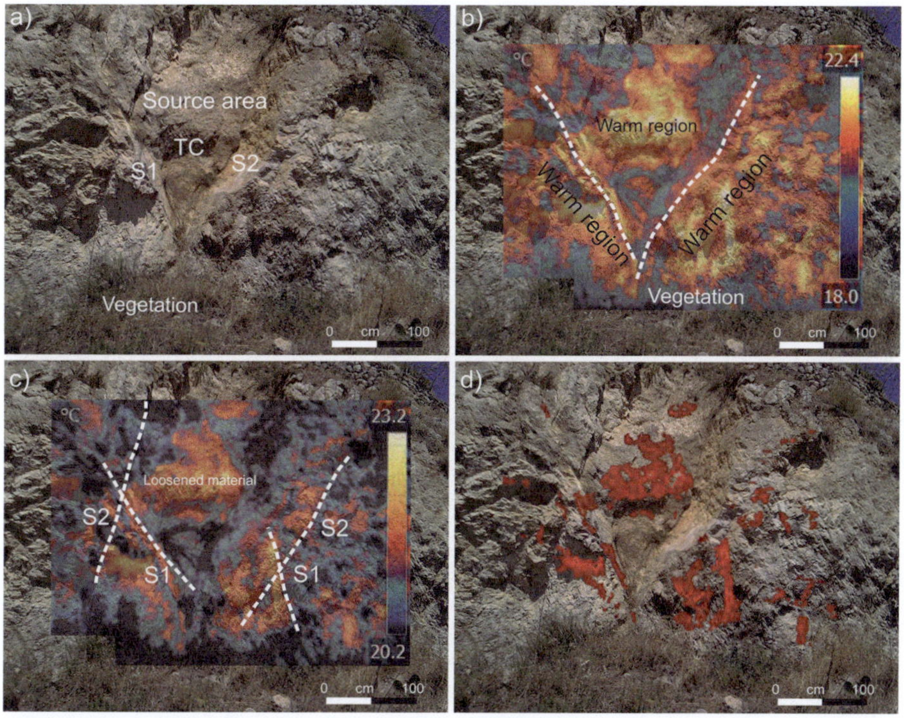

Fig. 5.12 (**a**) Digital photograph of the surveyed slope; (**b**) 18–22.4 °C thermogram overlapped to a digital photo of the slope; (**c**) 20.2–23.2 °C thermogram overlapped to a digital photo of the slope; (**d**) portions with surface temperature >21 °C highlighted on the slope

IRT images were taken in nighttime, when the heat is released by the rock masses through its discontinued systems. In this view, positive anomalies are related to fractures, caves and hollow parts, as reported before. Here, negative anomalies indicate low fractured planes as well as the weathered rock and jutting portions.

5.5.1 Thermal Outcomes

Acquired thermograms show the alternation of cold and warm areas, apparently without any specific shape, where the two main discontinuities bordering the source area of the wedge area can be individuated (Fig. 5.12b). In fact, S1 and S2 are highlighted by warm colors, while TC represents the internal face of the source area, where the highest surface temperatures were recorded.

By processing the IRT image, so to highlight the highest temperature values related to the most fractured rock, S1 and S2 traces become clearly visible

(Fig. 5.12c). In particular, S2 appears as a smooth surface externally bordered by intensely fractured rock.

Moreover, the traces of other discontinuities occurring next to the source area run parallel to the main discontinuity sets and are affected by close spacing. Such discontinuities border positive anomalies, highlighting a heavy fractured rock around the source area. This is relevant information obtainable by the analysis of a thermogram; in fact, these areas could represent potential enlargement sectors of the wedge, thus suggesting possible reactivation of the landslide movement.

A further element pointed out by Pappalardo et al. [27] is represented by the warm region occurring at TC plane, within the source area (Fig. 5.12d): such anomaly can be related to fractured and loosened rock, which is probably underwent decompression phenomena after the initial wedge sliding. This is a further hint provided by the IRT analysis, which suggests the presence of a potential source area for future failures, which would lead to retreat phenomenon of the slope.

5.6 Conclusions

This review paper reports an innovative approach for the remote survey of rock masses through InfraRed Thermography. Interesting results were achieved with reference to three different applications, whose outcomes are summarized by the following points, aimed at evaluating peculiar geomechanical features along rock slopes.

1. The first discussed case is related to a daily IRT shooting campaign commented to highlight the utility of such survey methodology applied to intensely jointed rock masses of different lithologies. The main discontinuity systems, as well as crushed portions of the rock are identified by positive thermal anomalies (highest surface temperature), while smooth and low fractured surfaces keep a lower temperature. Such an output allows mapping heavily fractured sector of the surveyed outcrop along with its open and persistent discontinuities, which are responsible for instability. Indeed, the maximum temperature was found at the most open and persistent fractures, while the lowest values were proper of closed or filled discontinuities.

   The use of IRT to highlight the different degree of fracturing along a rock slope proved a suitable tool, which could give a great contribution for the geostructural characterization of hardly-reachable rock masses, such as coastal cliffs or higher sectors.

   Furthermore, a new index describing the cooling attitude of rock masses was introduced. It was named as Cooling Rate Index (CRI) [10] and it is the expression of the cooling velocity of the rock, which is strongly conditioned by its

degree of fracturing. CRI proved well correlated with two geomechanical parameters used for the classification of the rock quality, i.e., RQD and Jv. Their statistical correlation with CRI returned satisfactory positive and negative linear relationships for CRI-Jv and CRI-RQD respectively, demonstrating that the cooling speed of a rock mass is proportional to its degree of fracturing.
2. The second application discussed herein is focused on the IRT imaging analysis carried out to detect unstable portions along a weathered rock slope, which had already suffered numerous rockfalls and debris-falls. In this case, the utility of IRT resides in the possibility of distinguishing elements with a different thermal attitude along the slope by highlighting the contrast between vegetation, debris/weathered rock and bare rock. Shooting was carried out in nighttime, because the difference between the above mentioned elements is not always clear in daylight conditions, due to the effects of light and shadows caused by the irregularity of the rock face [11]. In this way, the potential source areas of future events, including also possible reactivation sectors of previous landslides, were mapped and 11 critical portions were identified. The potentially unstable sectors are placed both at the crown of already occurred failures and between two or more denudation areas [12].

   Achieved outcomes confirm the utility of IRT as a support methodology during the modeling of the slope stability, even in highly weathered rock masses.
3. According to the afore mentioned outcomes, the application of IRT for the survey of potential reactivation areas along a rock mass has been discussed, highlighting how thermal images proved useful in the detection of relevant criticalities related to such cases. In detail, IRT was aimed at analyzing the source area of a wedge failure some years ago [27]. Such methodology provided interesting information on the poor quality of the rock mass, especially around and within the source area itself. In fact, the analysis of thermograms highlighted the presence of potential enlargement sectors of the main wedge, suggesting potential reactivation of the failure, involving new volumes of rock.

Results achieved by the researches reviewed herein demonstrate the high potentiality owned by IRT when applied to the geomechanics. The detection of unstable areas along a slope is a precious hint, especially when complex rock masses are surveyed.

It is clear that such application is at its pioneering stage and that further tests in different settings are needed to refine the outcomes. Nevertheless, this paper can be considered a starting point for possible classification systems and survey methods based on the thermal behavior of rock masses. Further studies will follow approaching to the analysis of thermograms not only from a qualitative point of view, but also considering the thermal behavior of the rock over time, at any scale [28].

References

1. Mineo S, Pappalardo G (2016) The use of infrared thermography for porosity assessment of intact rock. Rock Mech Rock Eng 49(8):3027–3039. https://doi.org/10.1007/s00603-016-0992-2
2. Mineo S, Pappalardo G (2016) Preliminary results on the estimation of porosity in intact rock through InfraRed Thermography. Rend Online Soc Geol It 41:317–320., 2016. https://doi.org/10.3301/ROL.2016.157
3. Pappalardo G, Mineo S (2017) Investigation on the mechanical attitude of basaltic rocks from Mount Etna through InfraRed Thermography and laboratory tests. Constr Build Mater 134:228–235., 2017. https://doi.org/10.1016/j.conbuildmat.2016.12.146
4. Rinker JN (1975) Airborne infrared thermal detection of caves and crevasses. Photogramm Eng Remote Sens 41:1391–1400
5. Wu JH, Lin HM, Lee DH et al (2005) Integrity assessment of rock mass behind the shotcreted slope using thermography. Eng Geol 80:164–173
6. Baron I, Beckovský D, Míca L (2012) Application of infrared thermography for mapping open fractures in deep-seated rockslides and unstable cliffs. Landslides 11:15–27. https://doi.org/10.1007/s10346-012-0367-z
7. Gigli G, Frodella W, Garfagnoli F et al (2014) 3-D geomechanical rock mass characterization for the evaluation of rockslide susceptibility scenarios. Landslides 11:131–140. https://doi.org/10.1007/s10346-013-0424-2
8. Casagli N, Frodella W, Morelli S et al (2017) Spaceborne, UAV and ground-based remote sensing techniques for landslide mapping, monitoring and early warning. Geoenviron Disasters 4:9. https://doi.org/10.1186/s40677-017-0073-1
9. Mineo S, Calcaterra D, Perriello Zampelli S et al (2015) Application of infrared thermography for the survey of intensely jointed rock slopes. Rend Online Soc Geol It 35:212–215. https://doi.org/10.3301/ROL.2015.103
10. Pappalardo G, Mineo S, Perriello Zampelli S et al (2016) InfraRed Thermography proposed for the estimation of the cooling rate index in the remote survey of rock masses. Int J Rock Mech Min Sci 83:182–196
11. Mineo S, Pappalardo G, Rapisarda F et al (2015) Integrated geostructural, seismic and infrared thermography surveys for the study of an unstable rock slope in the Peloritani Chain (NE Sicily). Eng Geol 195:225–235. https://doi.org/10.1016/j.enggeo.2015.06.010
12. Mineo S, Pappalardo G (2015) Infrared Thermography for the detection of potentially unstable areas along a slope. Special Publication, J of EG October 2015, 1330–1338
13. DeWitt N (1988) Theory and practice of radiation thermometry. Wiley, New York
14. Wolfe WL, Zissis GJ (eds) (1993) The infrared handbook. Office of Naval Research, Washington, DC
15. Hillel D (1998) Environmental soil physics. Academic, New York, p 771
16. Shannon HR, Sigda JM, Van Dam RL et al (2005) Thermal camera imaging of rock piles at the Questa Molybdenum Mine, Questa, New Mexico. Proc. 2005 National Meeting of the American Society of Mining and Reclamation, ASMR, 1015–1028 June 19–23, 2005
17. Prendes-Gero MB, Suárez-Domínguez FJ, González-Nicieza C et al (2013) Infrared thermography methodology applied to detect localized rockfalls in self-supporting underground mines. In: Kwaśniewski A, Lydżba D (eds) Rock mechanics for resources, energy and environment. Taylor & Francis Group, London, pp 825–829
18. Mineo S, Pappalardo G, D'Urso A et al (2017) Event tree analysis for rockfall risk assessment along a strategic mountainous transportation route. Environ Earth Sci 76(620):2017. https://doi.org/10.1007/s12665-017-6958-1
19. ISRM (2007) The complete ISRM suggested methods for rock characterization, testing and monitoring: 1974–2006. In: Ulusay R, Hudson JA (eds) Suggested methods prepared by the Commission on Testing Methods. International Society for Rock Mechanics, compilation arranged by the ISRM Turkish National Group, Kozan Ofset, Ankara, p 628

20. Pappalardo G, Mineo S, Rapisarda F (2014) Rockfall hazard assessment along a road on the Peloritani Mountains (northeastern Sicily, Italy). Nat Hazards Earth Syst Sci 14:2735–2748. https://doi.org/10.5194/nhess-14-2735-2014
21. Pappalardo G, Mineo S (2015) Rockfall hazard and risk assessment: the promontory of the Pre-Hellenic Village Castelmola Case, North-Eastern Sicily (Italy). In Lollino G et al (eds) The engineering geology for society and territory, vol 2, pp 1989–1993. doi:https://doi.org/10.1007/978-3-319-09057-3_353
22. Pappalardo G (2015) Correlation between P-wave velocity and physical–mechanical properties of intensely jointed dolostones, Peloritani mounts, NE Sicily. Rock Mech Rock Eng 48:1711–1721. https://doi.org/10.1007/s00603-014-0607-8
23. Deere DU (1963) Technical description of rock cores for engineering purposes. Felsmechanik und Ingenieurgeologie (Rock Mech Eng Geol) 1(1):16–22
24. Palmström A (1974) Characterization of jointing density and the quality of rock masses (in Norwegian).Internal report. A.B. Berdal, Norway. 1974, p 26
25. WP/WLI — The International Geotechnical Societies' UNESCO Working Party on World Landslide Inventory (1993) Multilingual landslide glossary. BiTech Publishers, Richmond
26. Cruden DM, Varnes DJ (1996) Landslide types and processes. In: Turner AK, Schuster RL (eds) Landslides, Investigation and Mitigation: Transportation Research Board. U.S. National Research Council, Special report 247, Washington, DC, pp 36–75
27. Pappalardo G, Mineo S, Calcaterra D (2017) Geomechanical analysis of unstable rock wedges by means of geostructural and infrared thermography surveys. Ital J Eng Geol Environ, Special Issue (2017), 93–101, doi:https://doi.org/10.4408/IJEGE.2017-01.S-09
28. Pappalardo G (2017) First results of infrared thermography applied to the evaluation of hydraulic conductivity in rock masses. Hydrogeol J. https://doi.org/10.1007/s10040-017-1670-5

Chapter 6
Ground Based Real Time Monitoring System Using Wireless Instrumentation for Landslide Prediction

D. P. Kanungo

Abstract Despite of our increasing knowledge on the subject, the damage tolls due to landslides are on rise during monsoon in hilly terrain. Hence, landslide prediction on temporal scale is a viable option for risk reduction. Prediction of shallow landslides developing rainfall thresholds using information on landslide occurrences and precipitation will be a cost effective risk reduction measure and may be applicable at a regional/catchment/district/tehsil/village/road corridor level in hilly terrain. Further, the installation of a real-time monitoring system can also be an alternate effective risk mitigation measure for perennial severe landslides and will be useful for community and traffic control on roads and railway tracks in hilly terrain. A Landslide Observatory with wireless instrumentation for real time monitoring of ground deformation and hydrologic parameters has been established at Pakhi Landslide in Garhwal Himalayas, India. The measurement sensors include in-place inclinometers (IPI), piezometers, wire-line extensometers and an automatic weather station (AWS). The real time data is being monitored to establish warning thresholds. The annual cumulative rainfall during 2015 was 1388 mm with cumulative monsoon period (June to September 2015) rainfall of 825 mm. At the crown of landslide beyond main scarp, there is negligible displacement being the stable part. Within the main body of the landslide, it could be inferred that the colluvium, greatly weathered bedrock and their interface experience somehow greater extent of movement at different depths in comparison to the interface between greatly weathered bedrock and unweathered bedrock. A correlation between higher intensity rainfall events and displacement pattern across the inclinometer sensors is also witnessed. However, these inferences can only be established with further data analysis of later periods. The principal aim of this chapter is to discuss the processes involved in establishment of a ground based real time monitoring system for landslides in hilly regions, in particular Indian Himalayas. Apart from establishing a landslide observatory in one of the severe landslide, the data acquisition and analysis for one monsoon season is also discussed.

D. P. Kanungo (✉)
Geotechnical Engineering Division, CSIR-Central Building Research Institute, Roorkee, India

S. P. Pradhan et al. (eds.), *Landslides: Theory, Practice and Modelling*, Advances in Natural and Technological Hazards Research 50,
https://doi.org/10.1007/978-3-319-77377-3_6

6.1 Introduction

Landslide prediction on a temporal scale is a viable option for landslide risk reduction. Rainfall thresholds such as intensity-duration and antecedent thresholds can be established for prediction of shallow landslides using information on landslide occurrences and precipitation records in a hilly region and may be applicable at a regional/catchment/district/tehsil/village/road corridor scale. Further, the installation of a real-time monitoring system can also be an alternate effective risk mitigation measure for perennial severe landslides and will be useful for community and traffic control on roads and railway tracks in hilly terrain.

Execution of a reliable landslide warning system depends on a detailed field investigation and understanding of the landslide characteristics along with a systematic quantitative analysis of the real time monitoring data obtained through a systematic instrumentation. A real time systematic landslide monitoring system can provide high quality data sets on the landslide movement dynamics in relation to the existing ground characteristics and climatic conditions. Initial investigation, instrumentation and monitoring of landslide can also guide for further augmentation of sensors for better understanding and monitoring of the real field scenario. The real time movement dynamics through real time monitoring can serve as a vital input for better geotechnical designs and also for issuing fore-warning aimed at mitigating the landslide disaster.

Real-time monitoring systems have been developed and executed in field by many countries to detect or forecast landslide occurrences. Regional level fore-warning systems based on real-time rainfall observations are already operational in Hong Kong, USA and Brazil to forewarn shallow landslides [1–4] and also have been developed for mountainous regions of Italy, New Zealand, and Taiwan [5–7]. Regional early warning systems using real-time monitoring network have also been developed and used by U.S. Geological Survey (USGS) for recording the dynamics of hazardous active landslides or landslide-prone hill slopes and fore-warning [8–11]]. Such real time monitoring systems are also popular in many countries to monitor vital infrastructures, such as dams, or landslides [12–16]. Deformation behavior of rainfall triggered landslides has been studied using GPS/GNSS, Acoustic Emission (AE) technique and other conventional sensors like inclinometer, extensometer or tilt meter [17–24]. Modern techniques including radar interferometry, LiDAR, GPS and photogrammetric tools have enormous potential to detect the activity prior to a land sliding phenomenon with high resolution [25–31]. In India, there are few attempts on real time monitoring of landslides using wired sensor network in Northwestern Himalayas [32–35] and wireless sensor networks in Western Ghat region of South India [36, 37].

In this chapter, we will discuss the establishment of a Landslide Observatory based on landslide instrumentation and real time monitoring of Pakhi landslide along Alaknanda Valley of Garhwal Himalayas using wireless sensors network to develop an early warning system.

6.2 Study Area

Earlier studies [38, 39] have revealed that the Pakhi Landslide is a potential debris slide. It is located at a distance of 9 km from Pipalkoti on National Highway 58 of Garhwal Himalayas in Uttarakhand State of India. The geology of the area is complex, consisting of Precambrian lithological units of NW Lesser Himalaya. The main geological formations belong to the Garhwal Group and Central Crystalline Zone of rocks and comprise of quartzites, dolomites with slates, metavolcanics, gneisses and schists. The rocks in the landslide area comprise mainly of dolomites. Physiographically, the area lies in a region of tectonic or folded and overthrust mountain chain with strata structurally marked by complex folds, reverse faults, overthrusts and nappes of great dimensions [40]. Main Central Thrust lies in the vicinity of the study area and is responsible for crushing and shearing of rocks. The area is traversed by the river Alaknanda and its tributaries.

Two natural streams (one on the extreme left flank boundary and the other at the centre) flow within the main body of the Pakhi landslide (Fig. 6.1b) and join the Alaknanda River at the downhill side. The topographic map showing elevation contours along with the locations of instrumentation is given in Fig. 6.1a. The portion of the landslide left to the central stream (Fig. 6.1b) is 112 m in length at the road level with two active stretches of about 57 m and 15 m spread lengths respectively. The landslide slope is dipping at 56° towards N320°. The unfavorable discontinuity with a dip-slope relationship has a dip of about 45°-55° towards N330°. The main scar is clearly visible on the uphill slope and a number of minor (secondary) scars have been developed along the radial transverse tension cracks on the main landslide body. There is a clear indication of detachment and displacement of the order of 0.5 m–1.0 m along these tension cracks.

The other portion of the debris slide right to the central stream (Fig. 6.1) occurred on a sub-ridge slope encompassing the up and down hill sides of the national highway. This landslide has a spread length of 90–100 m at road level. The landslide slope is dipping at about 40°–45° towards N245°. The unfavorable discontinuity in the dolomitic limestone rock has a dip of about 35°–40° towards slope direction itself. This landslide portion has the main scar along the sub-ridge line and many longitudinal tension cracks of about 30 cm wide and 40 cm deep are developed on the slide body.

The sub-surface material on the landslide slope comprises of three distinct layers such as colluvium, greatly weathered bedrock (dolomitic limestone) and unweathered bedrock. The depth of the top colluvium layer varies from 4.5 m to a maximum of 13.5 m from the ground level. The thickness of the next layer of greatly weathered bedrock varies from 4.5 m–10.5 m beyond colluvium at places. Beyond the weathered bedrock, in-situ unweathered bedrock is present as observed from the borehole investigation. At places, the unweathered bedrock is directly underlain by colluvium as observed in boreholes 1, 4, 5 and 6.

From the past records and precipitation history of the study area, it can be inferred that the initiation of this debris slide can be attributed to road excavation process and its further expansion and activation may be due to heavy monsoon precipitation in this area (generally from June to September months).

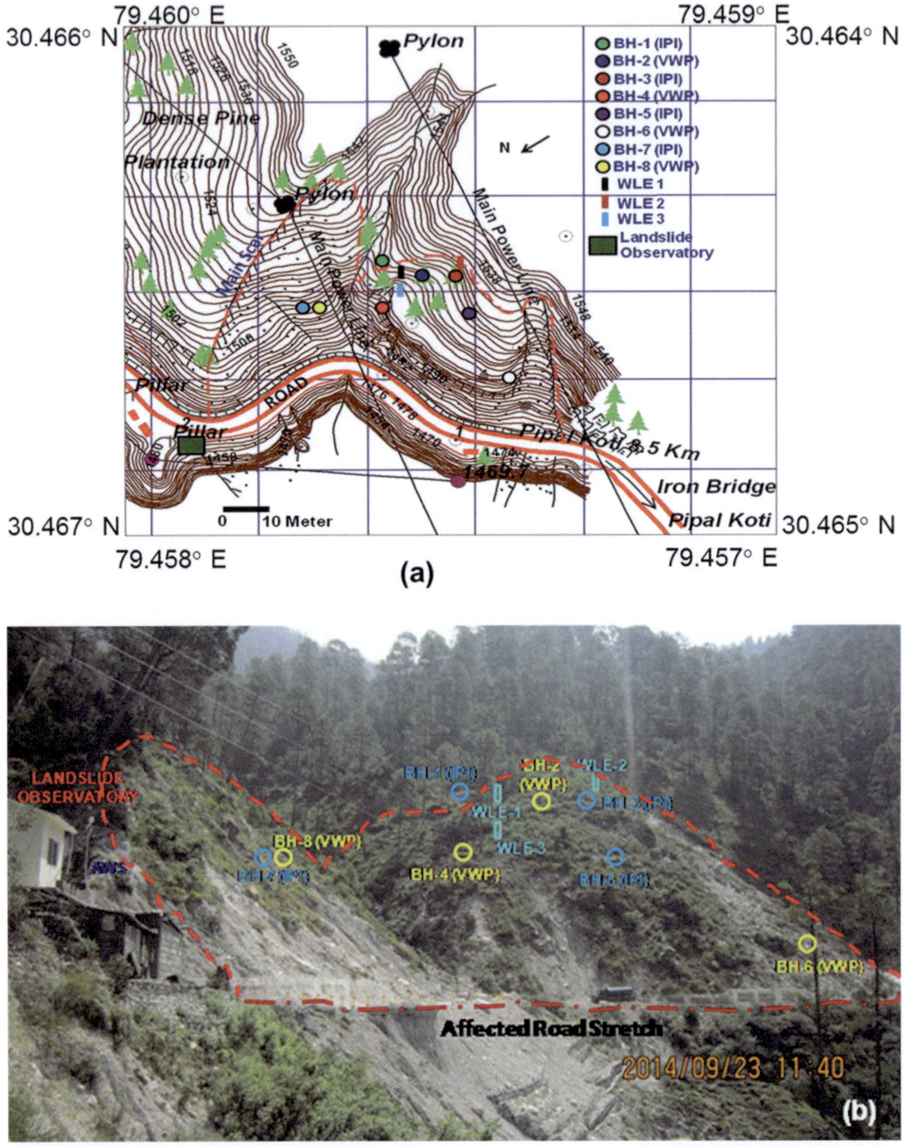

Fig. 6.1 (**a**) Topographic map (Source: Kanungo et al., 2013b) and (**b**) field photograph of Pakhi Landslide selected for instrumentation and real-time monitoring (*BH* – Bore Hole, *IPI* – Inplace Inclinometer, *VWP* – Vibrating Wire Piezometer, *WLE* – Wire Line Extensometer, *AWS* – Automatic Weather Station)

6.3 Landslide Hazard Scenarios

Landslide modeling is carried out to understand the failure mechanism and to assess the hazard scenario of the hill slope. 2D finite element modeling of the potential debris slide has been carried out under dry conditions. The analysis has shown that the slope is completely unstable under dry condition. The introduction of water table or consideration of pore water pressure in the analysis will further reduce the factor of safety. The analysis results validate and confirm the already initiated failure zones and expected progressive failure zones as also observed in the field. It may be inferred that the geometry of the slope, material profile and characteristics along the slope are the most crucial data for accurate modeling and assessing the hazard of the potential landslide. Though the field observation depicted two different soil/ debris layers above the in-situ rock on the slope, the shear strength properties indicated these two layers as homogeneous. Instead of adopting the limit equilibrium analysis method for the homogeneous material, finite element modeling approach has been adopted for better understanding of the displacement pattern and shear strain distribution along the slope profile. The analysis results indicated that the type of failure is more translational in nature along the interface between weathered disintegrated product of the dolostone & the competent in-situ dolostone. This failure mode has also been observed from the field. In a nutshell, finite element modeling has helped in identifying stress accumulation zones and also the extent of displacement expected to occur for accurately locating the sites for placement of sensors for instrumentation and monitoring of actual ground movements in a landslide. The details of the stability assessment of this potential landslide slope can be found in [38, 39] and, therefore, has not been provided here for brevity. The drilling at different locations on the landslide area and their logging will help in unfolding the real geological variation depth wise and will also help in accurately modeling the landslide behaviour.

6.4 Landslide Instrumentation and Real Time Monitoring Scheme

Ground-based instrumentation and real-time monitoring systems for landslide could be realized in recent years because of rapid development of various common components such as wireless sensors for parametric measurements in field, wireless networking technology for these sensors, web server based communication system from field to control stations and software for real time data monitoring and analysis. However, there is no unique system developed for monitoring of all types of landslides. The components of real time monitoring system for landslides in terms of type and number of sensors, location of sensor placement both on surface and sub-surface, design and configuration of sensor networking and communication system vary from site to site depending on types of landslides. This depends mainly on the following: (i) purpose of the systems such as understanding the dynamics of a landslide from scientific research point of view, monitoring the historical landslides for societal safety etc. and (ii) the types of landslides such as mainly debris/soil slides and rock slides.

A network of sensors is required to be placed both on the ground surface and in the subsurface strata of the landslide for measuring the activities related to its dynamics [41]. For this study, commercially available sensors [42, 43] are selected, procured and placed at the site. Special considerations have been kept in mind in selecting the sensors in terms of their ruggedness, weather resistant and portability for field implementation along with low power consumption and can be powered by solar energy through battery. Sensors having proper compatibility with the data acquisition systems and with adequate sensitivity and resolution have been selected. Also, sensors network with optimum spatial coverage, both on the ground and in the subsurface has been taken into consideration. A scheme of wireless landslide instrumentation and real-time monitoring system already implemented in the selected landslide site is demonstrated in Figs. 6.2 and 6.3.

In this attempt, both surface and sub-surface sensors were installed on the selected landslide area (Fig. 6.1). Two types of information such as actual displacements/ movements at different locations in the landslide area and environmental/weather conditions that affect the sliding activity have been targeted with the combination of surface and sub-surface mounted monitoring sensors. Surface sensors include wire-line extensometers (WLE) and automatic weather station (AWS) as shown in Fig. 6.4(a) and (b). Three numbers of wire-line extensometers have been installed

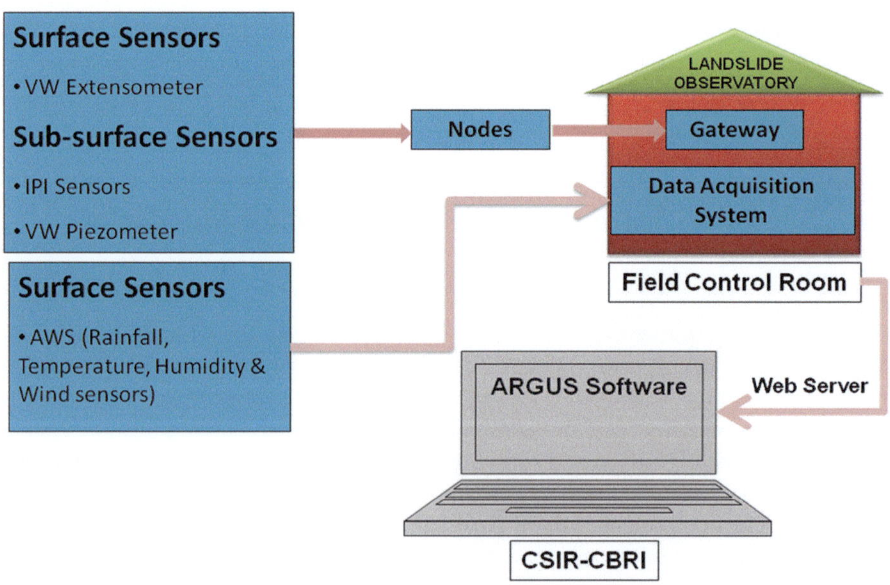

Fig. 6.2 Scheme of wireless landslide instrumentation and real-time monitoring system

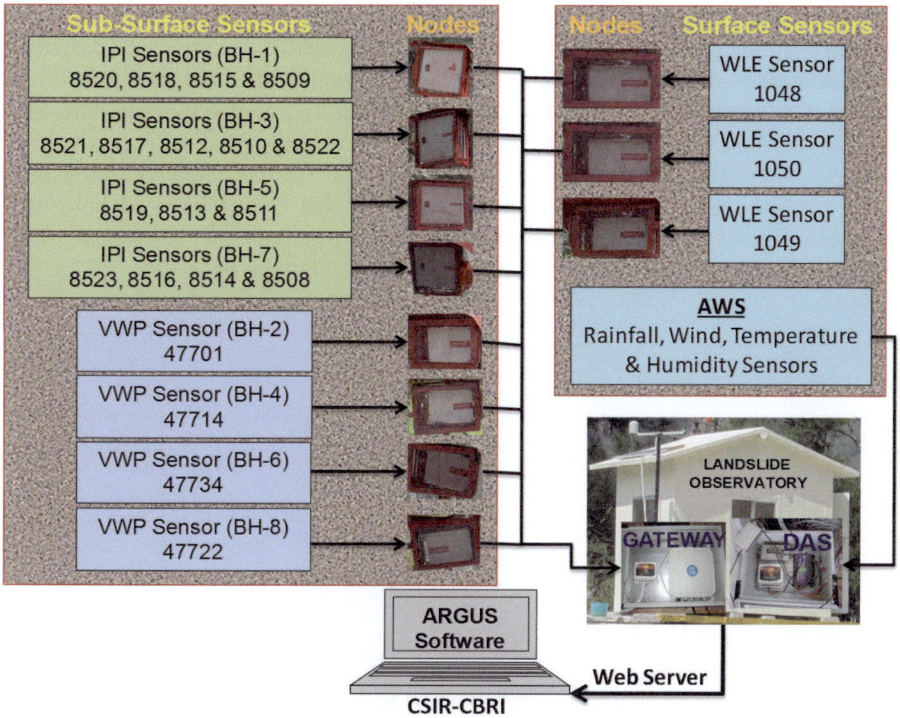

Fig. 6.3 Real-time monitoring system implemented in Pakhi landslide site

Fig. 6.4 (**a**) Wire-line extensometer (WLE); (**b**) Automatic Weather Station (AWS)

across the radial tension cracks developed on the landslide body to monitor the detachment/displacement of the downhill block from the uphill block of the landslide mass. The surface sensors are particularly subject to disturbance by animals; theft etc. and hence, are protected by providing wire mesh cages around the sensors at the site.

Automatic weather station includes different instruments/sensors to measure rainfall, air temperature, relative humidity and wind velocity and direction. Sub-surface sensors include biaxial in-place inclinometers (IPI) installed at different depths within a particular material and in the interface zones down the bore hole (BH) to measure the sub-surface displacements/movements (Fig. 6.5) and vibrating wire piezometers (VWP) in the bore holes (Fig. 6.6) to measure the variation in pore water pressure. Depths of these sensors were decided on the basis of bore hole geological logging information. In total, 16 in-place inclinometer sensors and four piezometric sensors were installed in total eight different bore holes (Table 6.1). The

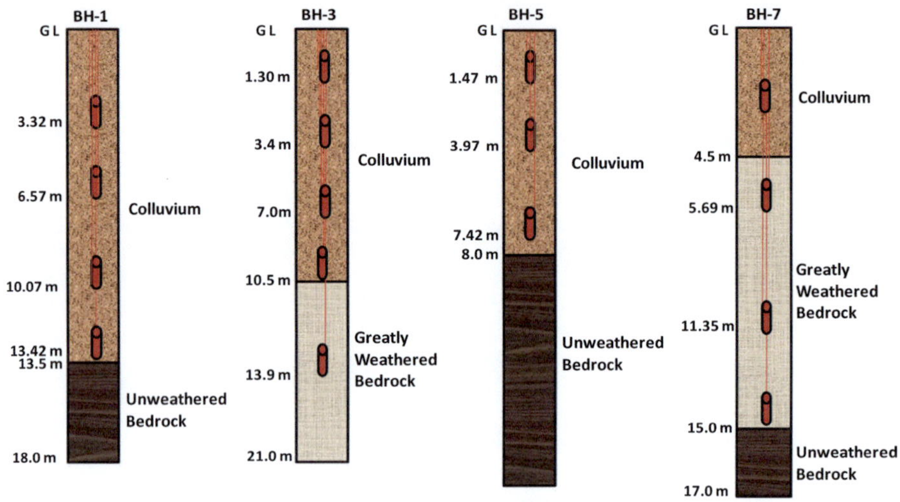

Fig. 6.5 In-place inclinometer sensors installed in different boreholes (*BH* Borehole, *GL* Ground level)

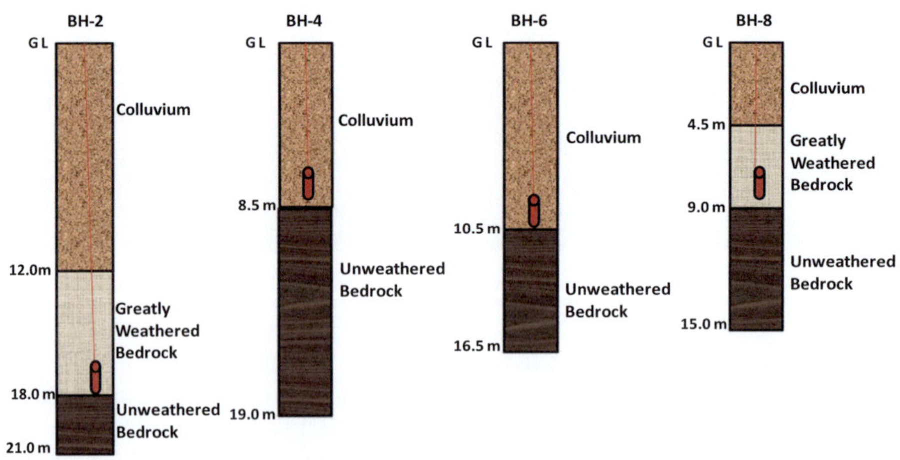

Fig. 6.6 Vibrating wire piezometric sensors installed in different boreholes (*BH* Borehole, *GL* Ground level)

Table 6.1 Sensor details in different boreholes

Borehole no.	Type of sensor	No. of sensor	Sensor depth (m)	Material at sensor depth	Displacement (mm)	
					A-axis	B-axis
1	Vertical biaxial in-place inclinometer	4	2.92–3.32	Colluvium	0	0
			6.17–6.57	Colluvium	0	0
			9.67–10.07	Colluvium	1	0
			13.02–13.42	Interface between colluvium and unweathered bedrock	2	0
2	Vibrating wire piezometer	1	17.60–18.0	Interface between greatly weathered bedrock and unweathered bedrock	–	–
3	Vertical biaxial in-place inclinometer	5	0.90–1.30	Colluvium	0	0
			3.00–3.40	Colluvium	0	0
			6.60–7.00	Colluvium	1	0
			10.05–10.45	Interface between colluvium and greatly weathered bedrock	3–4	3–4
			13.50–13.90	Greatly weathered bedrock	6	1
4	Vibrating wire piezometer	1	7.6–8.0	Interface between greatly weathered bedrock and unweathered bedrock	–	–
5	Vertical biaxial in-place inclinometer	3	1.07–1.47	Colluvium	1–2	2–10
			3.57–3.97	Colluvium	5–6	2–4
			7.02–7.42	Interface between colluvium and unweathered bedrock	2–4	1–4
6	Vibrating wire piezometer	1	10.1–10.5	Interface between greatly weathered bedrock and unweathered bedrock	–	–
7	Vertical biaxial in-place inclinometer	4	2.30–2.70	Colluvium	0	0
			5.29–5.69	Greatly weathered bedrock	2–4	1
			10.95–11.35	Greatly weathered bedrock	6–8	8–9
			14.45–14.85	Interface between greatly weathered bedrock and unweathered bedrock	2–3	3–4
8	Vibrating wire piezometer	1	8.1–8.5	Interface between greatly weathered bedrock and unweathered bedrock	–	–

suitable casings for installation of IPI sensors are placed in all the boreholes beyond the interface of greatly weathered bedrock and unweathered bedrock for accurate measurements except borehole 3 due to site constraints. The surface sensors are particularly subject to disturbance by animals; theft etc. and hence, are protected by providing wire mesh cages around the sensors at the site.

All these surface and sub-surface sensors except AWS are connected through wire to the specific nodes placed in close proximity to the sensors. These nodes are communicating wirelessly with the gateway placed in the field control station. AWS is connected to the data acquisition system (DAS) also placed in the field control station to store the data.

The data from the field control station are being transferred on real-time to the control computer at CSIR-CBRI, Roorkee through web server using ARGUS monitoring software. The commercially available ARGUS software designed for landslide monitoring is used for data analysis and visualization. The reference reading for all the sensors corresponds to 29th September 2014 (i.e., the date on which the monitoring system made operational).

6.4.1 Results and Discussion

The annual cumulative rainfall at the landslide site during 2015 was recorded 1388 mm with the cumulative monsoon period (June to September 2015) rainfall of 823.5 mm. During monsoon period, there were two occasions with somewhat higher intensities of rainfall during 25th June to 12th July having around 350 mm and during 6th August to 14th August having about 175 mm of rainfall (Fig. 6.7).

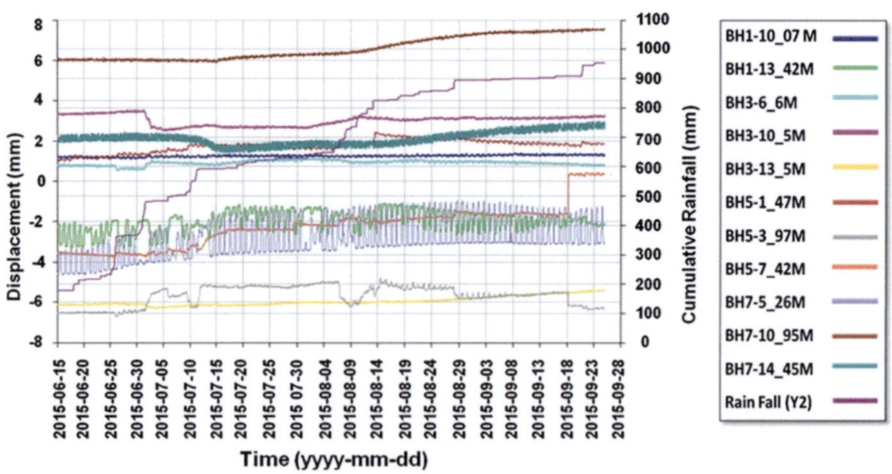

Fig. 6.7 Cumulative displacement (mm) as observed from IPI sensors (A-axis) in boreholes during monsoon period

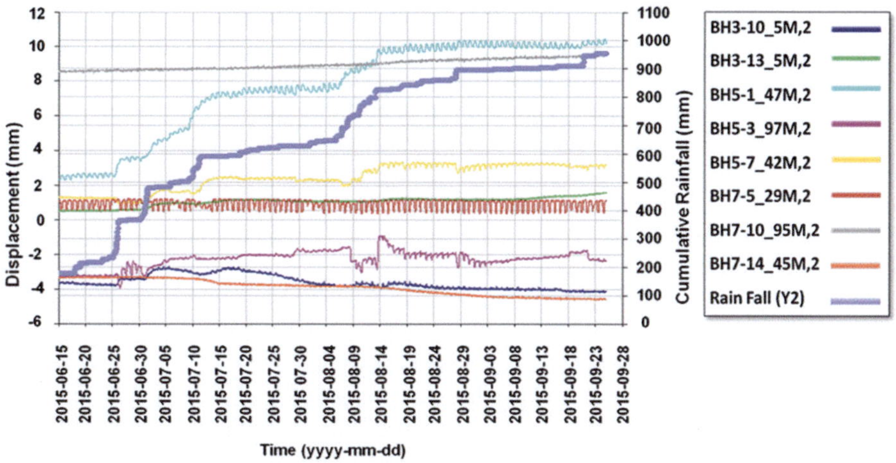

Fig. 6.8 Cumulative displacement (mm) as observed from IPI sensors (B-axis) in boreholes during monsoon period

The extent of displacements in IPI sensors along different boreholes on both A-axis (aligned in the slope direction) and B-axis (perpendicular to A-axis) are also given in Table 6.1. Figures 6.7 and 6.8 demonstrate the displacements measured by IPI sensors in different boreholes along both A-axis and B-axis respectively except those measurements with zero displacement (refer to Table 6.1). From the IPI data (BH-1) at the crown of the landslide beyond the main scar, it is observed that there is negligible displacement on sub-surface sensors along both the axes (Table 6.1 and Figs. 6.7 and 6.8) which was as expected being the stable part of the slope.

On the uphill slope at the boundary of main scar, displacements of the order of 3-4 mm in both axes at a depth of 10.45 m (i.e., interface between colluvium and greatly weathered bedrock) and of about 6 mm along A-axis at a depth of 13.9 m (i.e., within greatly weathered bedrock) in borehole 3 (BH-3) are observed (Figs. 6.7 and 6.8). At the middle portion of the landslide left to the central drain in borehole 5 (BH-5), maximum cumulative displacements of about 10 mm along B-axis at a depth of 1.47 m (Fig. 6.8) and about 5-6 mm along A-axis at a depth of 3.97 m (Fig. 6.7) within the colluvium are observed. At this location, the interface between colluvium and unweathered bedrock has observed a displacement of the order of 2-4 mm along both axes (Figs. 6.7 and 6.8). In borehole 7 (BH-7), the interface between greatly weathered bedrock and unweathered bedrock (at a depth of 14.85 m) witnessed a displacement of about 3-4 mm along both axes; whereas within the greatly weathered bedrock a maximum cumulative displacement of the order of 6–9 mm is experienced along both axes at a depth of 11.35 m (Figs. 6.7 and 6.8). There exists some ambiguity in the displacement data of borehole 3 as it could not be drilled and casings are not laid beyond the interface of greatly weathered and unweathered bedrock. However, the extreme end of the casings has been grouted properly.

Comparing rainfall events with the displacement patterns of all IPI sensors (Figs. 6.7 and 6.8), it can be observed that during 25th June to 12th July and 6th August to 14th August with higher intensities of rainfall there are increase in displacement rates at borehole 3 (at 10.5 m depth), borehole 5 (at 1.47 m, 3.97 m and 7.42 m depths) and borehole 7 (at 14.45 m depth). These displacements are observed to be higher along B-axis as compared to the A-axis of the sensors. This indicates a bearing of rainfall on the displacement in different sub-surface strata. However, such pattern and extent of movement activities can only be ascertained with future datasets of monsoon seasons.

It is further observed from the vibrating wire piezometric sensors installed in the bore holes (BH-2,4 & 6) at different depths at the interface between greatly weathered bedrock and unweathered bedrock that negative pressures of the order of −20 to −22 kPa during monsoon season have been developed (Fig. 6.9). This indicates that the strata have not attained the saturation level during monsoon. It may be inferred that the rain water could not infiltrate up to the unweathered bedrock level percolating through the greatly weathered bedrock. Hence, the installation of piezometric sensor at the interface between colluvium and greatly weathered bedrock would have been more useful to capture the pore pressure.

An initial displacement of the order of 2-4 mm has been observed from the surface installed wire line extensometer (WLE-1) till 13th September 2015. A

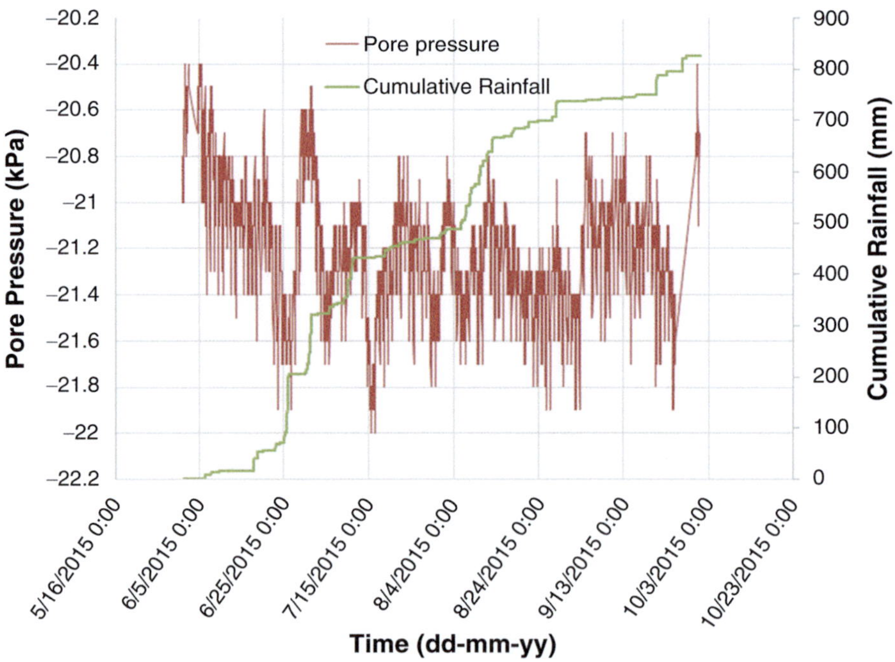

Fig. 6.9 Pore pressure (kPa) variations as observed from VW piezometric sensor in BH-2 during monsoon period

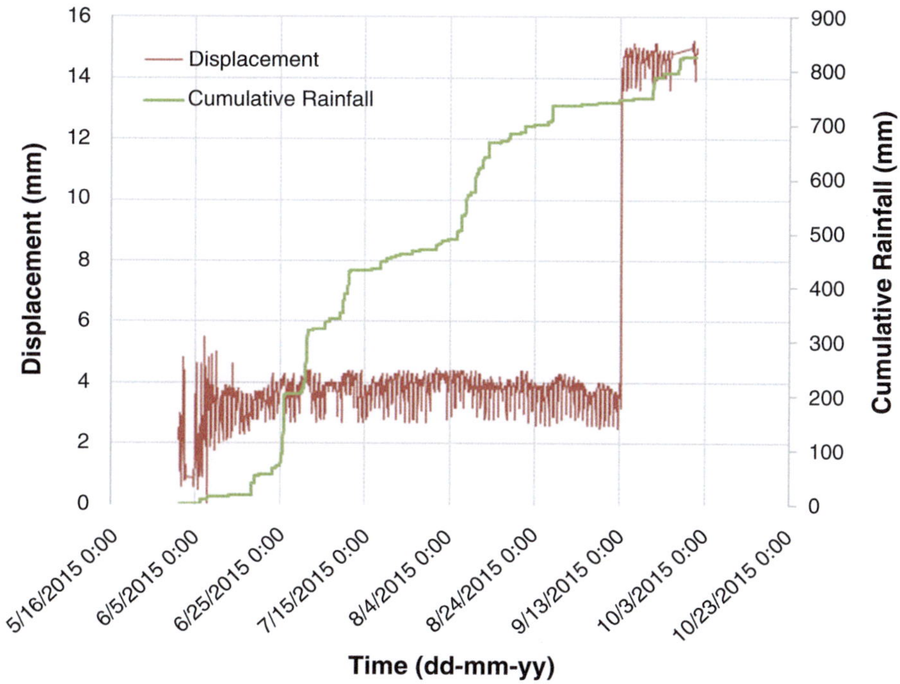

Fig. 6.10 Cumulative displacement from the surface installed wire line extensometer (WLE 1) during monsoon period

sudden increase in the displacement up to 14 mm has been observed on 13th September 2015 which cannot be correlated with the intensity of rainfall on this particular day (Fig. 6.10). Such type of sudden surficial movement may be attributed to the antecedent rainfall prior to the event which needs further in-depth study.

It can further be inferred from these displacement measurements that the strata within colluvium and greatly weathered bedrock have experienced higher displacements compared to the interface between greatly weathered and unweathered bedrock. Though the displacement at the interface between greatly weathered and unweathered bedrock was presumed and expected to be high, only a nominal displacement of maximum 2-4 mm is observed across the boreholes (BH-1, 5 & 7). This may be attributed to lack of sufficient data at this interface over the landslide area and specifically borehole 3 could not be drilled beyond this interface due to unforeseen site constraints. Such negligible or less displacement at the interface may further be attributed to lack of infiltration of rain water up to this depth which restrained mobilisation of movement (also complemented by the piezometric sensor data with negative pore pressures). Hence, future measurements during monsoon seasons with sufficient displacement data will help in addressing such issues with more clarity.

From the above observations, it can be deciphered that local slip surfaces exist within the colluvium and greatly weathered bedrock. However, this can be ascertained only after further data interpretation of later periods. But, it can be stated from the ground measurements that the Pakhi landslide is a very slow moving landslide.

6.5 Summary

In this chapter, a scheme of ground based real time monitoring system with wireless instrumentation is presented. The surface and sub-surface instruments/sensors were implemented in a test area of Pakhi Landslide in Alaknanda valley of the Garhwal Himalayas, India establishing a landslide observatory and the data acquired for one monsoon season was analyzed to throw some understanding on the dynamics of the particular landslide for its future prediction.

The landslide observatory with real-time landslide monitoring system is established at Pakhi Landslide in Garhwal Himalayas, India with an objective to understand the dynamics of landslide movement. Initial dataset indicates a slow movement of surface and sub-surface strata at certain depths across the landslide body and also its correlation with rainfall intensity. However, it needs to be established with more datasets.

Such type of real time monitoring of landslides through intensive instrumentation is somewhat a costly affair and may not be advisable to replicate in each landslide to issue early warning. Hence, this type of extensive instrumentation for establishing early warning can be planned and designed for perennial active landslides only. Further, establishing a warning threshold based on rainfall in relation to landslide occurrences in this terrain and then using this threshold for early warning to save lives and property and to control the traffic on hill roads is the only suitable, feasible and cost effective option. On establishing a rainfall threshold for a particular terrain having a specific lithotectonic and geomorphological setup, a number of AWS units in the region can be wirelessly networked at a central station and an early warning can be issued.

Acknowledgments Authors are grateful to the Director, CSIR-Central Building Research Institute, Roorkee, Uttarakhand (India) for granting permission to publish this paper.

References

1. Finlay PJ, Fell R, Maguire PK (1997) The relationship between the probability of landslide occurrence and rainfall. Can Geotech J 34(6):811–824
2. Ortigao B, Justi MG (2004) Rio-Watch: the Rio de Janeiro landslide alarm system. Geotech News 22(3):28–31

3. Wilson RC (2005) The rise and fall of debris-flow warning system for the San Francisco Bay region, California. In: Glade T, Anderson M, Crozier MJ (eds) Landslide hazard and risk. Wiley, West Sussex, pp 493–516
4. Chleborad AF, Baum RL, Godt JP (2006) Rainfall thresholds for forecasting landslides in the Seattle, Washington, Area - exceedance and probability. US Geol Surv Open-File Report:2006–1064
5. Aleotti P (2004) A warning system for rainfall-induced shallow failures. Eng Geol 73 (3–4):247–265
6. Chien-Yuan C, Tien-Chien C, Fan-Chieh Y, Wen-Hui Y, Chun-Chieh T (2005) Rainfall duration and debris-flow initiated studies for real-time monitoring. Environ Geol 47:715–724
7. Schmidt J, Turek G, Clark M, Uddstrom M (2007) Real-time forecasting of shallow, rainfall-triggered landslides in New Zealand. Geophys Res Abstr 9:5778
8. Keefer DK, Wilson RC, Mark RK, Brabb EE, Brown WM, Ellen SD, Harp EL, Wieczorek GF, Alger CS, Zatkin RS (1987) Real-time landslide warning during heavy rainfall. Science 238 (4829):921–925
9. NOAA-USGS Debris Flow Task Force (2005) NOAAUSGS debris-flow warning system—final report. U.S. Geological Survey Circular 1283
10. Reid ME, LaHusen RL (1998) Real-time monitoring of active landsides along Highway 50, El Dorado County. Calif Geol 51(3):17–20
11. Baum RL, Godt JP, Harp EL, McKenna JW, McMullen SR (2005) Early warning of landslides for rail traffic between Seattle and Everett, Washington, USA. Proceedings of 2005 International Conference on Landslide Risk Management, A.A. Balkema, New York, pp 731–740
12. Angeli MG, Gasparetto P, Menotti RM, Pasuto A, Silvano S (1994) A system of monitoring and warning in a complex landslide in northeastern Italy. Landslide News 8:12–15
13. Berti M, Genevois R, LaHusen RL, Simoni A, Tecca PR (2000) Debris flow monitoring in the Acquabona watershed on the Dolomites (Italian Alps). Phys Chem Earth Part B Hydrol Oceans Atmos 26(9):707–715
14. Husaini O, Ratnasamy M (2001) An early warning system for active landslides. Quar J Eng Geol Hydrogeol 34:299–305
15. Froese CR, Moreno F (2007) Turtle Mountain Field Laboratory (TMFL): Part 1—overview and activities.: In: Schaefer VR, Schuster RL Turner AK (eds) , Conference presentations: 1st North American Landslide Conference, AEG Special Publication 23. Association of Engineering Geologists, Vail, pp 971–980
16. Frigerio S, Schenato L, Bossi G, Cavalli M, Crema S, Mantovani M, Marcato G, Pasuto A (2014) Landslide monitoring with an integrated platform: methodology, design and case study. Rend. Online Societa Geologica Italiana 30:24–27
17. Malet JP, Maquaire O, Calais E (2002) The use of global positioning system techniques for the continuous monitoring of l and slides: application to the Super-Sauze earthflow (Alpes-de-Haute-Provence, France). Geomorphology 43:33–54
18. Corsini A, Pasuto A, Soldati M, Zannoni A (2005) Field monitoring of the Corvara landslide (Dolomites, Italy) and its relevance for hazard assessment. Geomorphology 66:149–165
19. Garcia A, Hordt A, Fabian M (2010) Landslide monitoring with high resolution tilt measurements at the Dollendorfer Hardt landslide, Germany. Geomorphology 120:16–25
20. Brückl E, Brunner FK, Lang E, Mertl S, Müller M, Stary U (2013) The Gradenbach observatory-monitoring deep-seated gravitational slope deformation by geodetic, hydrological, and seismological methods. Landslides 10:815–829
21. Dixon N, Spriggs MP, Smith A, Meldrum P, Haslam E (2015) Quantification of reactivated landslide behaviour using acoustic emission monitoring. Landslides 12(3):549–560
22. Kumsar H, Aydan O, Tano H, Celik SB, Ulusay R (2016) An integrated geomechanical investigation, multi-parameter monitoring and analyses of Babadag-Gundogdu Creep-like Landslide. Rock Mech Rock Eng 49(6):2277–2299
23. Macciotta R, Hendry M, Martin CD (2016) Developing an early warning system for a very slow landslide based on displacement monitoring. Nat Hazards 81(2):887–907

24. Uhlemann S, Smith A, Chambers J, Dixon N, Dijkstra T, Haslam E, Meldrum P, Merritt A, Gunn D, Mackay J (2016) Assessment of ground-based monitoring techniques applied to landslide investigations. Geomorphology 253:438–451
25. Teza G, Galgaro A, Zaltron N, Genevois R (2007) Terrestrial laser scanner to detect landslide displacement fields: a new approach. Int J Remote Sens 28(16):3425–3446
26. Monserrat O, Crosetto M (2008) Deformation measurement using terrestrial laser scanning data and least squares 3D surface matching. ISPRS J Photogramm Remote Sens 63:142–154
27. Abellán A, Jaboyedoff M, Oppikofer T, Vilaplana JM (2009) Detection of millimetric deformation using a terrestrial laser scanner: experiment and application to a rockfall event. Nat Hazards Earth Syst Sci 9:365–372
28. Barla G, Antolini F, Barla M, Mensi E, Piovano G (2010) Monitoring of the Beauregard landslide (Aosta Valley, Italy) using advanced and conventional techniques. Eng Geol 116:218–235
29. Casagli N, Catani F, Del Ventisette C, Luzi G (2010) Monitoring, prediction, and early warning using ground-based radar interferometry. Landslides 7(3):291–301
30. Intrieri E, Gigli G, Mugnai F, Fanti R, Casagli N (2012) Design and implementation of a landslide early warning system. Eng Geol 147–148:124–136
31. Barla M, Antolini F (2016) An integrated methodology for landslides early warning systems. Landslides 13(2):215–228
32. Mittal SK, Singh M, Kapur P, Sharma BK, Shamshi MA (2008) Design and development of instrumentation network for landslide monitoring and issue an early warning. J Sci Ind Res 67:361–365
33. Chaturvedi P, Jaiswal B, Sharma S, Tyagi N (2014) Instrumentation Based Dynamics Study of Tangni Landslide near Chamoli, Uttrakhand. Int J Res Advent Technol 2(10):127–132
34. Chaturvedi P, Srivastava S, Kaur B (2016) Landslide Early Warning System Development using Statistical Analysis of Sensors' Data at Tangni Landslide, Uttarakhand, India. 6th International Conference on "Soft Computing for Problem Solving" (SocProS 2016), Patiala, India, December 2016
35. Mishra PK, Chaturvedi P, Tyagi N, Joglekar PN (2015) Real time landslide monitoring and estimation of its movement velocity. IEEE Int Conf on Research in Computational Intelligence and Communication Networks (ICRCICN), Kolkata, Nov 2015, doi:https://doi.org/10.1109/ICRCICN.2015.7434260
36. Ramesh MV, Vasudevan N (2012) The deployment of deep-earth sensor probes for landslide detection. Landslides 9(4):457–474
37. Ramesh MV (2014) Design, development, and deployment of a wireless sensor network for detection of landslides. Ad Hoc Netw 13:2–18
38. Kanungo DP, Pain A, Sharma S (2013a) Stability assessment of a potential debris slide in Garhwal Himalayas, India. Indian Landslides 6(2):9–20
39. Kanungo DP, Pain A, Sharma S (2013b) Finite element modeling approach to assess the stability of debris and rock slopes: a case study from the Indian Himalayas. Nat Hazards 69:1–24
40. Krishnan MS (1982) Geology of India and Burma, 6th edn. CBS Publishers and Distributors, Delhi, 536p
41. Reid M E, Baum R L, LaHusen R G, Ellis W L, (2008) Landslides and Engineered Slopes. Chen et al.. Taylor & Francis Group, London, (ISBN 978-0-415-41196-7) 179–191
42. Dunnicliff J (1993) Geotechnical instrumentation for monitoring field performance. Wiley, New York
43. Mikkelsen PE (1996) Field Instrumentation. In: Turner AK, Schuster RL (eds) Landslides: investigation and mitigation, Special report 247. National Academy Press, Washington, DC, pp 278–316

Part III
Landslide Risk and Hazard Mitigation

Chapter 7
Presentation of RFFR New Ensemble Model for Landslide Susceptibility Assessment in Iran

Aiding Kornejady, Hamid Reza Pourghasemi, and Sayed Fakhreddin Afzali

Abstract The current study is focused on landslide susceptibility mapping over a critical mountainous watershed, Chehel-Chai, located in the Golestan Province, Iran. An integrated data mining new ensemble model, comprised of random forest and frequency ratio (RFFR), was proposed and employed as a robust computational algorithm in the study area. Landslide inventory map was prepared in Geographic Information System (GIS) by using several field surveys, local information, and available organizational resources. In this study, using different literature review and data availability, 12 landslide conditioning factors including proximity from fault (PFF), proximity from stream/river (PFS), proximity from road (PFR), lithological units, soil texture, land use/land cover (LU/LC), slope degree, slope aspect, altitude, plan curvature (PlanC), profile curvature (ProfC), and topographic wetness index were chosen and the corresponding maps were produced in the ArcGIS 10.2. For modeling, the FR values were calculated and then used for implementing RF in R 3.0.2 statistical software by "randomForest" package. In order to validate the built model, the receiver operating characteristic (ROC) curve using 30% of cast-off landslide was considered. The results revealed that the RFFR new ensemble model with the AUC value of 0.831 had a good performance (AUC = 83.10%) for landslide susceptibility zonation over the study area. Based on the RFFR model, about 42.27% of the Chehel-Chai Watershed has high (24.18%) and very high (18.09%) susceptibility to landslide occurrence. Hence, the proposed new algorithm was found to be suitable for landslide susceptibility modeling in the study area and, accordingly, for land use planning and landslide hazard management.

A. Kornejady
Department of Watershed Management Engineering, Gorgan University of Agricultural Sciences and Natural Resources, Gorgan, Iran

H. R. Pourghasemi (✉) · S. F. Afzali
Department of Natural Resources and Environmental Engineering, College of Agriculture, Shiraz University, Shiraz, Iran
e-mail: hr.pourghasemi@shirazu.ac.ir

S. P. Pradhan et al. (eds.), *Landslides: Theory, Practice and Modelling*,
Advances in Natural and Technological Hazards Research 50,
https://doi.org/10.1007/978-3-319-77377-3_7

7.1 Introduction

Varnes defined a landslide as all downward mass movement on slopes, including natural cliffs, soil, and artificial deposits or a mixture of them under influence of gravity [1]. Landslides are among the most significant natural damaging events occurring in different types, frequency, and intensity and controlled by different factors. About 4900 big notable landslides have occurred and mapped in Iran that cause too many damages until the end of September 2007 [2, 3].

While tens of numerical models with different weights, rates, parameters, and computational algorithm have been developed in the various conditions and studies have been devoted to map landslide susceptibility, there is a general consensus that these methods can be grouped into four different categories [4]: (1) inventory-based approaches which portray spatial and temporal patterns of landslide distribution, types and rates of movement, and types of displaced material [5, 6]; (2) statistical- probabilistic data driven approaches which minimize subjectivity in weightage assignment procedure and produce more objective and reproducible results [7, 8] including multivariate statistical approach such as logistic regression [9–11] or bivariate approaches such as Landslide Nominal Risk Factor (LNRF) [12–14], information value [15–17], weights of evidence [18–23], frequency ratio [24–28], fuzzy logic [29–32], Shannon's entropy [33–35, 26], artificial neural network [29, 36–38] and Certainty Factor [28, 39–41]; (3) heuristic approaches which estimate landslide potential from data on preparatory variables based on expert knowledge and opinions [12], such as Bureau of Indian standards (BIS) based landslide hazard evaluation factor (LHEF) method [42–48]; (4) deterministic approaches which deal with slope stability analysis and they are commonly used in site-specific scale such as factor of safety [49, 50] model.

Nowadays, machine learning models in terms of estimation of distribution algorithms (EDAs), data mining nature (DMN), and iterative modeling process (IMP), have presented high ability to detect the natural behavior and spatial occurrence of landslides. Several studies have been devoted to this field using different data mining model such as maximum entropy [23, 51–59], boosted regression tree (BRT) [60–62], classification and regression tree (CART) [63–66], support vector machine (SVM) [67–72], general linear model (GLM) [73, 74] and random forest (RF) [75–82]. These models have shown better performance rather than other bivariate or multivariate statistical methods [78, 83–87].

The Golestan Province, due to its semi-temperate climate and presence of highly susceptible lithological units to landslide occurrence, is known as a critical landslide prone area in Iran. The Chehel-Chai Watershed, in the mentioned province, shows the same symptoms of landslide occurrences. Furthermore, based on field surveys, unconsidered road constructions with poor foundations, human-induced land use changes and over housing features are the most tangible signs of mankind balance breaking behavior at the study area. Hence, preparation of landslide susceptibility map as the first stage of decision making process and land use planning is important in the study area. So, the aim of current study is to propose a new ensemble data mining technique namely RFFR for landslide spatial modeling in the Chehel-Chai Watershed, Iran.

7.2 Study Area

The study area (Chehel-Chai Watershed) is one of the mountainous basins in the Golestan Province, Iran, with an area of 255 km^2. It is located between 55° 23′ E to 55° 38′ E longitudes and 36° 59′ N to 37° 13′ N latitudes, UTM (Universal Transverse Mercator) Zone 40, with the minimum altitude of 190 m and maximum altitude of 2527 m above sea level (Fig. 7.1). Streams of the basin mainly have the general direction of south-north. Average annual precipitation is equal to 766 mm. According to Central Office of Natural Resources and Watershed Management of Golestan Province (hereafter CONRWMGP) [88], maximum and minimum precipitation occurs in February (99 mm) and May (28 mm), respectively. The basin area is mostly covered by forest and agriculture land use types. The Khosh-Yeylagh lithological formation, presented as Dkh in the lithological map, covers 42% of the study area consisting of dark grey shale, limestone, dolomite, sandstone, marl, and diabase (Table 7.1).

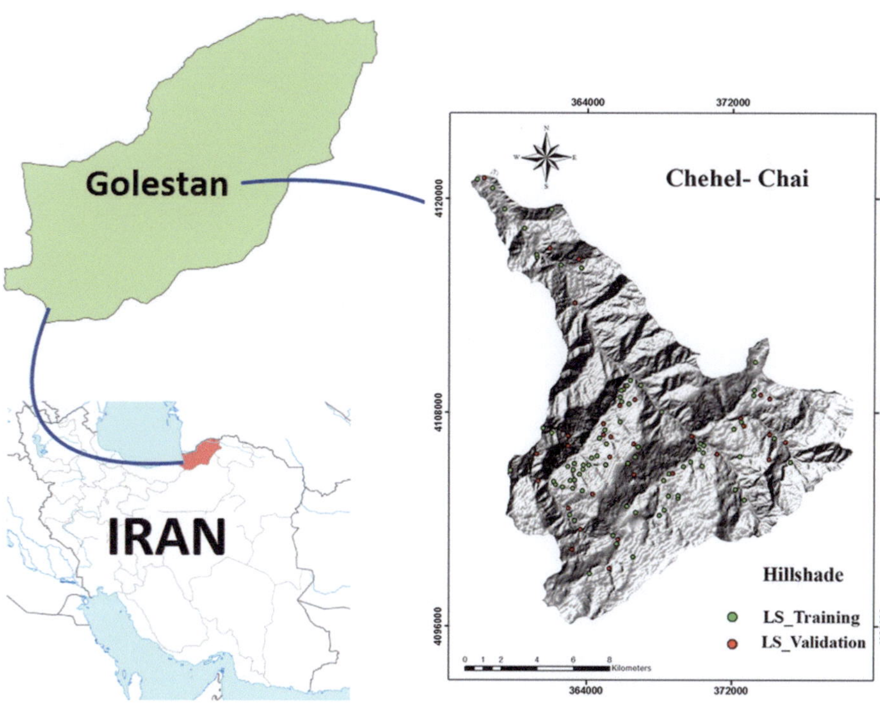

Fig. 7.1 Hillshade and landslide distribution over the study area

Table 7.1 Description of lithological units in the study area

Abbreviation	Formation	Geological system (period)	Material description
Qt	–	Quaternary	Young alluvial terraces
Jcb	Chaman-Bid	Jurassic	Limestone, marl, and shale
Jmz	Mozduran	Jurassic	Limestone, shale and shaly limestone
Jk	Kashaf-rud	Jurassic	Marl, dark and silty shale
Pr	Ruteh	Permian	Medium- bedded to massive limestone with intercalations of marl
Cm	Mobarak	Carboniferous	Thin to medium- bedded limestone, calcareous shale, and marl in some localities
Dkh	Khosh-Yeilagh	Devonian	Dark grey shale, limestone, dolomite, sandstone, marl, and diabase

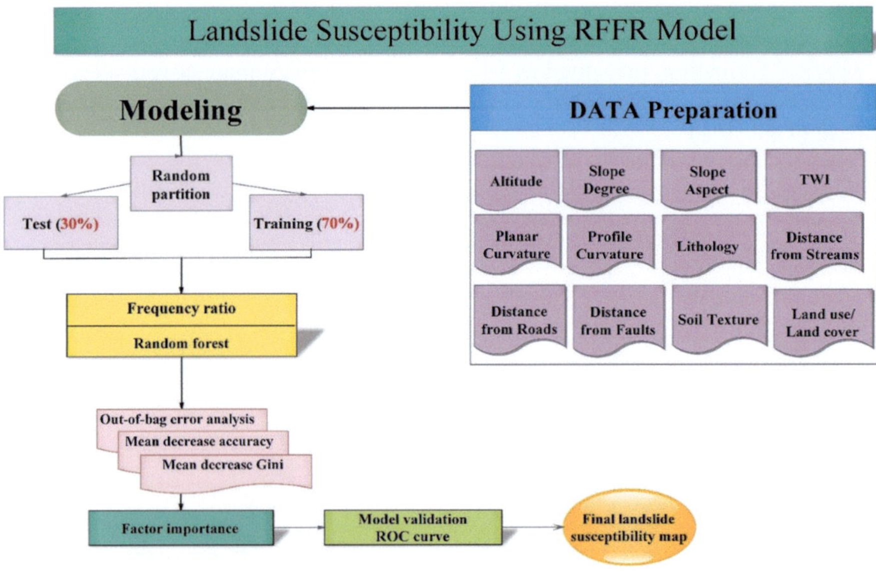

Fig. 7.2 Flowchart of the methodology applied in the study area

7.3 Methodology

The flowchart of current study is illustrated in Fig. 7.2. It consisted of following steps:

Data preparation

Landslides locations in the Chehel-Chai Watershed were recorded through field surveys using GPS (Global Positioning System) device, available organizational

Table 7.2 Conditioning factors' classes in the study area

Factor	Class	Factor	Class
TWI	2.17–5.64	Proximity from fault (m)	0–500
	5.64–7.78		500–1000
	7.78–12.14		1000–1500
	12.14–24.96		1500–2000
Slope degree	0–5		2000–6783.17
	5–15	Lithology	Cm
	15–30		Dkh
	30–65.54		Jcb
Proximity from road (m)	0–250		Jmz
	250–500		Jk
	500–750		Pr
	750–1000		Qt
	1000–4944.95	Land use/ land cover	Forest
Proximity from stream (m)	0–50		Farmland
	50–100	Slope aspect	F
	100–150		N
	150–200		NE
	200–584.15		NW
Plan curvature	Concave		S
	Neutral		SE
	Convex		SW
Profile curvature	Convex		E
	Neutral		W
	Concave	Soil texture	1.1.1
Altitude (m)	189.4–728.29		1.1.2
	728.29–1093.64		1.1.3
	1093.64–1140.72		1.1.4
	1440.72–1833.47		1.2.1
	1833.47–2527.64		1.2.2
			1.2.3
			9.3.4

data [88], and local information. A total 111 landslides with an area of 1192 ha have been recorded in this basin which covers about 5% of the study area. The built landslide inventory map divided into two sets of training (70%) and test data (30%) based on random sampling method [33, 89] (Fig. 7.1).

Based on data availability and literature review, 12 landslide conditioning factors including PFF, PFS, PFR, lithological units, soil texture, LU/LC, slope degree, slope aspect, altitude, PlanC, ProfC, and topographic wetness index (TWI) were mapped and classified in ArcGIS 10. 2 (Table 7.2). In order to create altitude map, the digital elevation model (DEM) with a cell size of 10×10 m was used. The DEM was created from topographic contour map (1:25,000-scale) and provided by the

Table 7.3 Description of soil texture codes in the study area

Code	Soil texture	Maximum depth (cm)	Additional description
1.1.1	Loamy sand	35	Light to semi light
1.1.2	Silty clay	100	Heavy
1.1.3	Silt loam, silty clay, silty clay loam	More than 100	Medium-textured
1.1.4	Silt loam	110	Medium-textured
1.2.1	Silt loam	40	Medium-textured
1.2.2	Silt loam, Silty clay loam	100	Medium-textured
1.2.3	Silty clay loam	100	Clay concentration increases in the depths of 45–90 cm
9.3.4	Sand, loamy sand	More than 100	Good aeration, highly susceptible to water erosion

National Cartographic Center (NCC). The road and stream maps were also obtained from the topographic contour map and then the corresponding proximity (Buffer) maps were created. The fault and lithological units' maps of the study area were obtained from Geological Survey Department of Iran at scale of 1: 100,000 [90]. The land use/ land cover and soil texture maps were provided from CONRWMGP [88]. The details of soil texture map are given in Table 7.3. The maps of slope degree and slope aspect were obtained from DEM in ArcGIS 10. 2. Slope aspect map was classified into eight primary and secondary directions and flat areas with no direction. Presence of open or closed fractures is directly correlated with the slope curvature [91]. So, profile and plan curvatures maps were obtained from DEM. For this, DEM Surface Tools Extension was used in ArcGIS 10. 2 based on Zevenbergen-Thorne method [92]. Topographic wetness index was made employing two maps of upslope area and slope degree based on Beven and Kirkby according to Eq. 7.1 [93]:

$$TWI = \alpha/\tan\beta, \qquad \alpha = A/L \tag{7.1}$$

7.3.1 *Landslide Susceptibility Modeling*

7.3.1.1 Ensemble of FR and RF (RFFR)

The frequency ratio (FR) is one of the most simple and applicable model among bivariate statistical techniques. FR relies on the basic rule of bivariate statistical analysis in which areal density of landslides within factor classes is calculated. Although, not considering interactions among factors is an important drawback of the mentioned method (unlike multivariate models), but simplicity in identifying probabilistic relationship between conditioning factors as independent variables and landslides as dependent variable [94, 95] could be considered as a great advantage. In this study, following equation (Eq. 7.2) is adopted for calculation of FR [94]:

$$FR = \frac{b/B}{a/A} \tag{7.2}$$

where, b is the number of pixels of landslides within a conditioning factor class, B is the number of pixels of total landslide in the basin/area, a is the number of pixels of each factor class and A is the number of total pixels of the study area. After rating the factor classes based on the above equation, the achieved weighting were applied to be used within RF model.

Random forests (RF) are in fact the novel extension of classification and regression trees (CARTs), introduced and developed by Breiman for the first time [96]. It has been used extensive in environmental modeling [75–77] and in some cases for landslide susceptibility mapping [78, 80]. Random forest is a machine-learning algorithm for non-parametric multivariate classification [79], based on averaging results of all decision trees and presents highly accurate classification for different data sets. The algorithm employs a technique called bootstrapping to use a subset of observations as training subset in which it takes advantage of random binary trees [79, 80] and the not included data are set aside as out-of-bag (OOB) [96]. The ensemble error of classification of data into landslide and non-landslide categories is computed by out-of-bag (OOB) error via comparing the out-of-bag predicted responses against the true responses. The most exclusive feature of this technique is using decrease of the classification accuracy which is suitable especially when variable values in a node of a tree switch randomly [97]. Factor importance, to answer the question which factor matters most, in random forests technique is analyzed through two types of errors, namely mean decrease in accuracy (MDA) and mean decrease in Gini (MDG) [96, 98]. Nicodemus states that mean decrease in accuracy makes better and more stable results, and consequently more reliable factor prioritizing, particularly when conditioning factors interact with each other [99]. In this study, RF was implemented in R 3.0.2 software by "randomForest" package based on FR weights.

Finally, for landslide spatial modeling by ensemble of random forest and frequency ratio approaches, the achieved weight values of FR were calculated and applied for running RF in R 3.0.2 software.

7.4 Results and Discussion

7.4.1 Using FR Model for Weighting Factor Classes

For weighting the conditioning factor classes, the FR values have been calculated which is briefly illustrated in Fig. 7.3. The FR values less than 1 represent low correlation between the factor classes and landslide occurrences, and in contrast values more than 1 point out high correlation [100]. So, high values of FR represent high probability of landslide occurrence. Based on the achieved results, the TWI class of 12.14–24.96 (saturated zone) has the highest FR value (2.21) and comparatively the class of 2.17–5.64 has the lowest FR value (0.7) with the lowest occurrence probability. The highest and lowest FR values for slope degree factor, can be found in the class of 15–30 (1.36) and 0–5 (0.18), respectively. The proximity

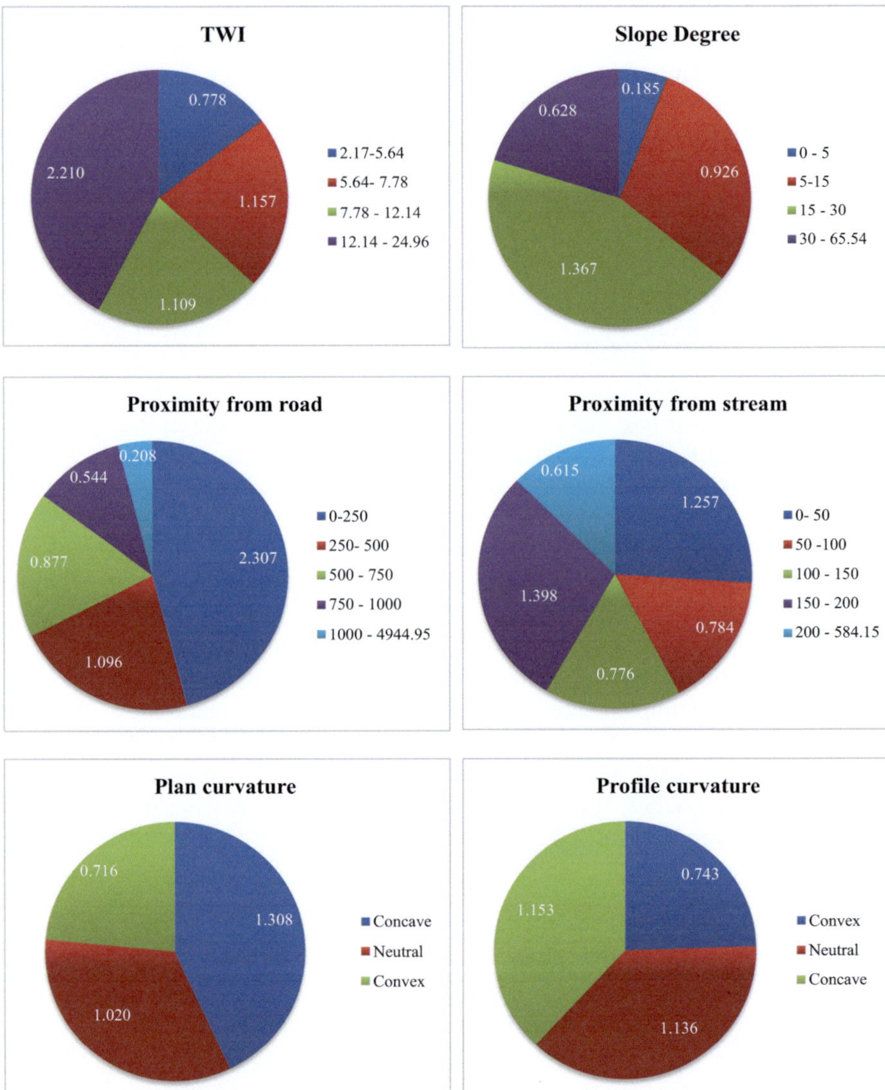

Fig. 7.3 FR values of conditioning factors' classes in the study area (Classes which are not shown in circle plot, have zero value)

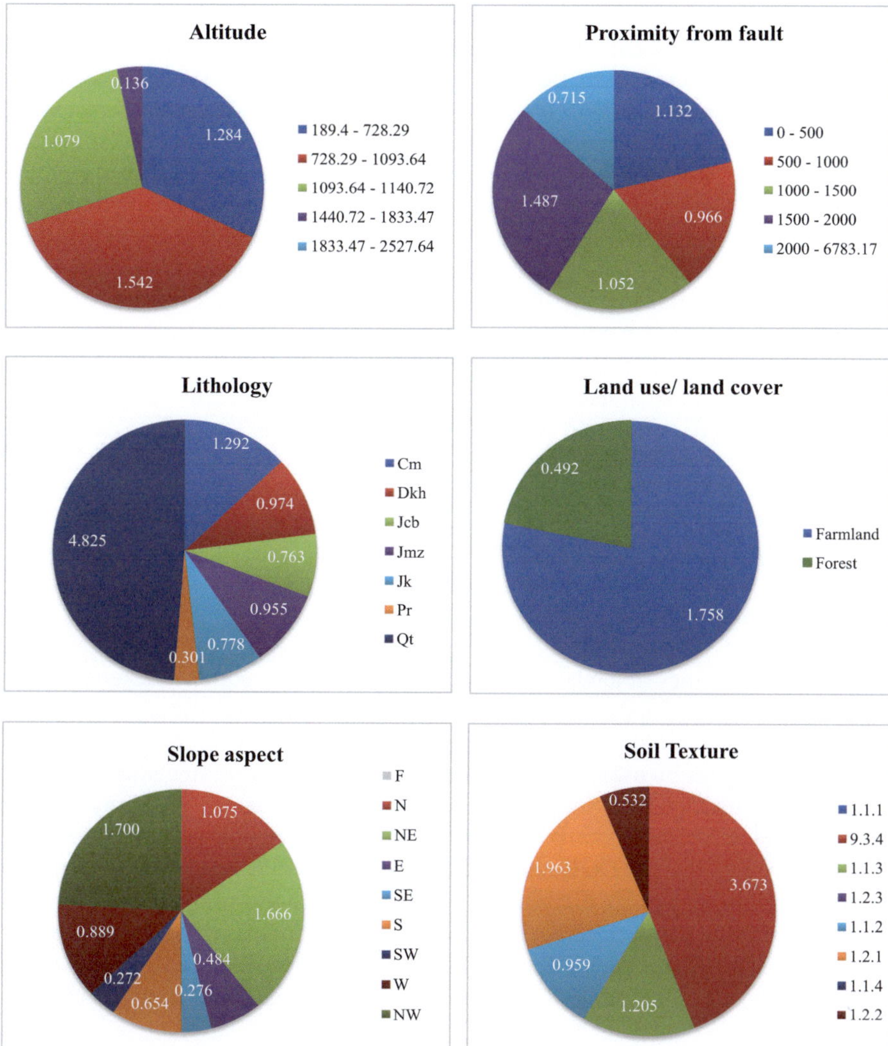

Fig. 7.3 (continued)

of 0–250 m from road has the highest FR values (2.3) and constantly decreases; so that, the lowest value is assigned to the class of 1000–4944.95 m (0.2). The proximity of 150–200 m from streams and rivers has the highest FR values (1.39), followed by the proximity of 0–50 (1.25) and the lowest value is calculated for the class of 200–584.15 m. The FR values of plan and profile curvatures have the same results, in which concave planar and profile curvatures have the highest FR values (1.3 and 1.15, respectively), followed by neutral and convex plan and profile curvatures. FR values of altitude factor showed that class of 1833.47–2527.64 m, has no (zero) correlation with landslide occurrence. In contrast, the altitudes of

728.29–1093.64 m has the highest values (1.54). The FR values of proximity/distance from fault show disarray orders (like to stream and road), in which 1500–2000 m distance from fault has the highest values (1.48) and the lowest value is assigned to the class of 2000–6783.17 m and hasn't any characteristic trend as increasing or decreasing in values. Reviewing the FR values of lithological units shows the Qt (young alluvial terraces) unit has the highest value (4.82), the highest level of correlation, by a wide difference compared to other lithological units. Comparatively, the Pr unit (Ruteh formation) has the lowest FR value (0.3). More information about lithological units is presented in Table. 7.1. Farmland land use type is gained the highest FR value (1.75) in comparison with forests (0.5). Northwest slope faces have the highest FR value (1.7), followed by northeast (1.6) and north (1) aspects. Flat areas have no correlation with landslide occurrence. The soil texture coded as 9.3.4 has the highest FR values (3.7) and classes of 1.1.1, 1.1.4, and 1.2.3 have zero correlation with landslide occurrence in the study area.

7.4.2 *Application of Random Forest Technique*

7.4.2.1 Parameter Assignment

Besides all difficulties in landslide susceptibility modeling, enriched results including manifold plots, figures, and tables play an important role in arriving at a suitable conclusion. The first issue that should be taken care of, is the number of trees (T#) which is usually set by users. There is no predefined number for tree numbers and it should be defined by user. So, by increasing number of trees we can't say that the model accuracy increases too (the role of choosing optimal tree numbers) [79]. One way to solve this issue is OOB error value as an unbiased estimator which helps to choose an optimum number of trees [23, 96, 101, 102]. In this study, to be sure about adequacy of T#, 1000 trees (ntree) and three predictor variables (mtry) as split points in each node, were considered. The results of OOB rate showed, the average error of 16.9% with accuracy equals to 83.1%, indicating a good performance of RF model.

7.4.2.2 Confusion Matrix

Basically, confusion matrix is a table that describes the performance of a classifier, RFFR here. Despite the summarized form, many results can be extracted from it which is added to the table to explore more features of the results (Table 7.4). Consequently, Table 7.4 can be interpreted as the following results:

Out of 156 observations, RFFR classifier predicted landslide locations 81 times and non-landslides 75 times. In the study area (reality), 75 actual landslides and 75 non-landslide observations exist. To dig more in the results, other variables such as true positive (TP), true negative (TN), false positive (FP), false negative (FN), accuracy, misclassification rate, sensitivity, specificity, precision and prevalence

Table 7.4 Confusion matrix obtained from RF model

Nr=156		Predicted		Sum	Class error	TP	TN	FP	FN	Accuracy	Mis-classification rate	Sensitivity	FPR	Specificity	Precision	Prevalence
		No	Yes													
Actual	No	56	22	78	0.282	59	56	22	19	0.737	0.263	0.756	0.282	0.718	0.728	0.5
	Yes	19	59	78	0.243											
	Sum	75	81													

Yes: landslide, No: landslide absence

were calculated and added to the Table. 7.4. According to TP and TN, there were 59 and 56 respective landslide and non-landslide locations that model predicted it right. On the other hand, according to FP and FN, there were 22 and 19 actual non-landslide and landslide locations that model predicted it wrongly as landslide and non-landslide, respectively. The accuracy value ((TP+TN)/Nr) indicates that overall, about 74% of observations were correctly predicted by RF. The misclassification rate ((FP+FN)/Nr) known as error rate implies that about 26% of observations were predicted wrong by RFFR model. Sensitivity (TP/sum of actual Yes) also known as true positive rate or recall points out that about 76% of landslide observations were predicted right by RFFR. According to specificity (TN/sum of actual No), about 72% of non-landslide observations were predicted right by RFFR. False positive rate (1-specificity or FP/sum of actual No) shows that about 28% of non-landslide observations were predicted wrong (landslide) by RFFR. Precision (TP/sum of predicted Yes) deals with the "Yes" situations which indicate when model predicts yes, how often it is correct (about 73% here). So, accuracy is more complete version of precision in which model success in identification of non-landslide observation are also taken into account. In fact, this is why precise models may not have a high accuracy. But sometimes predictive models with certain accuracy may have grater predictive power than other models with higher level of accuracy. In these cases, researchers may not consider the accuracy metric in favor of precision or sensitivity (recall) [103]. But here, both accuracy and precision meet satisfying values together. Finally, prevalence (sum of actual Yes/Nr) shows how often "Yes" situations occur or is predominant in our data. When the classes (landslide and non-landslide) are perfectly balanced, prevalence is equal to 50% which is the case here.

7.4.2.3 Factor Importance

The mean decrease accuracy (MDA) and mean decrease Gini (MDG) analyses were used to identify the most contributed conditioning factors which RFFR technique employed to constitute the optimal configuration of the susceptibility assessment. These results are summarized in Figs. 7.4 and 7.5, in which higher values represent higher importance and vice versa. According to MDA values in Fig. 7.4, distance from road has the highest importance, followed by altitude, aspect, plan curvature, and lithology. Also profile curvature is placed at the last level of importance of MDA analysis method. Also, based on Fig. 7.5, MDG analysis implies that distance from road achieved the highest importance similar to MDA results, followed by aspect, altitude, distance from stream and distance from fault. Results about distance from road support aforementioned statements about unconsidered road construction in the study area.

The final landslide susceptibility map made from RFFR assembled technique (Fig. 7.6) was classified into low, moderate, high, and very high classes (Fig. 7.7) based on natural break classifier [23, 28, 32, 80]. Consequently, high and very high susceptible areas to landslides cover about 42% of the study which is an issue of major concern.

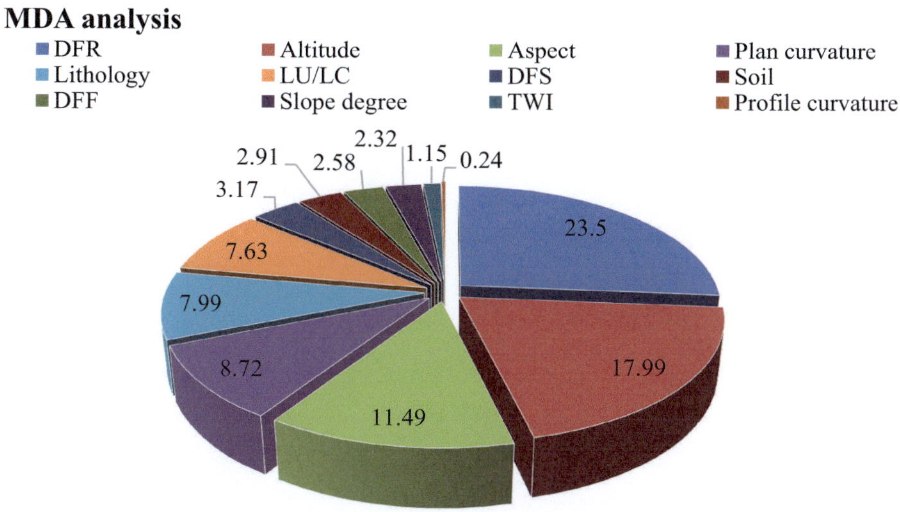

Fig. 7.4 Factor importance obtained from mean decrease accuracy analysis in RFFR model (*DFR* distance from road, *LU/LC* land use/land cover, *DFS* distance from stream, *DFF* distance from fault)

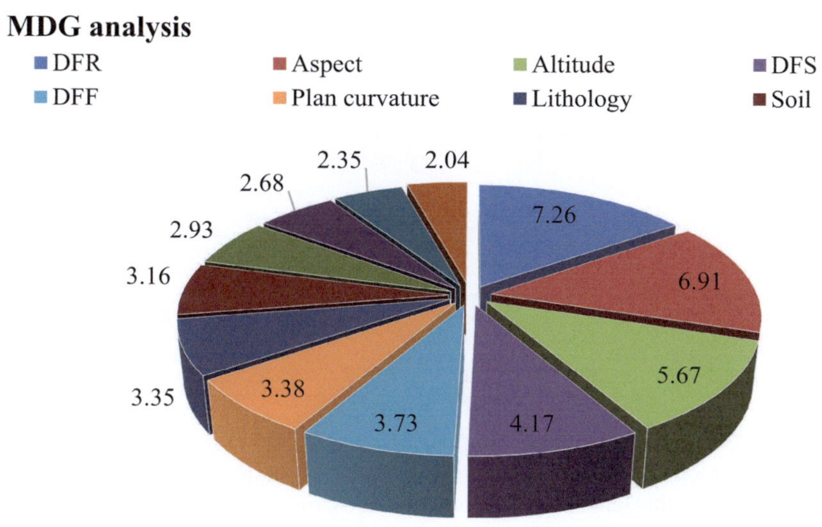

Fig. 7.5 Factor importance obtained from mean decrease Gini analysis in RF mode

7.4.3 Model Validation: ROC Curve

In order to test the performance of models' results, the receiver operating characteristic (ROC) curve was plotted (Fig. 7.8). ROC is a graphical form of model testing in

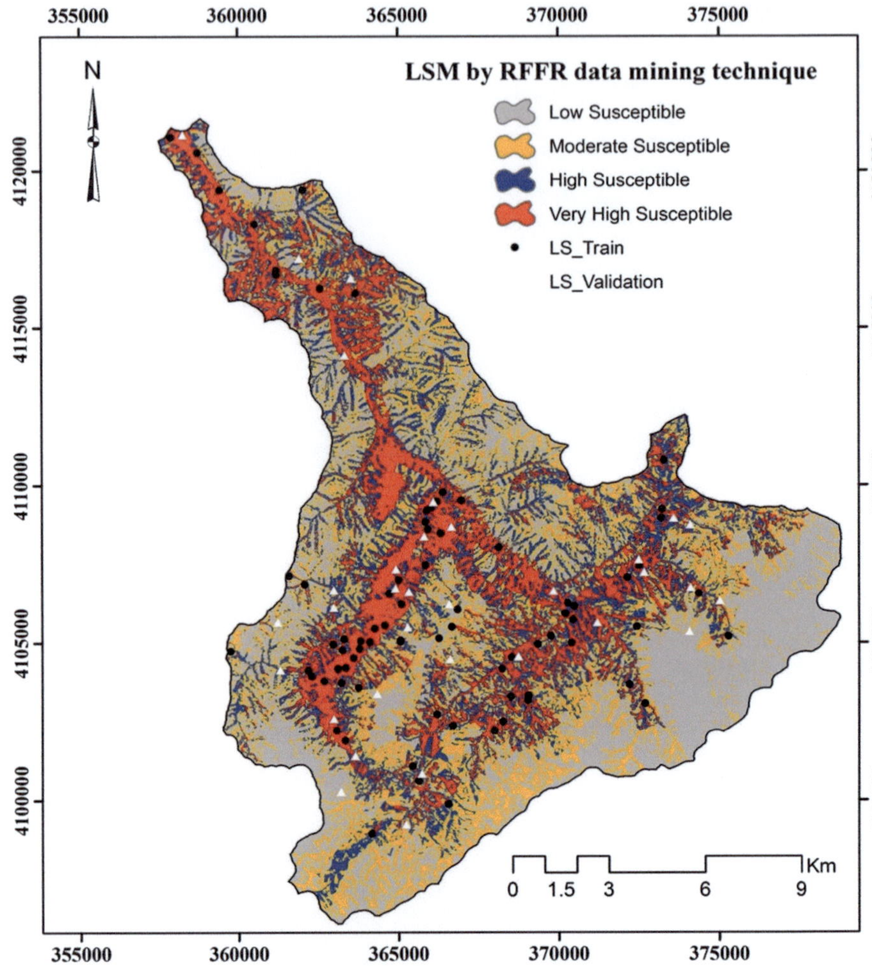

Fig. 7.6 Landslide susceptibility map made from RFFR model for the study area

which "sensitivity" of the test data (30%) (correct identification of the landslide presence locations by the model on y-axis) is plotted against "1- specificity" (wrong identification of the landslide absence locations by the model on x-axis) as user varies the threshold for assigning observations to a given class [104]. The area under the curve (AUC) is a measure of accuracy of the model. AUC of a model with a very low predictive power is equal to 0.5 and it will be equal to 1 for a complete model with the highest predictive power [105]. According to the Fig. 7.8, The AUC value was 0.831 with standard error of 0.05, and lower and upper bound values of 0.732 and 0.930, respectively; indicates high predictive power of the RFFR model in the study area.

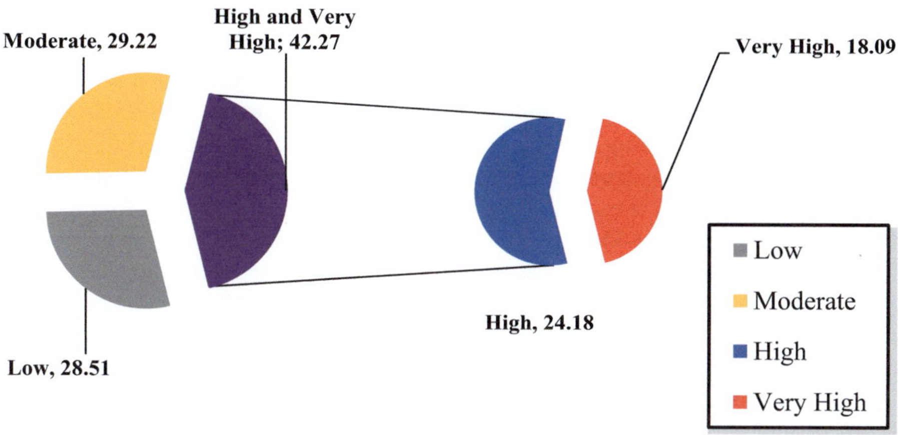

Fig. 7.7 Percentage of landslide susceptibility classes obtained from RFFR model in the study area

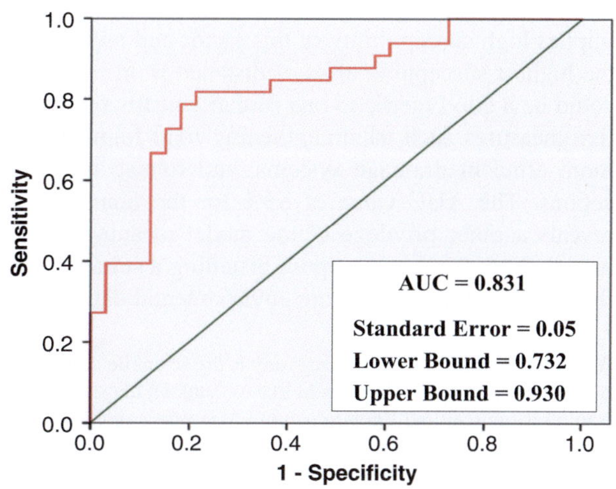

Fig. 7.8 Receiver operating characteristic (ROC) curve for the RFFR model

7.5 Conclusion

Nowadays, to address landslide susceptibility, several models have been proposed using different combination of conditioning factors, different scale, and different formats of landslide inventories. In the Golestan Province, these studies indicate high susceptibility to landsliding straight forward. Massive budgets are dedicating to land use planning projects, still in action less improvement have happened and in some cases we are faced with retrograding situations. The current study, covers one of the critical watersheds namely Chehel-Chi in the Golestan Province, again addressing high percentage of susceptibility, about half of the area (42%), to landsliding. A novel bivariate statistical – data mining method to wit random forest-frequency ratio (RFFR) is proposed using 111 landslide inventories and different conditioning factors including PFF, PFS, PFR, lithological units, soil texture, LU/LC, slope degree, slope aspect, altitude, PlanC, ProfC, and TWI. Factors have been weighted via FR computational algorithm. According to FR values, Cm lithological unit with value of 4.82, code 9.3.4 soil texture with value of 3.67, and 0–250 m distance from road with the value of 2.3 have the highest weight. Three variables as split points at each node and 1000 trees have been set to run RF model in R3.0.2 environment. Out-Of-Bag (OOB) error indicated the accuracy value of 83% for the model which shows the best accuracy for the proposed model in Chehel-Chi Watershed. Confusion matrix brought us several results regarding accuracy (0.74), misclassification rate (0.26), sensitivity (0.76), specificity (0.72), and precision (0.73). Two types of error namely mean decrease accuracy (MDA) and mean decrease Gini (MDG) had notable results about factor importance. According to MDA, distance from road, altitude, aspect, plan curvature, and lithology had the highest importance and correlation with landslide occurrence. According to MDG, distance from road, aspect, altitude, distance from stream, and distance from fault had the highest importance. Distance from road at the top of two above analyses implies high susceptibility of this factor and requires immediate intervention. Also, the highest susceptible class of distance from road based on FR values (0–250 m) could be a good metric to ban further construction activities and plan more preventive measures such as strengthening road foundation, designing and constructing more efficient drainage systems, and robust and flexible land use planning and actions. The AUC value of 83% for the final map obtained from RFFR model reveals another privilege of the model meaning the high predictive power of the model. At the end, we propose installing a suitable landslide monitoring station for the study area and upgrading environmental data to compensate scarcity of data.

Acknowledgement The present study is the scientific outcome of a six month sabbatical leave at Shiraz University, Iran. We would like to thank all the staff and the members of the Department of Natural Resources and Environmental Engineering and the College of Agriculture who supported our work and helped us to get results of better quality.

References

1. Varnes DJ, Radbruch-Hall DH (1976) Landslides cause and effect. Bull Int Assoc Eng Geol 13 (1):205–216
2. Iranian Landslide Working Party (ILWP) (2007) Iranian landslides list. Forest, Rangeland and Watershed Association, Iran, p 60
3. Pourghasemi HR, Pradhan B, Gokceoglu C, Mohammadi M, Moradi HR (2013) Application of weights-of-evidence and certainty factor models and their comparison in landslide susceptibility mapping at Haraz watershed, Iran. Arab J Geosci 6(7):2351–2365
4. Van Westen CJ, Van Asch TW, Soeters R (2006) Landslide hazard and risk zonation—why is it still so difficult? Bull Eng Geol Environ 65(2):167–184
5. Cruden DM (1991) A simple definition of a landslide. Bull Eng Geol Environ 43(1):27–29

6. Guzzetti F, Reichenbach P, Cardinali M, Galli M, Ardizzone F (2005) Probabilistic landslide hazard assessment at the basin scale. Geomorphology 72(1):272–299
7. Kanungo DP, Arora MK, Sarkar S, Gupta RP (2009) Landslide susceptibility zonation (LSZ) mapping—a review. J South Asia Disaster Stud 2(1):81–105
8. Pardeshi SD, Autade SE, Pardeshi SS (2013) Landslide hazard assessment: recent trends and techniques. Springer Plus 2(1):523
9. Van Den Eeckhaut M, Vanwalleghem T, Poesen J, Govers G, Verstraeten G, Vandekerckhove L (2006) Prediction of landslide susceptibility using rare events logistic regression: a case-study in the Flemish Ardennes (Belgium). Geomorphology 76(3):392–410
10. García-Rodríguez MJ, Malpica JA, Benito B, Díaz M (2008) Susceptibility assessment of earthquake-triggered landslides in El Salvador using logistic regression. Geomorphology 95 (3):172–191
11. Kornejady A, Heidari K, Nakhavali M (2015) Assessment of landslide susceptibility, semi-quantitative risk and management in the Ilam dam basin, Ilam, Iran. Environ Resour Res 3 (1):85–109
12. Gupta RP, Joshi BC (1990) Landslide hazard zoning using the GIS approach—a case study from the Ramganga catchment, Himalayas. Eng Geol 28(1–2):119–131
13. Mohammadi A, Heshmatpoor A, Mosaedi A (2004) Study on efficiency of an Iranian method for landslide hazard zonation in Golestan Province. EGU- 1st General Assembly, Nice, France
14. Kornejady A, Heidary K, Sarparast M, Khosravi G, Mombeini M (2014) Performance assessment of two "LNRF" and "AHP-area density" models in landslide susceptibility zonation. J Life Sci Biomed 4(3):169–176
15. Akbar TA, Ha SR (2011) Landslide hazard zoning along Himalayan Kaghan Valley of Pakistan—by integration of GPS, GIS, and remote sensing technology. Landslides 8 (4):527–540
16. Sharma LP, Patel N, Ghose MK, Debnath P (2015) Development and application of Shannon's entropy integrated information value model for landslide susceptibility assessment and zonation in Sikkim Himalayas in India. Nat Hazards 75(2):1555–1576
17. Chen T, Niu R, Jia X (2016) A comparison of information value and logistic regression models in landslide susceptibility mapping by using GIS. Environ Earth Sci 75(10):1–16
18. Sterlacchini S, Ballabio C, Blahut J, Masetti M, Sorichetta A (2011) Spatial agreement of predicted patterns in landslide susceptibility maps. Geomorphology 125(1):51–61
19. Mohammadi M, Pourghasemi HR, Pradhan B (2012) Landslide susceptibility mapping at Golestan Province, Iran: a comparison between frequency ratio, Dempster–Shafer, and weights-of-evidence models. J Asian Earth Sci 61:221–236
20. Pourghasemi HR, Moradi HR, Mohammadi M, Pradhan B, Mostafazadeh R, Goli Jirandeh A (2012) Landslide hazard assessment using remote sensing data, GIS and weights-of-evidence model (South of Golestan Province, Iran). In Asia Pacific Conference on Environmental Science and Technology (APEST 2012). Adv Biomed Eng 6:30–36
21. Martha TR, van Westen CJ, Kerle N, Jetten V, Kumar KV (2013) Landslide hazard and risk assessment using semi-automatically created landslide inventories. Geomorphology 184:139–150
22. Sumaryono DM, Sulaksana N, DasaTriana Y (2015) Weights of evidence method for landslide susceptibility mapping in Tandikek and Damar Bancah, West Sumatra, Indonesia. IJSR 4 (10):1283–1290
23. Rahmati O, Pourghasemi HR, Melesse AM (2016) Application of GIS-based data driven random forest and maximum entropy models for groundwater potential mapping: a case study at Mehran Region, Iran. Catena 137:360–372
24. Ozdemir A, Altural T (2013) A comparative study of frequency ratio, weights of evidence and logistic regression methods for landslide susceptibility mapping: Sultan Mountains, SW Turkey. J Asian Earth Sci 64:180–197
25. Pourtaghi ZS, Pourghasemi HR (2014) GIS-based groundwater spring potential assessment and mapping in the Birjand Township, southern Khorasan Province, Iran. Hydrogeol J 22 (3):643–662

26. Naghibi SA, Pourghasemi HR, Pourtaghi ZS, Rezaei A (2015) Groundwater qanat potential mapping using frequency ratio and Shannon's entropy models in the Moghan watershed, Iran. Earth Sci Inform 8(1):171–186
27. Demir G, Aytekin M, Akgun A (2015) Landslide susceptibility mapping by frequency ratio and logistic regression methods: an example from Niksar–Resadiye (Tokat, Turkey). Arab J Geosci 8(3):1801–1812
28. Hong H, Chen W, Xu C, Youssef AM, Pradhan B, Tien Bui D (2016) Rainfall-induced landslide susceptibility assessment at the Chongren area (China) using frequency ratio, certainty factor, and index of entropy. Geocarto Int 32:1–16
29. Tien Bui D, Pradhan B, Lofman O, Revhaug I, Dick OB (2012) Landslide susceptibility assessment in the Hoa Binh province of Vietnam: a comparison of the Levenberg–Marquardt and Bayesian regularized neural networks. Geomorphology 171:12–29
30. Shahabi H, Hashim M, Ahmad B (2015) Remote sensing and GIS-based landslide susceptibility mapping using frequency ratio, logistic regression, and fuzzy logic methods at the central Zab basin, Iran. Environ Earth Sci 73(12):8647–8668
31. Kumar R, Anbalagan R (2015) Landslide susceptibility zonation in part of Tehri reservoir region using frequency ratio, fuzzy logic and GIS. J Earth Syst Sci 124(2):431–448
32. Pourghasemi HR, Beheshtirad M, Pradhan B (2016) A comparative assessment of prediction capabilities of modified analytical hierarchy process (M-AHP) and Mamdani fuzzy logic models using Netcad-GIS for forest fire susceptibility mapping. Geomatics Nat Hazards Risk 7(2):861–885
33. Pourghasemi HR, Mohammady M, Pradhan B (2012) Landslide susceptibility mapping using index of entropy and conditional probability models in GIS: Safarood Basin, Iran. Catena 97:71–84
34. Devkota KC, Regmi AD, Pourghasemi HR, Yoshida K, Pradhan B, Ryu IC, Althuwaynee OF (2013) Landslide susceptibility mapping using certainty factor, index of entropy and logistic regression models in GIS and their comparison at Mugling–Narayanghat road section in Nepal Himalaya. Nat Hazards 65(1):135–165
35. Jaafari A, Najafi A, Pourghasemi HR, Rezaeian J, Sattarian A (2014) GIS-based frequency ratio and index of entropy models for landslide susceptibility assessment in the Caspian forest, northern Iran. Int J Environ Sci Technol 11(4):909–926
36. Ermini L, Catani F, Casagli N (2005) Artificial neural networks applied to landslide susceptibility assessment. Geomorphology 66(1):327–343
37. Ercanoglu M (2005) Landslide susceptibility assessment of SE Bartin (West Black Sea region, Turkey) by artificial neural networks. Nat Hazards Earth Syst Sci 5(6):979–992
38. Tsangaratos P, Benardos A (2014) Estimating landslide susceptibility through an artificial neural network classifier. Nat Hazards 74(3):1489–1516
39. Dou J, Oguchi T, Hayakawa YS, Uchiyama S, Saito H, Paudel U (2014) GIS-based landslide susceptibility mapping using a certainty factor model and its validation in the Chuetsu Area, Central Japan. In: Landslide science for a safer geoenvironment. Springer, Cham, pp 419–424
40. Pourghasemi HR, Moradi HR, Fatemi Aghda SM, Gokceoglu C, Pradhan B (2014) GIS-based landslide susceptibility mapping with probabilistic likelihood ratio and spatial multi-criteria evaluation models (North of Tehran, Iran). Arab J Geosci 7(5):1857–1878
41. Chen W, Li W, Chai H, Hou E, Li X, Ding X (2016) GIS-based landslide susceptibility mapping using analytical hierarchy process (AHP) and certainty factor (CF) models for the Baozhong region of Baoji City, China. Environ Earth Sci 75(1):1–14
42. Dimri S, Lakhera RC, Sati S (2007) Fuzzy-based method for landslide hazard assessment in active seismic zone of Himalaya. Landslides 4(2):101–111
43. Naithani AK (2007) Macro landslide hazard zonation mapping using univariate statistical analysis in a part of Garhwal Himalaya. J Geol Soc India 70(2):353–368
44. Anbalagan R, Chakraborty D, Kohli A (2008) Landslide hazard zonation (LHZ) mapping on meso-scale for systematic town planning in mountainous terrain. J Sci Ind Res 67:486–497

45. Ghosh S, Van Westen CJ, Carranza EJ, Ghoshal TB, Sarkar NK, Surendranath M (2009) A quantitative approach for improving the BIS (Indian) method of medium-scale landslide susceptibility. J Geol Soc India 74(5):625–638
46. Singh CD, Behera KK, Rocky WS (2011) Landslide susceptibility along NH-39 between Karong and Mao, Senapati district, Manipur. J Geol Soc India 78(6):559–570
47. Kannan M, Saranathan E, Anbalagan R (2011) Macro landslide hazard zonation mapping-case study from Bodi–Bodimettu Ghats Section, Theni District, Tamil Nadu-India. J Indian Soc Remote Sens 39(4):485–496
48. Kannan M, Saranathan E, Anbalagan R (2015) Comparative analysis in GIS-based landslide hazard zonation—a case study in Bodi-Bodimettu Ghat section, Theni District, Tamil Nadu, India. Arab J Geosci 8(2):691–699
49. Motamedi M, Liang RY (2014) Probabilistic landslide hazard assessment using Copula modeling technique. Landslides 11(4):565–573
50. Hindayar JN, Dasarwar P, Srivastava SP, Kumar NT, Mohan M, Som SK (2016) Dynamicity of the Himalayan landslide—A tectono-geotechnical appraisal of the 13th mile landslide, Sikkim. J Geol Soc India 88(2):197–205
51. Phillips SJ, Anderson RP, Schapire RE (2006) Maximum entropy modeling of species geographic distributions. Ecol Model 190(3):231–259
52. Dudík M, Phillips SJ, Schapire RE (2007) Maximum entropy density estimation with generalized regularization and an application to species distribution modeling. J Mach Learn Res 8 (6):1217–1260
53. Phillips SJ, Dudík M (2008) Modeling of species distributions with Maxent: new extensions and a comprehensive evaluation. Ecography 31(2):161–175
54. Elith J, Phillips S, Hastie T, Dudík M, Chee Y, Yates C (2010) A statistical explanation of MaxEnt for ecologists. Divers Distrib 17:43–47
55. Kleidon A, Malhi Y, Cox PM (2010) Maximum entropy production in environmental and ecological systems. Philos T Roy Soc B 365(1545):1297–1302
56. Nieves V, Wang J, Bras RL (2011) Statistics of multi-fractal processes using the maximum entropy method. Geophys Res Lett 38:17
57. Convertino M, Troccoli A, Catani F (2013) Detecting fingerprints of landslide drivers: a MaxEnt model. J Geophys Res Earth Surf 118(3):1367–1386
58. Kim HG, Lee DK, Park C, Kil S, Son Y, Park JH (2015) Evaluating landslide hazards using RCP 4.5 and 8.5 scenarios. Environ Earth Sci 73(3):1385–1400
59. Park NW (2015) Using maximum entropy modeling for landslide susceptibility mapping with multiple geoenvironmental data sets. Environ Earth Sci 73(3):937–949
60. Friedman JH (2001) Greedy function approximation: a gradient boosting machine. Ann Stat 29:1189–1232
61. Elith J, Graham CH, Anderson RP, Dudik M, Ferrier S, Guisan A, Hijmans RJ, Huettmann F, Leathwick J, Lehmann A, Li J, Lohmann LG, Loiselle BA, Manion G, Moritz C, Nakamura M, Nakazawa Y, Overton JMM, Peterson AT, Phillips SJ, Richardson K, Scachetti-Pereira R, Schapire RE, Soberón J, Williams S, Wisz MS, Zimmermann NE (2006) Novel methods improve prediction of species' distributions from occurrence data. Ecography 29:129–151
62. Elith J, Leathwick JR, Hastie T (2008) A working guide to boosted regression trees. J Anim Ecol 77:802–813
63. Breiman L, Friedman JH, Olshen R, Stone CJ (1984) Classification and regression trees. Wadsworth, Belmont, California, USA. Nucl Instrum Meth 543:57
64. Aertsen W, Kint V, Van Orshoven J, Özkan K, Muys B (2010) Comparison and ranking of different modeling techniques for prediction of site index in Mediterranean mountain forests. Ecol Model 221:1119–1130
65. Westreich D, Lessler J, Funk MJ (2010) Propensity score estimation: neural networks, support vector machines, decision trees (CART), and meta-classifiers as alternatives to logistic regression. J Clin Epidemiol 63(8):826–833

66. Naghibi SA, Pourghasemi HR (2015) A comparative assessment between three machine learning models and their performance comparison by bivariate and multivariate statistical methods in groundwater potential mapping. Water Resour Manag 29(14):5217–5236
67. Brenning A (2005) Spatial prediction models for landslide hazards: review, comparison and evaluation. Nat Hazards Earth Syst Sci 5(6):853–862
68. Yao X, Tham LG, Dai FC (2008) Landslide susceptibility mapping based on support vector machine: a case study on natural slopes of Hong Kong, China. Geomorphology 101 (4):572–582
69. Pourghasemi HR, Goly Jirandeh A, Pradhan B, Xu C, Gokceoglu C (2013) Landslide susceptibility mapping using support vector machine and GIS at the Golestan Province, Iran. J Earth Syst Sci 122(2):349–369
70. Jebur MN, Pradhan B, Tehrany MS (2015) Manifestation of LiDAR-derived parameters in the spatial prediction of landslides using novel ensemble evidential belief functions and support vector machine models in GIS. IEEE J Sel Top Appl Earth Obs Remote Sens 8(2):674–690
71. Ren F, Wu X, Zhang K, Niu R (2015) Application of wavelet analysis and a particle swarm-optimized support vector machine to predict the displacement of the Shuping landslide in the Three Gorges, China. Environ Earth Sci 73(8):4791–4804
72. Tien Bui D, Tuan TA, Klempe H, Pradhan B, Revhaug I (2016) Spatial prediction models for shallow landslide hazards: a comparative assessment of the efficacy of support vector machines, artificial neural networks, kernel logistic regression, and logistic model tree. Landslides 13(2):361–378
73. Brenning A, Schwinn M, Muenchow J (2015) Landslide susceptibility near highways is increased by 1 order of magnitude in the Andes of southern Ecuador, Loja province. Nat Hazards Earth Syst Sci 15(1):45–57
74. Pourghasemi HR, Rossi M (2016) Landslide susceptibility modeling in a landslide prone area in Mazandarn Province, north of Iran: a comparison between GLM, GAM, MARS, and M-AHP methods. Theor Appl Climatol:1–25. https://doi.org/10.1007/s00704-016-1919-2
75. Prasad AM, Iverson LR, Liaw A (2006) Newer classification and regression tree techniques: bagging and random forests for ecological prediction. Ecosystems 9(2):181–199
76. Strobl C, Boulesteix AL, Kneib T, Augustin T, Zeileis A (2008) Conditional variable importance for random forests. BMC Bioinforma 9:307
77. Bachmair S, Weiler M (2012) Hillslope characteristics as controls of subsurface flow variability. Hydrol Earth Syst Sci 16(10):3699–3715
78. Vorpahl P, Elsenbeer H, Märker M, Schröder B (2012) How can statistical models help to determine driving factors of landslides? Ecol Model 239:27–39
79. Catani F, Lagomarsino D, Segoni S, Tofani V (2013) Landslide susceptibility estimation by random forests technique: sensitivity and scaling issues. Nat Hazards Earth Syst Sci 13 (11):2815–2831
80. Pourghasemi HR, Kerle N (2016) Random forests and evidential belief function-based landslide susceptibility assessment in Western Mazandaran Province, Iran. Environ Earth Sci 75(3):1–17
81. Naghibi SA, Pourghasemi HR, Dixon B (2016) GIS-based groundwater potential mapping using boosted regression tree, classification and regression tree, and random forest machine learning models in Iran. Environ Monit Assess 188(1):1–27
82. Pourtaghi ZS, Pourghasemi HR, Aretano R, Semeraro T (2016) Investigation of general indicators influencing on forest fire and its susceptibility modeling using different data mining techniques. Ecol Indic 64:72–84
83. Stumpf A, Kerle N (2011) Object-oriented mapping of landslides using random forests. Remote Sens Environ 115(10):2564–2577
84. Stumpf A, Kerle N (2011) Combining Random Forests and object-oriented analysis for landslide mapping from very high resolution imagery. Procedia Environ Sci 3:123–129
85. Lee S, Hwang J, Park I (2013) Application of data-driven evidential belief functions to landslide susceptibility mapping in Jinbu, Korea. Catena 100:15–30

86. Trigila A, Frattini P, Casagli N, Catani F, Crosta G, Esposito C, Iadanza C, Lagomarsino D, Mugnozza GS, Segoni S, Spizzichino D (2013) Landslide susceptibility mapping at national scale: the Italian case study. Landslide Sci Prac 1:287–295
87. Youssef AM, Pourghasemi HR, Pourtaghi ZS, Al-Katheeri MM (2015) Landslide susceptibility mapping using random forest, boosted regression tree, classification and regression tree, and general linear models and comparison of their performance at Wadi Tayyah Basin, Asir Region, Saudi Arabia. Landslides 13:1–18. https://doi.org/10.1007/s10346-015-0614-1
88. Central Office of Natural Resources and Watershed Management in Golestan Province (2007) Detailed action plan of Ziarat watershed. Pazhouhab Sharq Consulting Engineers Co., Iran
89. Hussin HY, Zumpano V, Reichenbach P, Sterlacchini S, Micu M, van Westen C, Bălteanu D (2016) Different landslide sampling strategies in a grid-based bivariate statistical susceptibility model. Geomorphology 253:508–523
90. Geological Survey Department of Iran (GSDI) (1997) http://www.gsi.ir/Main/Lang_en/index.html
91. Jenness J (2013) DEM surface tools for ArcGIS (surface_area. exe). Jenness Enterprises. http://www.jennessent.com/arcgis/surface_area.htm
92. Zevenbergen LW, Thorne CR (1987) Quantitative analysis of land surface topography. Earth Surf Proc Landf 12:47–56
93. Beven KJ, Kirkby MJ (1979) A physically based, variable contributing area model of basin hydrology/Un modèle à base physique de zone d'appel variable de l'hydrologie du bassin versant. Hydrol Sci J 24(1):43–69
94. Bonham-Carter GF (1994) Geographic information systems for geoscientists: modelling with GIS. Computer msethamphetamine geos, vol 13. Pergamon, New York, p 398
95. Tehrany MS, Pradhan B, Jebur MN (2013) Spatial prediction of flood susceptible areas using rule based decision tree (DT) and a novel ensemble bivariate and multivariate statistical models in GIS. J Hydrol 504:69–79
96. Breiman L (2001) Random forests. Mach Learn 45(l):5–32
97. Bureau A, Dupuis J, Hayward B, Falls K, Van Eerdewegh P (2003) Mapping complex traits using Random Forests. BMC Genet 4:64
98. Calle ML, Urrea V (2011) Letter to the editor: stability of random forest importance measures. Brief Bioinform 12(1):86–89
99. Nicodemus KK (2011) Letter to the editor: on the stability and ranking of predictors from random forest variable importance measures. Brief Bioinform 12:369–373. bbr016
100. Oh HJ, Lee S (2010) Assessment of ground subsidence using GIS and the weights of evidence model. Eng Geol 115:36–48
101. Goetz JN, Brenning A, Petschko H, Leopold P (2015) Evaluating machine learning and statistical prediction techniques for landslide susceptibility modelling. Comput Geosci 81:1–11
102. Trigila A, Iadanza C, Esposito C, Scarascia-Mugnozza G (2015) Comparison of Logistic Regression and Random Forests techniques for shallow landslide susceptibility assessment in Giampilieri (NE Sicily, Italy). Geomorphology 249:119–136
103. Bruckhaus T (2007) The business impact of predictive analytics. Knowledge discovery and data mining. Challenges and Realities 114–138
104. Yesilnacar E K (2005) The application of computational intelligence to landslide susceptibility mapping in Turkey, PhD Thesis. Department of Geomatics the University of Melbourne, p 423
105. Pontius RG, Schneider LC (2001) Land-cover change model validation by an ROC method for the Ipswich watershed, Massachusetts, USA. Agric Ecosyst Environ 85(1):239–248

Chapter 8
Landslide Susceptibility Mapping, Vulnerability and Risk Assessment for Development of Early Warning Systems in India

Sudesh Kumar Wadhawan

Abstract Landslide or the landmass movement is a geomorphic hill slope physical process of mass-wasting resulting in downslope rolling of large mass of debris, regolith and soil under influence of gravity. It is caused by a combination of particular geo-factors that are region or territory specific. Landslides are generally triggered and activated by substantial precipitation and/or earthquake tremors and other anthropogenic interventions such as over the top cutting of slant for development of mountainous roads/streets and other excavations for civil structures, etc. The relatively young entire Himalayan hilly tract, mountainous steep slopes in sub-Himalayan landscape of North-east India, Western Ghats, the Nilgiris in Tamil Nadu and Konkan ranges are susceptible to landslides or debris flow.

In order to formulate strategies to minimize societal impacts of landslides, a systematic approach would entail preparation of Landslide Susceptibility Maps linked to landslide incidence inventory and making them available to the concerned stakeholders for necessary preparatory and mitigation measures. Geological Survey of India (GSI) being the nodal agency for landslides studies in India formally launched on February 05, 2014 the National Landslide Susceptibility Mapping (NLSM) programme which has been a geoscientific exercise on 1:50,000 scale on GIS platform in making both quantitative or qualitative estimates of spatial distribution of landslides which either exists or has the potential to occur in a given area. GSI has formulated a set of standard operating procedures that emphasize on geo-parametric data collection (as per standard and devised formats) for landslide inventory. These data sets are synthesized with relevant spatially-distributed causative thematic maps into susceptibility zonation which represents geospatial information indicating intensity and propensity of landslides. Such baseline data will ultimately lead to the collation and evaluation of landslide hazard and risk and mitigation plans. It will also help in disaster preparedness of the country and to indicate areas critical for landslide monitoring and developing early warning system

---

S. K. Wadhawan (✉)
Geological Survey of India, Jaipur, India

S. P. Pradhan et al. (eds.), *Landslides: Theory, Practice and Modelling*, Advances in Natural and Technological Hazards Research 50,
https://doi.org/10.1007/978-3-319-77377-3_8

(EWS). It is aimed to demarcate and facilitate prioritization of areas for further detailed studies (Meso- and Micro-scales) and help in Regional Land Use Planning and provide the scientific basis for framing the Land Use Zoning Regulations. Several lessons were learnt from Uttarakhand disaster of June 2013 in India that compelled re-evaluation of the existing methodology of conducting geosurveys of macro scale landslide susceptibility maps. Additional geofactors that also need to be considered include: effect of toe erosion by higher order streams; effect of long run-outs of the debris flows and drainage morphometry; nature and size of clastic components, etc.

It is intended to elaborate here a synthesis of various approaches and constraints on continuing research on such country-wide programmes on landslides related geohazards characterization and its implications on evolving EWS for the societal preparedness and resilience for mitigating impending disasters. However, any method of predicting landslide susceptibility needs validation which sometimes may be difficult in areas having no land sliding history. Besides, EWS need also to highlight mitigation efforts/remedial measures through geotechnical and engineering solutions as suited to Indian conditions on case to case basis, delineation of safe escape routes in the event of a landslide/debris flow/flash floods, and for optimum utilization of available resources.

8.1 Introduction

Landmass movement or the landslide is a geomorphic hill slope physical process of mass-wasting resulting in downslope rolling of large mass of debris, regolith and soil owing to influence of gravity. It is caused by a combination of particular geo factors that are region or territory specific (e.g., geomorphology, slant and slope aspects, geology: bed-rock and litho-structures, nature of chemically altered or physically disintegrated weathered material, land-cover and land-use and so on). Such hill slope processes are generally activated by substantial precipitation or earthquake tremors. In Indian landscape, landslide or slope failure occasions are for the most part activated by monsoonal precipitation, however cases of seismic tremor activated landfalls and avalanche are likewise normal in India, for instance, the Uttarkashi Earthquake, Chamoli Earthquake, Kashmir Earthquake, Sikkim Earthquake, and so forth, produced several landslides in recent years [1, 2]. The whole Himalayan tract, mountains slopes in sub-Himalayan landscape of North-east India, Western Ghats, the Nilgiris in Tamil Nadu and Konkan ranges are susceptible to debris flow or landslides (~0.42 million km^2; including the permafrost locales makes it around 0.49 million km^2). Other than the precipitation and earth shaking tremors, the toe disintegration by stream and unscientific or informal anthropogenic activities, for example, over the top cutting of slant for development of mountainous roads/streets and other excavations for civil structures can likewise trigger downslope movement of loose/poorly consolidated rock-mass or debris.

India is amongst the worst affected countries in the global scenario where landslides are perennial natural hazards causing irreparable losses to properties and infrastructure. Even, fatality due to landslides is quite high in India compared to the rest of the landslide-affected nations in the world. In India, the vulnerability to landslides mainly owes to the presence of complicated geo-environmental setting, long stretch of active Fold-Thrust-Belt (FTB) in the steep mountainous tract of the Himalayas in the north, long near coastal steep mountains alongside the Arabian Sea and unsustainable anthropogenic activity in hilly regions (Fig. 8.1). More prominently, the landslide-prone terrains in India also receive very high amount of monsoon rainfall for about 5 months (June–October) in a row every year besides severe storms, cyclones and depressions, which act as a vital triggering factor for landslides. Moreover, the vast extra-peninsular tract in India which hosts mainly the landslide-prone areas of the country (0.42 million km^2) is also equally prone to earthquakes of moderate to high/very high magnitudes, which also act as another potential triggering factor. Therefore, the risk to landslides of all types remain to be quite substantial in India.

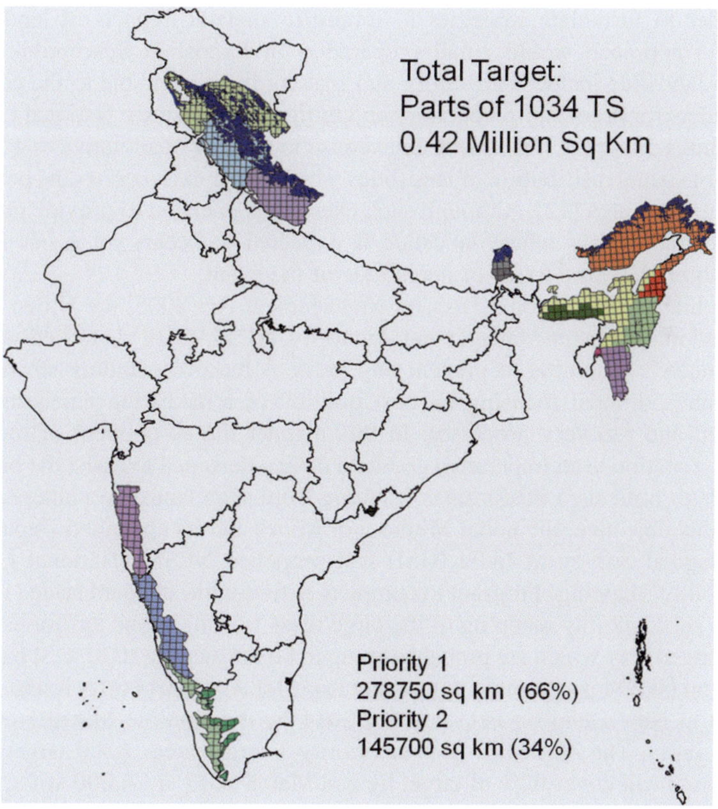

Fig. 8.1 Landslide-prone terrains in India (Source: GSI portal <http://www.portal.gsi.gov.in>, [32])

It is evaluated that landmass movements have caused more harm to properties than some other unstable topographical risk [3–5]. The reasons ascribed to this include: over exploitation of characteristic natural assets, extensive deforestation, change in atmospheric conditions, increment in slope-dwelling populace and uncontrolled strip mining thereby bringing about a higher vulnerability of surface soil to creep/flow under gravity and higher instability of the occupied human settlements [6–9]. As indicated by the National Disaster Management Act (2005) [64], the Landslide Disaster Management Plan should give utmost weightage on readiness and mitigation as opposed to adopting more consumptive temporary courses of action for rescue, help and recovery measures. In this manner, the onus of successful overseeing landslide related risks turn into a challenging undertaking because preventing or minimizing losses owing to an inevitable and consequential geomorphic phenomenon like landslide needs proper geo-scientific appraisal and studies which need to be an integral part of any legitimate geo-scientific examination and studies which ought to be the fundamental piece of any detailed point by point disaster management plan or the debacle administration design leading to an advancement of early cautioning framework or Early Warning System [10, 11].

In order to formulate strategies to minimize societal impacts of landslides, a systematic approach would entail preparation of Landslide Susceptibility Maps linked to landslide incident inventory and making them available to the concerned stake holders for necessary preparatory and mitigation measures. National Landslide Susceptibility Mapping (NLSM) is an exercise in making quantitative or qualitative estimate of spatial distribution of landslides which either exists or has the potential to occur in a given area [12]. Although such exercise is intended to provide potentially most vulnerable areas where landslide is expected to occur, yet it has not been feasible to predict time frame or magnitude of the event.

According to the National Disaster Management Act 2005, the Union Government (National Disaster Management Authority, NDMA) [13] is tasked with fortifying country's readiness to prevent any perilous hazard including slope failures rather than additional focusing on cost prohibitive remediation amelioration and alleviation and recovery processes. In this manner the assignment of multi-scale landslide zonation is an imperative geo-data device designed to assist the organizers and heads to limit such misfortunes resulting from such landslide vulnerability. To achieve this objective, the nodal Ministry of Mines and its appended Department – the Geological Survey of India (GSI) had propelled NLSM (National Landslide Susceptibility Mapping) Program to complete nation-wide standard scale (1:50,000) landslide susceptibility mapping of the large mass wasting prone territories of India in an arranged way which are probably going to be finished by 2020. GSI had started working on NLSM programme with effect from 1st April 2014 after it was formally launched in New Delhi on February 05, 2014 by the Hon'ble Minister of Mines, Govt. of India. The NLSM project in Priority-1 target areas (total target – 0.282 million km^2) will cover 62% of target by end March 2017 (1,76,000 km^2). One can view the output maps of Uttarakhand at:- http://bhukosh.gsi.gov.in/Bhukosh/MapViewer.aspx

Another major concern is the maintenance of the transportation corridors or the so called "Life Line" in hill areas. Frequent landslides along transportation lines not only result in the direct loss to properties but also indirect loss by blocking the road and rail and communication links [14, 15]. Therefore, a timely forecasting of landslides along the transportation corridors and issuance of early warning to alert the traffic is extremely important for the benefit of the society. Early warning of landslide is also important in the context of the 'residual risk'. Humans deliberately thrive in such areas and become adopted to the impending risks owing to certain inherent advantages such as running businesses, proximity to workplace, etc. Studies have shown that people residing in such areas for many generations do not want to leave their native and ancestral places in spite of having witnessed landslide disasters [16]. Under such circumstances, the only opportunity to minimize the landslide loss is to issue timely alert to the community about the impending hazard through the use of an early warning system, EWS in India. Official Nodal Agency for landslide investigations in India is the Geological Survey of India. It realized the importance of landslide forecasting in the aftermath of 2003 Varuna Parvat Landslide in Uttarakashi, Uttarakhand. Landslide forecasting and risk assessment is particularly significant in countries with emerging economies where spatial planning is not yet fully appreciated while attempting land use planning, leading to civic and infrastructural developments in vulnerable and unstable hill slopes and areas prone to landslide disasters. The recent research in risk quantification and threshold modeling for landslide initiation [16, 17] formed reasonable basis for GSI to step into a multidisciplinary science of landslide risk reduction through early warning system [11, 18].

8.2 Role of Geosciences and Engineering Geologists

Landslide, being a geomorphic phenomenon largely depends on various pre-disposing geofactors such as slope morphometry, lithology, structure, land use/cover, geohydrology, etc. and a variety of triggering factors such as rainfall, earthquake, etc. [19–22] Therefore, it is naive to consider that landslides and their spatial distribution to be a stochastic phenomenon, rather their spatial distribution is well controlled by various combinations of geofactors and their relative importance depends on different failure mechanisms [23–26], which can best be studied by application of different geoscientific tools [27]. Thus role of geosciences and application of its core knowledge in studying landslides remain the most important analytical and scientific approach to understand the causes and behaviour of landslide-related hazards [28].

The Geological Survey of India, being the premier geoscientific organization in India, is mandated to provide a gamut of necessary inputs on geological/geotechnical attributes of various slope-forming materials to meet the requirement of various modes of landslide-related studies and designing of disaster-resilient plans and protective structures for minimizing the losses caused by this particular hazard. However, being the nodal agency, the responsibilities of the GSI includes coordinating all activities related to landslide hazard mitigation and management, assisting and providing technical support to the National Disaster Management Authority (NDMA) towards the capacity building and skill development centres such as the National Institute of Disaster Management (NIDM) and the newly-formed Technical Advisory Committee (TAC) on landslide disaster mitigation and management – the topmost advisory body of the Government of India in addressing all matters of landslides and related issues.

8.2.1 Geoscience-Based Standard Operating Procedure (SOP)

The sequence of systematic geoscientific activities in investigating and managing landslide disasters can be specified and elaborated as a Standard Operating Procedure (SOP). GSI has formulated a set of operating procedures on large mass wasting investigations in India that emphasize on systematic geo-parametric data collection and compilation (as per standard and devised format) for landslide inventory. This data is synthesized with relevant spatially-distributed causative thematic maps into susceptibility zonation which represents integrated geospatial information indicating intensity and propensity of landslides. This baseline data should ultimately lead to the collation and evaluation of landslide hazard and risk and mitigation plans.

Geoscientific studies of landslides also provide necessary inputs on geotechnical aspects to meet the requirement of disaster-resilient structures. It engages in studying and ascertaining the causes, nature of various geohazards and associated disasters with an aim to provide input parameters to work out suitable corrective measures. Technical design and execution of relevant corrective measures are the responsibility of the concerned stakeholders. Geoscientific organizations like GSI and others, render geotechnical and engineering geological inputs to the structural designer/ geotechnical engineers, who in turn design the protective structures. The possible instability of slopes and seismic status of the area on regional scale are indicated beforehand in their reports so that the stakeholders can incorporate necessary corrective measures in design of the structure required.

8.2.2 Pre-hazard Stage Investigations

Macro Scale Landslide Susceptibility Zonation on 1:50,000 are followed to classify the landslide prone terrains of the country into different tracts according to their proneness to vulnerability to mass wasting for perspective project planning. It is a multi-thematic exercise taking into account the relevant causal geofactors such as: (i) slope morphometry (slope gradient, aspect, slope shape), (ii) lithology, (iii) structure, (iv) geomorphology, (v) land use/cover, (vi) geohydrology, etc., as separate causative themes and by establishing either through an expert-driven or data-driven interrelationships with different types of landslides, adopting a number of standard and well-established techniques [27, 29, 30]. The main objective of this susceptibility zonation is to facilitate the planners and inhabitants to understand the slope stability potential of the land parcels in an area in fragile hilly or treacherous mountainous terrains for use, further/future development and deciding on protection measures to ameliorate the slope stability conditions.

However, from the experiences of major event like that of the recent Uttarakhand disaster of June 2013, several lessons were learnt which compel re-evaluation of the existing methodology to prepare macro-scale landslide susceptibility maps. The additional geofactors that also need to be considered in light of the initiation of numerous landslides and colossal damages in Uttarakhand event are: (i) effect of toe erosion by higher order streams, (ii) effects of unplanned development of settlements on lower-level geomorphologic terraces and flat lands and proximal to trunk or higher order river/drainage systems, (iii) effect of long run- outs of the debris flows on drainage system, etc.

The existing landslide susceptibility maps exhibit zones having varying degree of susceptibility (e.g., high, low, moderate, etc.) but lack the basic information such as location of stable flat ground nearby which is to be delineated away from the river and the steep potentially failure-prone back slopes; roads/foot tracks leading to stable higher grounds that could act as safe escape routes in the event of a landslide/debris flow/flash flood, etc. It is thus relevant that after evaluating the scientific results of the landslide susceptibility maps, a user-friendly map depicting stable and unstable slopes along with all the available road/foot-tracks, rivers, higher order streams/drainages, etc. should also be prepared, marked on map and the same also to be supplied to the stakeholders and end users to facilitate them in planning disaster escape routes and relief and rehabilitation measures.

8.2.3 National Landslide Susceptibility Mapping (NLSM) Programme

National Landslide Susceptibility Mapping Programme aims to prepare seamless Landslide Inventory and Landslide Susceptibility maps in the mountainous areas of India on 1:50,000 scale which would act as the fundamental input to devise and compile maps on the macro-scale landslides geohazards and disaster prone areas of India. NLSM intends to provide Pan India macro-scale baseline information for the first time to accurately and quantitatively assess the spatial locations of landslide-prone areas in the country (Fig. 8.1). It will help in disaster preparedness of the country and will indicate areas critical for landslide monitoring and early warning. It will demarcate and facilitate prioritization of areas for further detailed studies (on Meso- and Micro-scales) and help in Regional Land Use Planning and provide the scientific basis for framing the Land Use Zoning Regulations.

Geo-parametric data sheet for collecting information and geodata on landslide inventory and details on consequent damage is compiled and documented at national

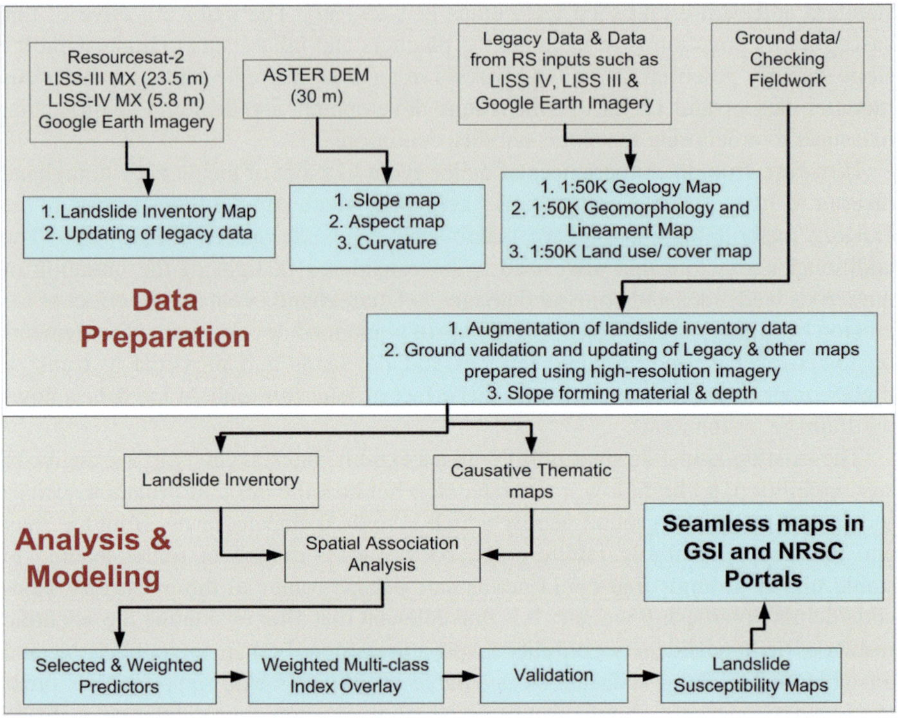

Fig. 8.2 Flow diagram is prepared to present various steps and stages of relevant data generation/compilation and synthesis (Source: GSI portal <http://www.portal.gsi.gov.in>, [12, 32])

level by GSI. A flow diagram is prepared to present various steps and stages of relevant data generation/compilation and synthesis (Fig. 8.2). As a first step, it is imperative to create and strengthen a Landslide Inventory which documents the specific spatial and temporal and other attribute of a landslide incidence on a GIS platform [12, 31, 32]. Thus a geo-referenced spatial database is created. GSI's endeavour in preparing such a national landslide database for the entire country in a GIS-based dynamic platform would be extremely beneficial in future and will also enable the Indian landslide scientists to work on predicting the temporal and magnitude component of such hazard. As a necessary logical follow-up, effective implementation of NLSM is facilitated which would lead finally to research and developement of EWS at the most vulnerable selected sites in India.

8.2.4 *Web-Based Landslide Incidence Inventory Map Service*

Geological Survey of India has developed a Web-Based Landslide Incidence Inventory Map Service where details of more than 1000 landslide incidences are currently available (as on 25/04/2017). Landslide incidence compilation and uploading in such web-based interactive platform is a dynamic exercise and needs updating on a regular basis as and when a new landslide occurs. In this endeavour, GSI has already taken up systematic field surveys and research programme, the first phase of which is scheduled to be completed by March 2017. GSI primarily gathers the relevant information for this purpose from two type of sources: (i) GSI's old landslide investigation reports through compilation, and (ii) from the on-going landslide investigation programmes of GSI such as NLSM Projects which are targeted to be completed on priority by 2020, and other site specific post-monsoon field-based landslide inventory mapping programmes taken up in all the landslide-prone areas each year. To access this data, one can visit GSI Portal (www.portal.gsi.gov.in) and navigate to "Recent Landslide Occurrences in India" under "Interactive Maps" within "Interactive GIS Maps". Or, the following link can be used to access the Beta-version of the map service directly at: <http://www.portal.gsi.gov.in/gismap/landslide/index.html>.

The spatial database of landslide incidences can be seen on an Indian base map with important roads, state boundaries and important places (towns, villages) marked on it. Landslide incidences are shown as point objects. Details about various attributes (maximum 41 attributes, as given in Table 8.1) per landslide incidence are also listed [11, 12], which can be observed using "identify" button in an interactive GIS Map Service environment. The metadata showing such textural attributes can also be exported as "pdf".

Benefits likely to accrue as a result of this project include the following:-

- Can give pan-Indian geospatial scenarios about past/historic landslide occurrences or hazard
- Can evaluate the vulnerability and risk to roads, buildings and other physical features etc.
- Such type of GIS-based landslide inventory is easier to update, retrieve and also to spatially evaluate its relations with other geofactors such as slope, geology, land use, geomorphology etc.
- Acts as the most crucial and fundamental input to any landslide susceptibility, hazard and risk analyses

8.2.5 Post-Hazard Stage Investigations

8.2.5.1 Immediate Response: Reconnaissance (Level-1)

The immediate response to any natural disaster is undoubtedly evacuation, relief and rehabilitation. At this crucial stage, consultation of the available macro-scale landslide susceptibility maps by the people engaged in relief and rehabilitation work (NDRF, Army, NGO, etc.) may become helpful to plan evacuation and relief routes, etc. Embarking on a field-based scientific investigation, by GSI and related agencies,

Table 8.1 Geo-parametric sheet for collecting information on landslide inventory and damage data

No.	Field	Description
1	Landslide no.	State/district/toposheet/year/serial number
2	State	Name
3	District	Name
4	Toposheet	Survey of India number
5	Name of the slide	Maximum 20 character
6	NH/SH/locality	National highway/state highway/name of nearest village or locality (maximum 50 character)
7	Latitude	Latitude in decimal degree (using WGS 1984 datum)
8	Longitude	Long in decimal degree (using WGS 1984 datum)
9	Length	Meters
10	Width	Meters
11	Height	Meters
12	Area	m^2
13	Depth	Meters
14	Volume	m^3
15	Run-out distance	Meters
16	Type of material	Rock/soil/debris/rock-cum-debris/debris-cum-rock
17	Type of movement	Falls/topple/slide/subsidence/flows/lateral spread/creep
18	Rate of movement	Exceedingly quick, severely quick, rapid, moderate speed, gradual, very gradual, exceedingly gradual speed
19	Activity	Discernible/rejuvenated/indiscreet/temporarily quite/residually firmed-up/inactive/firmed-up in place/fully anchored
20	Distribution	Active slope process/retrogressive/spreading/expanding/restricted/creeping
21	Style	Superimposed/successive/repetitive/distinctive event
22	Failure mechanism	Shallow planar failure/deep planar failure/shallow rotational failure/deep rotational failure
23	History	Date of initiation and subsequent reactivations (date/year)
24	Geomorphology	Topography, landform, processes (maximum 100 character)
25	Geology	Lithological description, stratigraphic disposition (100 character)
26	Structure	Details of discontinuity in whole circle (e.g.: J1: 50°:150°)
27	Land use/land cover	Barren/cultivated/forest /rural/urban/anthropogenic activity/any other (maximum 50 character)
28	Hydrological condition	Dry/wet/dripping/flowing
29	Triggering factor	Rainfall in mm/daily or hourly rainfall data/earthquake (intensity in Richter scale)/lake bursting (GLOF)/anthropogenic factors like road cutting, blasting, etc. (maximum 50 character)
30	Death of persons	Numbers
31	People affected	Numbers
32	Live stock loss	Cattle death
33	Communication	Road/rail/transmission line (blocked/damaged)

(continued)

Table 8.1 (continued)

No.	Field	Description
34	Infrastructure	Numbers of houses/building/dam/barrages damaged
35	Other land use	Agriculture/forest/barren (maximum 50 character)
36	Geo-scientific causes	Topple/planar failure/wedge failure / pore water pressure/piping/ reduction of strength on super-saturation/toe erosion by stream/gully erosion/head ward erosion, anthropogenic activities (unplanned cutting of slope, unplanned construction, loading at head region, afforestation, adverse cultivation pattern/combination of above factors/any other factors (maximum 200 character)
37	Remedial measures	(maximum 200 character)
38	Remarks, if any	Assessment remarks, if any (maximum 200 character)
39	Photographs, sketch	Photographs, sketch of plan affected by landslide – Quantity/type of cultivation and section of the slide send in jpg format
40	Summary/ abstract	(within 1000 character)
41	Pdf, soft copy	Soft copy of landslide report if data is collected/compiled from past data/maps, where feasible

immediately after any disaster may be difficult because utmost priority just after the disaster is given for relief and rehabilitation work. Moreover, inaccessibility due to damages in surface communication route hinders any immediate field inspections. Therefore, during that time, maximum stress is given for preliminary assessment by interpreting satellite based data products generated through the state of art satellite imageries and/or air surveillance, as has been carried out presently by Indian Space Research Organization (ISRO) through its National Remote Sensing Centre (NRSC), State Government officials, forest, army, etc. GSI has also earmarked trained resource personnel to be associated for any such endeavour for such rapid assessment either through remote sensing or ground truth surveys.

The field-based first-level ground appraisal is generally taken up within 15 days of the event by respective Regional HQ of GSI (where the hazard has taken place) by employing adequate number of trained personnel after due consultation with the State Government agencies, army, rescue party, etc., and as per the accessibility to the affected terrain. This reconnaissance is generally completed within 3 months after initiation of the fieldtrip along with submission of preliminary report to the concerned State Government authorities. For undertaking technical data in field per individual landslide/trouble spot, suitable geo-parametric data format is already developed (Table 8.1) which is focused on collection of all sorts of relevant information of any landslide following the international norms of classification and guidelines [3, 33–38]. It includes rapid assessment of damages, preliminary identification of landslides, its broad typology, identification of stretches of affected roads/accessibility corridors and if possible, tentative identification of probable safer slopes for temporary rehabilitation. Another main objective of this reconnaissance would be to delineate vulnerable tracts and to assess quantum of further detailed work to be taken up for the second level appraisal.

8.2.5.2 Response Study: Preliminary/Feasibility and Multi-thematic (Level-2)

At this level (to be initiated expectedly within three to 6 months of the disastrous event), and as per the suggestions of the first level response studies and according to the specific demands of the State Government, GSI can take up a rapid macro-scale (1:50,000/25,000) slope stability assessment (preferably GIS-enabled) of the affected area as an additional item of the investigation in that particular Field Season. The total time period of this study can range from 6 months to 1.5 years including submission of reports depending upon the extent of the affected area. Arrangements for provision of interim reports may also be kept to facilitate initiation and continuation of restoration and rehabilitation works.

8.2.5.3 Landslide Vulnerability Tracts on Macro Scale

Landslide Vulnerability Tracts on Macro Scale (1:50,000) is followed to classify the landslide prone terrains of the country into different risk prone categories according to their degree of propensity to landsliding for perspective project planning. It is a multi-thematic exercise dependant on several causal geofactors such as (i) *slope morphometry* (Slope gradient, aspect, slope shape), (ii) *lithology,* (iii) *structure,* (iv) *geomorphology,* (v) *land use/cover,* (vi*) geohydrology, etc.,* by establishing either through an expert-driven or data-driven interrelationships with different type of landslides, adopting a number of standard and well-established techniques [27, 29, 31, 39, 40, 63]. The main objective of this susceptibility zonation is to create awareness amongst the planners and inhabitants to become better equipped and prepared to cope with impending fragile and natural hilly terrain related dangers of such area and for making well informed decisions for protection, appropriate land use planning, civic development and improving resilience to address the slope stability issues.

However, based on the experiences of major event like that of the recent Uttarakhand disaster of June 2013, several lessons were learnt which compelled re-evaluation of the existing methodology to of conducting preparation of macro scale landslide susceptibility maps. Additional geofactors that also need to be considered in light of the initiation of numerous landslides and colossal damages in Uttarakhand event include: (i) effect of toe erosion by higher order streams, (ii) effects on settlements that are built on lower-level geomorphological terraces and flat lands and proximal to trunk or higher order river/drainage systems, (iii) effect of long run-outs of the debris flows and drainage morphometry, (iv) nature and size of clastic components, etc.

The landslide susceptibility maps need to exhibit zones having varying degree of susceptibility (e.g., high, low, moderate, etc.) and depict the basic information such as location of stable flat ground nearby, existing roads/foot tracks leading to stable higher grounds that could act as safe escape routes in the event of a landslide/debris flow/flash flood, etc.

Based on geoscientific evaluation a user-friendly landslide susceptibility map depicting stable and unstable slopes along with all the available infrastructure: foot-tracks/road, communications network, drainage lines and higher order streams, safe shelters, etc., need to be prepared to facilitate stakeholders and end-users in planning disaster escape routes and relief and rehabilitation strategies. A collaboration framework has already been finalized between NRSC and GSI regarding event-based landslide inventory mapping and landslide susceptibility/hazard zonation in the prioritized landslide vulnerable areas of India and integrated application of modern GIS based remote sensing data products and field surveys and processing. It is planned to get ready different geofactor layers on GIS stage, for example, satellite data products based interpretation and derivative information with respect to slope angles, aspects and bend, elevation and relief, landuse-landcover, site geomorphology and operative surface processes, tectonic features and structural highlights: Faults/Lineaments thrusts and discontinuities, waste support, upslope contributing range and field review surveys to synthesise geo-information in light of slope and configuration of material and its thickness, morpho-structural understanding and geoscientific reconciliation of every topical layer for relegating appropriate weightage and hazard appraisal for simulated and projected frequency-magnitude demonstrations. Notwithstanding any technique for foreseeing landslide geohazard vulnerability, it needs validation which once in a while might be troublesome in zones having no land sliding history. However, for all NLSM programs to be completed, a quantitative approval through interpolated successful achievement and related inferred forecast rate curves, following the globally acknowledged strategies proposed by Chung and Fabbri [41] and embraced by Ghosh et al. [22, 42] are firmly prescribed, which would likewise encourage grouping of crude mass movement vulnerability score maps into subjective ("High", "Moderate", "Low") landslide proneness maps, effectively justifiable and understood by all interested partners.

8.3 Vulnerability and Risk Assessment

Landslide susceptibility analysis is performed utilizing modern methods to predict sites vulnerable to disasters related to the future incidence of landslides. Such analyses reiterate that given the required combination of geofactors and hill slope process parameters, a future landslide would occur. However, this requires a prior knowledge about and reference to parametric data sets on the landslide that occurred in any area. This means that this predictive analysis would answer where, when and how large the future landslide could take place [19, 47, 48]. Therefore, it is imperative that for landslide hazard analysis, availability of historical information on past landslides is essential for desired levels of accuracy of predictions.

Landslide geohazards vulnerability (risk) analysis involves methods to fore see or estimate financial aspects of losses caused to material assets at risk (roads, buildings, telecommunications and power transmissions) or estimation of losses of population due to any mass wasting hazard [43]. This is the consequential aim of any slope-failure geohazards investigations as it serves useful purpose to planners and insurance agencies for realistic evaluation of the inherent vulnerabilities. Nevertheless, such analysis is extremely difficult due to non-availability of either spatial and temporal information of landslide incidences in particular or magnitude-frequency data of mass wasting occurrences in general. India has been having the similar problems to grapple with, like many landslide-prone countries in the world. India has geared itself to face these challenges and launched several programmes to build its own authentic data base to rely on and work for risk evaluation for long-term mitigation measures and disaster preparedness. In order to find dependable solutions to the above and to prepare upgradable and spatially-distributed national landslide repository, the Web-based National Landslide Incidence Inventory project of GSI has been launched that will finally prepare a substantial and quantitative nation-wide landslide parametric database.

Landslide Risk Management is the ultimate assemblage and syntheses of geofactor data generated and compiled on mass movement (landslides) investigations starting from identification of a landslide-related issues, knowing its vulnerability status (predictions of spatial locations), geohazards assessment (comprehensive interpretation of spatial, temporal and magnitude aspects) and estimation of risk (loss of life and material estimation), followed by total risk evaluation and process to define and implement ground-level action towards mitigation and reduction of the forecasted dangers. Therefore, GSI's endeavour in preparing such a national landslide database for the entire country on a GIS-based dynamic platform would be extremely beneficial in future and will also enable the Indian landslide scientists to work on predicting the temporal and magnitude component of such hazard more realistically.

8.4 Development of Early Warning Systems in India

In India, landslides mainly affect the Himalayan region, Western Ghats, Nilgiris and Konkan Hill ranges [10–12, 63]. It is assessed that 30% of world's landslides occur in the geologically youthful mountain framework: the Himalayas [44]. Geoscientists consider down slope land movement events for an assortment of targets inside the system of landslide vulnerability and chance occurrences. These objectives incorporate systematic and reliable compilation of hazard incidence data base, ground surveys and susceptibility mapping, site-specific and detailed geoscientific investigation of danger prone area, hazard estimation, incidence appraisal and likely rescue and relief preparedness, and so on [27]. The ultimate aim of such studies is to help in reduction of risk to the human population, and to make life and property safe from landslide disasters. Therefore, concerted and integrated efforts need to be made to reduce the societal risk, either by reducing the probability of failure or by reducing adverse effects of consequences. Crozier [45] listed nine different approaches, grouped into three strategies, to reduce landslide risk: "land use zoning, engineering solutions and emergency preparedness". These techniques significantly work towards either mitigation of the spatio-temporal probability of occurrence of landslide by using different remedial options or lessening the vulnerability of operative geofactors involved through enhancements of the geomorphologic built-up components or scenario by making them more safe or strong or resilient to the conceivable landslide impact as enumerated below:–

(i) In land use zoning, either building regulation or relocation of the elements at risk to hazard-free areas is often used to diminish landslide risk. However, practice has shown that relocation is not always viable as in any case, rehearse has demonstrated that migration isn't generally reasonable as individuals would prefer not to be moved unless the neighbourhood specialists/administration offer monetarily substantially greater and attractive options.
(ii) In designing engineering geological arrangements, slope adjustment works, development of waste water drainage, and so forth, are regularly used to diminish landsliding chances by limiting the likelihood of occurrence of such dangerous disaster. This procedure isn't just costly yet is reasonable for a couple of known insecure hill slope situations, as it is hard to portray the exact area of every single potential landmass movement at a catchment scale. However, technology infusion in due course will certainly make this approach cost-effective.
(iii) In crisis readiness technique or emergency preparedness strategy, chance/mishap risk lessening is accomplished through skill building, i.e., by sensitising public, expanding mindfulness and increasing awareness among the people in general and suitable training with an aim to imbibe appropriate response to the warning before or when the disaster strikes. Although studies have indicated that these three strategies, alone or in combination, are effective in mitigating landslide, yet the fact remains that since it is not possible to completely mitigate all landslide-prone slopes, the chances of 'residual' risk remains to exist.

Issues are more complicated in large countries such as India where landslide-prone areas are spread over 0.42 million km^2. Using engineering solutions or carrying out risk awareness programmes for mitigating landslides for such a large area with huge geodiversity remains a challenge. In the developing countries, risk management further gets complicated due to their societal and economic considerations. Here the financial issues (socio-economic problems) are frequently so substantial and genuine that adequate consideration is not feasible to potential mass movement issues. Besides, these geohazards mostly remain confined and influence a smaller segment of the general public. Under this circumstance, endeavours are expected to enhance the practical and pragmatic methodologies of developing landslide hazard diminishing systems. One such approach is the utilization of early-cautioning frameworks or the EWS. Value-added relevance of this strategy in India in reducing human deaths and destruction was acknowledged in the recent two natural disasters: the Odisha cyclone "Phailin" where numerous lives were saved due to the timely early warning; and the Uttarakhand landslide/flood disaster of June 2013 where numerous lives were lost due to the lack of an early warning.

8.4.1 *What Is an Early-Warning System?*

In context of landslide, an early-cautioning framework or early-warning system (EWS) is a technology which is capable of monitoring and modelling landslide initiation and issuance of advance warning to the concerned of impending danger. In order to develop reliable EWS, the geological/geotechnical processes and mechanisms need to be understood in addition to their quantified measurements and processing for modelling and prediction. Modelling for EWS can be done manually or through the use of automated systems, such as wireless sensor networks. Warning system may be of different types, depending on the type of landslide event and resulting landslides, spatial coverage of the warning area and the communities to be warned. It can be developed for:

(a) An individual site-specific mass movement prone environment and to warn people about the movement of the delineated landslide. The warning is based on site-specific study of the landslide using movement sensors, geodetic and piezometric monitoring, etc. In India, Amrita University, Kollam has been carrying out such study since 2008 in Munnar, Kerala and some parts of eastern Himalaya [18, 63].
(b) Communities or organizations for small areas for one or more landslides. The warning is based on simple rainfall threshold values estimated for landslide initiation.
(c) Large areas and extreme weather events (such as very high rainfall prediction that could be critical for initiating landslides).

In all these three types, the most important input parameter for landslide early-warning system is the projected threshold value. For rainfall-triggered landslides, the threshold is the amount of rainfall required to initiate a ground displacement that can prograde into a mass movement capable of causing damage to life and property.

In general, a warning system for landslide has four major components:

1. **Monitoring**, understanding of geofactors and comprehensive recording which includes data collection and compilation, archival and retrieval devises, ease of accessibility and upkeep of the gadgets and machinery;
2. **Synthesis, analysis and interpretation,** leading to predictions with riders, mainly arrived at with of a set of limiting criteria and use of modern processing tools and specialist inputs, etc.
3. **Cautioning or early alert,** i.e., the rapid communication of unambiguous understandable statements and instructions alerting public for the potential disaster and
4. **Resilience and adoptability,** i.e., emergency preparedness for safety, relief and rescue operations where people are able to willingly cooperate and show collective maturity on how they must respond to the warning.

In these above four components, the role of geoscientists is restricted to the first two, i.e., the monitoring of a landslide, and analysis of the data to forecast the initiation of a landslide. The last two components, i.e., warning and response are of multidisciplinary nature and involve many stake holders such as administrators, planners, disaster response force, etc.

8.4.2 Why Is It So Important to Have Early Warning for Landslide Events?

Dissimilar to other common perils and resultant disasters, for example, woodland fire, coastal wave surges and earthquakes, landslides occur fairly consistently in hilly territories and make significant harm to life and property, both regarding immediate tragedy and consequential misfortune [4, 43, 46, 63]. Since most recent couple of decades, mass movements have been seen as a genuine and persistent risk to humankind. It is generally assessed that all inclusive, every year; landslides owing to slope failures would become more disastrous and harmful to life and properties than some other land risk [3, 5]. The reasons ascribed to this assertion are the overexploitation of natural assets, fast and extensive deforestation, change in atmospheric conditions (impacts of global warming) and steady growth in slope dwelling populace and uncontrolled anthropogenic excavations along these inherently perilous terrains thus bringing about destabilization of slope and enhanced proclivity of surface soil to creep and erode to damage the human settlements and add to insecurity of the uncovered populace [6–8].

The situation is grimmer in India where the increase in hill tourism and population have not only resulted in inappropriately planned fast and enhanced developments of infrastructure, but also prompted encroachment in hazardous areas. This was one of the reasons attributed to huge unprecedented loss of life and property in the June 2013 Uttarakhand landslide disaster. Researchers have argued that the Uttarakhand disaster was the result of an extreme rainfall, of a 1000-year return period, and could have been averted if timely warnings were issued, similar to the 'Phailin' cyclone of Odisha. Studies have shown that people residing in such hazard prone areas for many generations are generally poor and not fully aware of such potential hazards. They do not want to leave their place of settlement in spite of having witnessed landslide disasters [14–16]. Under such circumstances, the only opportunity to minimize the landslide-related loss is to timely alert the community about the expected hazard through the use of an early-warning system. Maintenance of the transportation corridors and communications network is important pre-requisites in such naturally hazardous tracts for the benefit of society [17].

8.4.3 What Makes It So Difficult to Forecast and Warn a Landslide Event?

Development of landslide EWS essentially requires:-

(i) Establishing criteria for parameters to be monitored and working out their threshold values,
(ii) Acquiring/developing monitoring equipments and systems,

(iii) Coordinating satellite-based data/interferometric radar data (Synthetic Aperture Radar) with local monitoring stations, where needed,
(iv) Planning detailed monitoring programmes for high-risk areas, and
(v) Acquiring/developing computer-aided decision-making tools, information management strategies using GIS with remote sensing and 3D modelling, etc.

Unlike floods, which have defined inundation area and earthquake, which has defined source zone and follow defined pattern of ground acceleration, landslide is more complex in terms of magnitude, source location and run-out pattern, etc. [18, 19, 47, 48]. Landslide, being an individual phenomenon, each landslide is unique in terms of its morphometry (area and volume), magnitude (intensity) and spatial location. The uniqueness of landslide phenomenon and the fact that each triggering event can result in one or many landslides in an area, make this hazard comparatively difficult to predict both in space and time.

8.4.4 Difficulties Related to Spatial Prediction of a Landslide

The success of an early warning depends on prediction of the precise location of a landslide. The state-of the art model available in spatial mapping of landslide-prone zones provides only a qualitative or probability-based quantitative susceptibility maps showing the probable initiation zones (e.g., [41, 48, 49, 50]). Such maps do not point out 'precisely' where the landslide will actually initiate given the triggering condition and how it will behave down-slope. Useful links need to be established between ground movement, rainfall and pore-water pressure. The need for R&D on these issues was discussed at length during the recently concluded regional workshops on landslide disaster management in Shimla (for NW Himalaya), Shillong (for NE Himalayan region) and Wellington (for Western Ghats and south India), India [55–57]. However, the uncertainty in the absolute spatial prediction of a landslide makes early warning difficult to implement [21]. Further, absence of geoscientific data on the size of the landslide, its run-out distance, its inundation zone makes warning difficult. It has been observed that most of the landslide-related loss is not merely to vulnerable assets around locations on catchment source of landslide; rather it is more to establishments and contained elements located on the run-out zone [11, 63].

8.4.5 Difficulties Related to Temporal Prediction of a Landslide

The most important component of an early warning is the prediction of the precise time of a landslide. Till date, only a hand full of research has been carried out where temporal component of landslide initiation has been modelled for a large area (e.g.,

[17, 48, 51, 52, 60]). In this aspect, almost all research focuses on identified vulnerability with associated likelihood estimation on of occurrence of hill slope process leading to land mass movement. Uncertainty related to probability estimation is one of the hurdles in the temporal forecasting of a landslide.

8.4.6 Challenges Related to Communication of Landslide Warning

Advancement of a reasonable cautioning dialect and directions that are appropriate to the hazard prone areas and its populace are fundamental for the fruitful execution of an early-cautioning framework on ground (EWS). Although relevant Rules and guiding instructions are framed on local dialect and articulations mechanisms for various levels of caution are accessible (e.g., [16, 53, 64]), yet the issue is the means by which to educate and involve individuals or group about the approaching threat. Since, landslide distribution is localized and affects only a few individual or a certain segment of the society, communicating the warning, which are developed and issued from a place, located far away, to the concerned remains a challenge particularly in India. With hill area spreading over 0.42 million km^2 and dispersed settlements located at distant places accessible through track routes, communicating warning message within a short span has been a major challenge in real time ground alerts and execution of an early-cautioning framework.

8.4.7 Difficulties Related to Data Dissemination and Awareness

One important component of early-warning system is the availability of updated maps such as safe shelter map, alternate route map, susceptibility map, etc. In India, safe shelter and alternate route maps for landslide hazard are often not prepared or not available. These maps need to be prepared for individual settlement. People concerned should be aware of such maps and their physical representation on ground for timely response to the warning. In India, though susceptibility maps are available for important settlements, but they need to be updated periodically with respect to the changing land use condition. Another important aspect that makes early warning ineffective is the lack of public awareness. The mindfulness, prompting crisis readiness, means the capacity and skill levels of a group to put vigorously the set up action plans and methodology for vulnerability moderation [45]. This technique has a tendency to lessen the hazard by expanding awareness among the general population with a plan to convenient and suitable reaction to the notice when the perilous fiasco strikes. Albeit, Geological Survey of India, National Disaster Management Authority together with National Institute of Disaster Management and the

State Disaster Management Centres complete standard group based debacle administration (disaster management) program at a state/region level, yet at the same time more deliberate exertion is required to reach to individual and dispersed hilly settlements through the State Government, NGO and other stake-holders.

8.5 A Way Forward in Landslide Warning and GSI's Initiative

GSI had realized the importance of landslide forecasting in the aftermath of 2003 Varunavat Parvat landslide in Uttarakashi, Uttarakhand. It resulted in high expenses of alleviation and recuperation or relief and recovery after landslide disasters had taken place [63]. Adequate considerations on geoscientific planning for safe and cost effective land use practices need to be followed rigorously in such hazardous territories that need to be equipped with early cautioning mechanisms on impending catastrophes. The recent research in risk quantification and threshold modelling for landslide initiation [12, 16, 17] form a potential basis for any geoscientific organisation to step into a multi-disciplinary science of landslide risk reduction through early-warning system. Geoscientific observations based on detailed field- and laboratory based studies would help to formulate strategies on hazard characterization. These include frequency, magnitude, extent, onset and consequences of past mass-wasting events.

GSI has already built-up database structure with a provision of storing as many as 41 different attributes per landslide incidences. It is currently operational and any stake-holder can access upwards of 281 such data sets in GSI Portal. The web-based landslide incidence inventory will also have spatial locations of all the landslides that are being mapped in the on-going NLSM programme of GSI which is being carried out to collate and map several landslides in different mass wasting vulnerable tracts in the country on a dynamic GIS platform. Collation and syntheses of such dynamic data sets will enable formulations on Early Warning Systems including: Landslide Hazard Warning Plans, routines participation and operating procedures, monitoring using in-situ real-time monitoring devices (e.g., geophones, inclinometers, piezometers, extensometers, etc.) and monitoring using geodetic scanners (Terrestrial Laser Scanner). Geodetic monitoring using differential GPS systems and detailed-scale field mapping of permanent surface features using Total Station, remote monitoring systems for field measurements with automated data collection by web-enabled devices are needed for comprehensive determination of both empirical and deterministic threshold modelling of triggering factors (e.g., rainfall, earthquakes, etc.) for devising an empirical system of early warning based on triggers. Development of EWS is considered to be the next higher-level step towards forecast or pre-warn the essential elements-at-risks. Therefore, the general step for developing EWS is essentially to be preceded by detailed site-specific stability and hydrologic modelling, followed by sufficiently long-term monitoring of any particular landslide.

Developing such site-specific landslide early warning system is mainly instrument-based, time-consuming and costly and requires specialized knowledge on this particular subject.

One of the important components of landslide forecasting is the establishment of rainfall threshold value for rainfall-induced landslides. In 2009, GSI made an attempt to model rainfall threshold for shallow landslides in Nilgiri Hills using landslide events up to 2006 [17, 57]. The model was found accurate in forecasting 2009 landslide events in Nilgiri Hills [54]. Making one stride ahead of the 'threshold model', GSI had published a conceptual model and operating procedure of an early cautioning framework after evaluating risk perception of affected community in Nilgiri Hills [16]. However, the efficacy of the model is yet to be tested in the field and in genuine ground circumstance. The model includes the establishment of rainfall related thresholds triggering mass movement, region specific meteorological predictions, establishing a network of rainfall gauges for real time monitoring and an automated computing system for processing and evaluation of data sets. The system is essentially based on actual precipitation dependent forecast and using satellite data products linked to a large network of rain gauges that are equipped to issue alerting messages on surpassing tripping limit – the computed threshold. Such work practiced on the Nilgiri Hills in Tamil Nadu can be a good basis for the development of an early-warning system in the country [11, 16].

Recently, GSI has initiated a joint venture with Defense Terrain Research Laboratory (DTRL), New Delhi and Amrita University, Kollam for automated instrumentation-based monitoring of landslides in Himalayan region. The objective is to develop and test an early-warning system for a single landslide. DTRL and Amrita University are already in advance stage of data collection for the pilot study sites in Uttarakhand and Munnar, Kerala, respectively. GSI is also in advance stage of collaboration with Indian Meteorological Department (IMD) for rainfall threshold modelling in a catchment scale, with an aim to develop an early-cautioning framework for multiple landmass movements covering a large area in the Himalayas. GSI's national programme on the generation of seamless landslide susceptibility map (NLSM) for the entire country on 1:50,000 scale is nearing completion [10, 31]. These maps in GIS platform will form an important input in: (i) targeting potential hazardous areas for landslide monitoring (ii) evaluation of hazard potential and (iii) development of an early-cautioning framework – the EWS.

Besides, the recently created Geohazards Research Division in GSI has also taken up a 2-year long project in collaboration with Natural Resources, Canada (NRCan) for an Interferometric Synthetic Aperture (InSAR) based landslide monitoring and early warning research programme at two active landslides in Eastern Himalayas (14th Mile-Gayabari and Chibo landslides, Darjeeling district, West Bengal). The main aim of this collaboration is to develop a suitable InSAR-based technique to monitor activity of ground movements of these two landslides. After completion of the target of this collaborative research item, a time series of ground movement at two active landslide sites and a map showing active kinematic movement of slope will be prepared which can be used in developing landslide hazard scenarios in the study area and as a suitable output for monitoring of landslides and early warning in the Himalaya. For this particular collaborative investigation, NRCan is providing GSI the required technical advice, guidance, training and 20 scenes of RADARSAT-2 images of the study area for a period of 24 months. In addition, NRCan has also provided GSI five Trihedral Corner Reflectors for installation at site. Apart from the five Canadian-make Corner Reflectors, GSI's Geohazards Research & Management Centre Kolkata has indigenously manufactured five more Corner Reflectors at Kolkata following the similar design and installed all the ten Corner Reflectors at two landslide sites in Darjeeling Himalayas, India (five at **14th Mile** and five at **Chibo** sites). All these ten (10) Trihedral Corner Reflectors were successfully sighted by the RADARSAT satellite and the official acquisition of RADARSAT-2 image has started by NRCan since September 2016 and till March 2017, seven (07) RADARSAT-2 scenes were acquired by NRCan for further analysis in developing landslide hazard scenarios in the study area.

Owing to complex environmental conditioning and triggering processes that cause landslides, the extent and variability of their spatial and temporal scales imply that they are inherently difficult to forecast and mitigate at site, slope, catchment and regional spatial scales. However, various programmes of research across the world have mapped, catalogued and monitored landslides (e.g., field mapping, remote sensing) to better understand triggering mechanisms of landslide processes. As these processes are often triggered by local in-situ site conditions/ events (e.g., antecedent hydro-metrological conditions, land-use) it is difficult to develop forecasting and early-warning systems (EWS) that are equally-applicable and useful beyond a local site (<1 km^2) to larger area (≈ 10 to ≈ 100 km^2) on spatial scale. There are many projects nationally and internationally that monitors landslides and landscapes in real-time through instrumental sensor networks (e.g., rain, soil moisture, slope movement; now, many of these wireless) and investigate specific triggering mechanism associated with individual or sets of landslides. These networks provide local site to slope/catchment scale EWS for landslides and help individual communities at risk (e.g., Munnar, India; Hong Kong, China, and several sites in Japan, etc.). However, these studies provide triggering thresholds and warnings that are difficult to scale-up (to catchment and regional spatial scales) and apply to other areas/regions. This knowledge gap and challenge are being addressed in a 4-year long UK-funded Project called **LANDSLIP (Landslide Multi-Hazard Risk Assessment, Preparedness and Early Warning in South Asia: Integrating Meteorology, Landscape and Society)** in India, which aims to integrate landscape, climatic and social dynamics by a multi-institutional international consortium, where Geological Survey of India (GSI) has agreed to be one of the nine partner organisations of the LANDSLIP by signing an agreement with the British Geological Survey (BGS), the latter being one of the main co-leads of the LANDSLIP [3, 11, 12, 18, 60].

LANDSLIP will aim to reduce impacts of hydrologically related landslide multi-hazards (in terms of fatalities, livelihoods, assets) and build skill levels and capacity to mange disaster owing to mass wasting in hazard prone terrains (often remote) and vulnerable tracts in South Asia. LANDSLIP will address this issue by using two pilot study areas in north-eastern (Darjeeling/East Sikkim) and southern (Nilgiris) India and will enhance access and response to both landslide risk assessments and robust landslide related multi-hazard early warnings, on a slant/catchment and regional satellite data based aerial scale and diurnal to seasonal time series scale. This means that the proposed LANDSLIP project, where output of this item will be incorporated, aims at to provide simultaneous and better understanding of landslide risk for a smaller catchment area (~25 km^2) as well as for a larger regional area (~400 km^2) with the development of an early cautioning framework/mechanism of landslides for the same on differing temporal scales (~1–15 days to 1 month, etc.). In these particular study areas, 1:50,000 scale national landslide susceptibility mapping (NLSM) database are already available, which will be used as base input thematic maps for further up-scaling at 1:25,000/1:10,000 scales.

The above collaborative endeavour will ultimately prepare an improved rainfall threshold model by incorporating weather regime data, improved landslide susceptibility, hazard models at 1:25,000 and 1:10,000 scales to assess landslide risk at multiple scales. Ultimately, the final outputs of landscape and meteorological dynamics so developed will be assimilated with social dynamics (output generated by other LANDSLIP Collaborators) to achieve the final goal of LANDSLIP that is: improved understanding and risk evaluation on slope failure mass movement and advance alert mechanism.

8.6 Discussion and Conclusions

According to the National Disaster Management Act 2005 [64], the Union Government (National Disaster Management Authority, NDMA) [13] is occupied with reinforcing country's readiness to prevent and keep any danger at bay rather than focusing on allocating additional resources on alleviation and recovery from disaster. Similarly carrying out of multi-scale landslide hazards propensity zonation is a vital geo-data instrument to the organizers and overseers to limit such misfortunes because of such land movement risk. To fulfil this objective, the nodal Ministry of Mines and its attached Department – the Geological Survey of India embarked upon the NLSM Program to finish large scale (1:50,000) landslide susceptibility mapping of the whole mass wasting prone tracts of India in a prioritised way on GIS stage which is likely to be completed by 2020. Standard operating procedures for NLSM have been discussed and debated amongst the stake-holders during three major regional workshops on Landslide Disaster Management held in 2013–2014 respectively at Shimla, Shillong and Wellington (Nilgiris) and the Brain Storming Sessions held at GSI, Kolkata [55–57].

GSI started working on NLSM programme with effect from the Field Season 2014–2015, after it was formally launched on February 05, 2014 by the Hon'ble Minister of Mines, Govt. of India. The NLSM project aims to prepare and integrate seamless Landslide Incidence Inventory and Landslide Susceptibility Maps in the mountainous areas of India on 1:50,000 scale on GIS platform which would provide the fundamental inputs to compile the macro scale landmass movement (landslide) geohazards and spatially documented propensity scenarios of India. It will give pan India macro scale baseline information for the first time to accurately and quantitatively assess the spatial locations of landslide prone areas in the country. It will also help in disaster preparedness of the country and to indicate areas critical for landslide monitoring and developing early warning system.

Significance of documenting down slope landcover movement and forecasting was acknowledged in India since the repercussions of Varuna Parvat Landslide tragedy in Uttarakashi, Uttarakhand in 2003. It has been envisaged to get ready different geofactors layers on GIS stage, for example, satellite data products based interpretation and derivative information with respect to slope angles, aspects and bend, elevation and relief, landuse-landcover, site geomorphology and operative surface processes, tectonic features and structural highlights: Faults/Lineaments thrusts and discontinuities, waste support, upslope contributing range and field review surveys to synthesise geo-information in light of slope and configuration of material and its thickness, morpho-structural understanding and geoscientific reconciliation of every topical layer for relegating appropriate weightage and hazard appraisal for simulated and projected frequency-magnitude demonstrations. The recent multi-disciplinary and collaborative research in risk quantification and threshold modelling for landslide initiation will pave way for landslide risk reduction through development of Early Warning System [11, 18]. Notwithstanding, any technique for foreseeing landslide geohazards vulnerability, it needs validation which once in a while might be troublesome in zones having no land sliding history.

Himalayan landscape is relatively young and slopes are steeper, hence causative factors triggering landslide will be more complex. Besides, slopes in the Himalayas are covered under a very different type of material which is mostly transported fluvio-glacial and slope-wash from younger geological formations [61–63]. Therefore, geo-scientific considerations will play greater role in the Himalayas in Uttarkhand, Himachal and Kashmir as well as in Sikkim, Darjeeling in West Bengal, Arunanchal Pradesh, Nagaland and Mizoram in addition to rainfall and meteorological parameters and will have to be factored in for evolving appropriate models for EWS. These are the topics of advanced research aimed to creating reliable high quality geo-data devices to empower the organizers/managers for delineating landuse zoning regulations for sustainable and safe development with appropriate considerations, geoscientific planning and preparedness in handling such approaching geohazards and for potential disaster management. Therefore it is reasonable to assert that EWS innovative research work and development of instrumentation will take off from the robust products and outcomes of the NLSM programme [12]. EWS need also to highlight mitigation efforts/remedial measures through geotechnical and engineering solutions as suited to Indian conditions on case to case basis for optimum utilization of available resources.

Developing early-warning system for Indian conditions is an important tool for risk reduction, both for a single site as well as for large area. Since, the majority of the landslides in the nation are precipitation activated, advancement of an early-cautioning framework based on precipitation received versus landslide incidence relations, through the use of a rainfall limiting or triggering constraint (threshold), appears to be the best and most cost-effective solution. EWS for a single landslide, threshold can provide information on the precise loci of ground displacement and its acceleration, while for a catchment area it can provide information on when and where land sliding would occur. This in conjunction with presently prepared land sliding propensity maps (NLSM) can portray conceivably the perilous territories. In fact, in a few nations, precipitation limits have been utilized to gauge precipitation initiated land sliding, especially in the San Francisco Bay range, USA [53, 58, 59]. The capacities of early cautioning depend on the observational limit models which in turn depend on different precipitation estimations, for example, forerunner precipitation, precipitation intensity and duration, cumulative precipitation and areal spread and total pouring time, and standardized/normalised precipitation, and so on. The subtle elements of the sorts of limit models utilized everywhere throughout the world can be found in Guzzetti et al. [47, 48] and would require being adjusted to Indian condition.

Although, world over threshold-based early warning is considered to be the most reliable option, yet the lack of information on past landslide event dates and corresponding rainfall data for the desired locations make threshold definitions difficult to model. Since, landslide is a site specific phenomena and the fact that in hills rainfall vary considerably due to 'orographic effect', the use of rainfall data from distant stations with respect to the location of landslide is yet another factor inducing uncertainty to the threshold modelling [61–63].

Successful usage of any early caution framework requires critical assets and skilled man power. Such EWS methodology entails automated rain gauges, updated and refreshed landslide susceptibility maps, specialists to decipher outcomes about, logical and reliable precipitation conjectures, and so forth. In India, a large portion of these assets are not promptly accessible and subsequently the initial step is to set up the operational centralised infrastructural establishments, field based monitoring instruments and involvement of committed proficient staff for timely observing of threshold values and to provide estimated early cautioning on incidence of potential landslides. Besides, a micro early warning system can be developed at a community level, with community participation. This, however, would require organizing a regular Community Based Disaster Management Programme (CBDMP) for public awareness. Through participatory activities, trainings and dedicated information campaigns, it is conceivable to make the targeted community groups mindful of possibly dangerous territories and of potentially hazardous areas, to fortify their ability to be strongly prepared to face challenges pertaining to perilous fiascos and to build up an enlightened association of humans to shoulder the risk diminishment activities in affected areas [10, 11, 57, 63].

The launch of a national programme on Landslide Susceptibility Mapping on scale 1:50,000, as a joint venture between GSI and NRSC, on modern digital base will contribute significantly to develop national risk monitoring plans. It will be built on robust underpinning of landslide hazard studies and research and will lead to improved understanding and provide perspectives at multiple scales of vulnerability and resilience to ongoing adverse land and climate changes and impending hazards. Recognizing the importance and the need, GSI has taken important steps for adopting interdisciplinary and collaborative multi-institutional approach to generate and provide relevant, reliable geoscientific data needed for management of landslide hazard and risk, development of EWS and increase of public awareness to face the challenges.

Acknowledgement Fruitful discussions with and value added assistance received from Drs. T.B. Ghoshal, M.S. Bodas, Saibal Ghosh and Pankaj Jaiswal, Geoscientists at GSI, CHQ Kolkata are gratefully acknowledged.

References

1. Champati Ray PK, Parvaiz I, Jayangonda-perumal R et al (2009) Analysis of seismicity induced landslides due to the October 8, 2005 earthquake in Kashmir Himalaya. Curr Sci 97 (3):1742–1751
2. Ghosh S, Chakraborty I, Bhattacharya D et al (2012) Generating field-based inventory of earthquake-induced landslides in the Himalayas – an aftermath of the 18 September 2011 Sikkim earthquake. Indian J Geosci 66(1):27–38
3. Varnes DJ (1984) IAEG commission on landslides and other mass-movements. Landslide hazard zonation: a review of principles and practice. UNESCO Press, Darantiere. 61p
4. Schuster RL, Fleming RW (1986) Economic losses and fatalities due to landslides. Bull Assoc Eng Geol 23:11–28
5. Petley DN, Dunning SA, Rosser NJ (2005) The analysis of global landslide risk through the creation of a database of worldwide landslide fatalities. In: Hungr O, Fell R, Couture R, Eberhardt E (eds) Landslide risk management. Taylor and Francis, London, pp 367–374
6. Nadim F, Kjekstad O, Peduzzi P et al (2006) Global landslides and avalanche hotspots. Landslides 3:159–173
7. Hoyois P, Scheuren JM, Below R et al (2007) Annual disaster statistical review: numbers and trends 2006. Centre for research on the epidemiology of disasters (CRED), Brussels
8. Sarvothaman H, Kumar AKJ (2013) Disaster management – engineering and environmental aspects. Asiatech Pub. Inc, New Delhi, pp 73–83
9. Hemalatha T, Ramesh MV (2015) Indian landslide scenario: with special reference to landslide research methods. J Eng Geol XL(1):45–61. [Spl. Issue for Landslides: Management & Mitigation Strategies – 2015, Defence Terrain Research Laboratory, Defence Research and Development Organisation, New Delhi]
10. Wadhawan SK, Raju M, Ghosh S et al (2013) Geoscience considerations in formulation of National Landslide Disaster Management plan. Indian J Geosci 3&4:203–216
11. Wadhawan SK, Jaiswal P, Ghosh S (2013) Landslide early warning in India – prospects and constraints. Indian J Geosci 3&4:229–236
12. Wadhawan SK (2015) Implementation of landslide susceptibility mapping programme, vulnerability and risk assessment – a gateway to research and development of early warning systems in India. J Eng Geol XL(1):20–32
13. NDMA (2009) Management of landslides and snow avalanches, 2009. National Disaster Management Authority (NDMA), Government of India, New Delhi, p 144
14. Jaiswal P, van Westen CJ, Jetten V (2011) Quantitative estimation of landslide risk from rapid debris slides on natural slopes in the Nilgiri hills, India. Nat Hazards Earth Syst Sci 11:1723–1743
15. Jaiswal P, Srinivasan P, Venkataraman NV (2013) A data-guided heuristic approach for landslide susceptibility mapping along a transportation corridor in the Nilgiri Hills, Nilgiri District, Tamil Nadu. Indian J Geosci 3&4:273–288
16. Jaiswal P, van Westen CJ (2012) Use of quantitative landslide hazard and risk information for local disaster risk reduction along a transportation corridor: a case study from Nilgiri district, India. Nat Hazards. https://doi.org/10.1007/s11069-012-0404-1
17. Jaiswal P, van Westen CJ (2009) Estimating temporal probability for landslide initiation along transportation routes based on rainfall thresholds. Geomorphology 112(1–2):96–105
18. Hemalatha T, Ramesh MV (2015) Challenges in predicting landslides with space borne SAR technology. J Eng Geol XL(1):99–106

19. Guzzetti F, Carrara A, Cardinali M, Reichenbach P (1999) Landslide hazard evaluation: a review of current techniques and their application in a multi-scale study, Central Italy. Geomorphology 31(1–4):181–216
20. Cardinali M, Reichenbach P, Guzzetti F et al (2002) A geomorphological approach to the estimation of landslide hazards and risks in Umbria, Central Italy. Nat Hazards Earth Syst Sci 2:57–72
21. van Westen CJ, van Asch TWJ, Soeters R (2006) Landslide hazard and risk zonation—why is it still so difficult? Bull Eng Geol Environ 65(5):167–184
22. Ghosh S, Carranza EJM, van Westen CJ et al (2011) Selecting and weighting spatial predictors for empirical modeling of landslide susceptibility in the Darjeeling Himalayas (India). Geomorphology 131(1–2):35–56
23. Baeza C, Corominas J (2001) Assessment of shallow landslide susceptibility by means of multivariate statistical techniques. Earth Surf Process Landf 26:1251–1263
24. Popescu ME (2002) Landslide casual factors and landslide remedial options. Keynote lecture. In: Proceedings of the 3rd international conference on landslide, slope stability and safety of infrastructures, Singapore, pp 61–81
25. Castellanos Abella EA, van Westen CJ (2008) Qualitative landslide susceptibility assessment by multicriteria analysis: a case study from San Antonio del Sur, Guantánamo, Cuba. Geomorphology 94(3–4):453–466
26. Lee S, Hwang J, Park I (2013) Application of data-driven evidential belief functions to landslide susceptibility mapping in Jinbu, Korea. Catena 100:15–30
27. Fell R, Corominas J, Bonnard C et al (2008) Guidelines for landslide susceptibility, hazard and risk zoning for land use planning. Eng Geol 102(3–4):85–98
28. Crozier MJ (1989) Landslides: causes, consequences and environment. Routledge, London
29. Bureau of Indian Standards, BIS (1998) Preparation of landslide hazard zonation maps in mountainous terrains – guidelines. Bureau of Indian Standards (BIS) IS 14496 (Part – 2). Government of India Press, New Delhi
30. Aleotti P, Chowdhury R (1999) Landslide hazard assessment: summary review and new perspectives. Bull Eng Geol Environ 58:21–44
31. Saibal G, Das R, Goswami B (2013) Developing GIS-based technique for application of knowledge and data driven methods of landslide susceptibility mapping. Indian J Geosci 3&4:249–272
32. Ghosh S, Raju M, Jaiswal P (2014) Need to launch the national landslide susceptibility mapping (NSLM) programme in India – a roadmap to create a national database on landslide. ISEG News Bull 10(2):3–4
33. UNESCO-WP/WLI (1990) A suggested method for reporting a landslide. Bull Int Assoc Eng Geol 41:5–12
34. UNESCO-WP/WLI, Cruden DM (1991) A suggested method for a landslide summary. Bull Int Assoc Eng Geol 43:101–110
35. UNESCO-WP/WLI (1993) Multilingual landslide glossary. Biotech Publishers Ltd, Richmond. 34 p
36. UNESCO-WP/WLI (1993) A suggested method for describing the activity of a landslide. Bull Int Assoc Eng Geol 47:53–57
37. UNESCO-WP/WLI (1994) A suggested method for reporting landslide causes. Bull Int Assoc Eng Geol 50:71–74
38. Cruden D, Varnes DJ (1996) Landslide types and processes. In: Turner AK, Schuster RL (eds) Landslides investigation and mitigation, Special report 247. Transportation Research Board, National Academy of Sciences, Washington, DC, pp 36–75
39. Sarkar NK, Megotsohe C, Theophilius PK et al (2013) Debris flow characterization causes and consequences – a study from Kohima District Nagaland, India. Indian J Geosci 3&4:303–328
40. Sharma VK, Rawat PVS (2013) Post-disaster slope stability evaluation of catastrophic events in Uttarakhand. Indian J Geosci 3&4:337–346
41. Chung C-JF, Fabbri AG (1999) Probabilistic prediction models for landslide hazard mapping. Photogramm Eng Remote Sens 65(12):1389–1399

42. Ghosh S, van Westen CJ, Carranza EJM et al (2012) Generating event-based landslide maps in a data-scarce Himalayan environment for estimating temporal and magnitude probabilities. Eng Geol 128:49–62
43. Sterlacchini S, Frigerio S, Giacomelli P et al (2007) Landslide risk analysis: a multi-disciplinary methodological approach. Nat Hazards Earth Syst Sci 7:657–675
44. Government of India (2011) Disaster management in India. Ministry of Home Affairs, New Delhi. 233p
45. Crozier MJ (2005) Multiple occurrences of regional landslide events in New Zealand – hazard management issues. Landslides 2:247–256
46. Zezere JL, Oliveira SC, Garcia RAC et al (2007) Landslide risk analysis in the area North of Lisbon (Portugal) – evaluation of direct and indirect costs resulting from a motorway disruption by slope movement. Landslides 4:123–136
47. Guzzetti F, Mondini AC, Cardinali M et al (2012) Landslide inventory maps: new tools for an old problem. Earth Sci Rev 112(1–2):42–66
48. Guzzetti F, Reichenbach P, Cardinali M et al (2005) Probabilistic landslide hazard assessment at the basin scale. Geomorphology 72:272–299
49. Soeters R, van Westen CJ (1996) Economic losses and fatalities due to landslides. Bull Assoc Eng Geol 23:11–28
50. Lee S, Pradhan B (2006) Probabilistic landslide hazards and risk mapping on Penang Island, Malaysia. J Earth Syst Sci 115(6):661–672
51. Zezere JL, Reis E, Garcia R et al (2004) Integration of spatial and temporal data for definition of different landslide hazard scenarios in the area North of Lisbon (Portugal). Nat Hazards Earth Syst Sci 4:133–146
52. Harp EL, Reid ME, McKenna JP et al (2009) Mapping of hazard from rainfall-triggered landslides in developing countries: examples from Honduras and Micronesia. Eng Geol 104:295–311
53. Keefer DK, Wilson RC, Mark RK et al (1987) Real-time landslide warning during heavy rainfall. Science 238:921–925
54. Jaiswal P, van Westen CJ, Jetten V (2010) Quantitative assessment of direct and indirect landslide risk along transportation lines in southern India. Nat Hazards Earth Syst Sci 10:1253–1267
55. Geological Survey of India (2013) Proceedings of the regional workshop on landslide disaster management. 19–20 June 2013, Shimla, Himachal Pradesh, 112p
56. Geological Survey of India (2013) Proceedings of the regional workshop on landslide disaster management. 22–23 November 2013, Shillong, Meghalaya, 131p
57. Geological Survey of India, (2014) Proceedings of the regional workshop on landslide disaster management, 17–18 January 2014, Wellington, the Nilgiris, Tamil Nadu, 105p
58. Baum RL, Godt JW (2010) Early warning of rainfall induced shallow landslides and debris flows in the USA. Landslides 7(3):259–272
59. Chan R.K.S., Pang P.L.R., Pun, W.K., (2003) Recent developments in the landslip warning system in Hongkong. In: Proceedings of the 14th Southeast Asian geotechnical conference, Balkema, Lisse, The Netherlands
60. Ramesh MV, Vasudevan N (2012) The deployment of deep-earth sensor probes for detection of landslides. Landslides 9(4):457–474
61. Gabet EJ, Burbank DW, Putkonen JK et al (2004) Rainfall thresholds for landsliding in the Himalayas of Nepal. Geomorphology 63(3–4):131–143
62. Dahal RK, Hasegawa S (2008) Representative rainfall thresholds for landslides in the Nepal Himalaya. Geomorphology 100(3–4):429–443
63. Pachauri AK (2016) Disaster management of landslides in the Himalaya. J Indian Geol Congr 8 (1):27–48
64. National Disaster Management Act-2005 (2005) Ministry of Home Affairs, Government of India. New Delhi. The Gazette of India Extraordinary Part-II, Section 1[No.64]:1–29

Chapter 9
Soil Nailing: An Effective Slope Stabilization Technique

Mahesh Sharma, Manojit Samanta, and Shantanu Sarkar

Abstract The present chapter discusses the soil nailing technique as an effective stabilization measure for slopes, excavations, rail or road embankments, tunnels and retaining walls. Different aspects of the technique such as favorable ground conditions, advantages and limitations over other methods have been reported. Further, different installation process, failure modes of soil nailed structures, design philosophies, effects of various construction parameters on the design method has been discussed in detail. The pullout response of the soil nail is the critical parameter for the soil nail design. Analytical, numerical, field and lab testing procedures are usually used to determine the pullout capacity of the soil nail. A chronological literature review examines the influence of various parameters such as grouting pressure, overburden pressure, soil dilation, degree of saturation, roughness of the nail surface and borehole on pullout capacity of soil nail. A comparative study based on different types of experimental setup reported in the literature along with the innovative pullout system developed at CSIR-CBRI for determination of pullout capacity of soil nail has also been summarized. The last section briefly describes the recent advancements in the soil nail technique and its beneficial effects over the conventional soil nailing system.

Keywords Soil nail · Slope stabilization · Pullout test · Helical soil nail · Roughness · Overburden pressure

9.1 Introduction

Soil nailing is a top down in-situ reinforcement technique used for improving the stability of slopes, excavations, rail or road embankments, tunnels and retaining walls by insertion of closely spaced passive slender elements known as nails

---

M. Sharma · M. Samanta (✉) · S. Sarkar
Geotechnical Engineering Group, CSIR-Central Building Research Institute, Roorkee, Uttarakhand, India

S. P. Pradhan et al. (eds.), *Landslides: Theory, Practice and Modelling*, Advances in Natural and Technological Hazards Research 50,
https://doi.org/10.1007/978-3-319-77377-3_9

[1, 2]. The soil nails are generally of steel bars or steel bars surrounded by cement grout, carbon fiber reinforced polymers (CFRP) and glass fiber reinforced polymers (GFRP) materials and are primarily capable of resisting tensile stresses. Generally, the soil nails are installed at a horizontal and vertical spacing of 1–1.5 m. Whereas, the diameter and length of soil nails usually lie in the range of 25–40 mm and 4–20 m respectively. They are driven or drilled and grouted into the soil or soft weathered rock mass usually at an inclination angle of 10–20° with the horizontal and an initial shotcrete facing is provided. Subsequently, the final shotcrete facing is provided thereafter. The water-cement ratio of the cement grout for soil nails usually varies from 0.40 to 0.45. The cement grout enhances the bond between the soil and nail and also protects the nails against corrosion. Rotary or percussion rotary drill rigs with water/air flush are generally used for drilling of boreholes. The soil nails along with initial and final shotcrete facing contribute to the overall stability of the slope or excavations. The main difference in working principles of the soil nail and other soil reinforcement elements such as tie back anchors or geo-synthetics is in the stress transfer mechanism along the reinforcement. As the soil nails are grouted to the full length, stresses are developed along the full length of the nail. The development and mobilizations of the interface shear stress at the grout soil interface depends on the relative displacement between the soil nail/grout and surrounding soil, making soil nailing a passive soil reinforcement system. Whereas, tieback anchors are active soil reinforcements and are partially grouted. After the strength development of the grout material, the free zone (ungrouted portion) is pre-stressed to preload the tie back anchors for reducing the soil or wall deformations. This makes the tieback anchor an active soil reinforcement system where the stress development along the interface of tieback anchor and soil does not depend on the relative movements of the two. The geosynthetics reinforcement remains in stress free conditions during the applications. Geosynthetics strips derive their stabilizing actions through the stress mobilizations of the geosynthetics reinforcement due to the relative movement between geosynthetics and soil similar to the soil nails.

In the past few decades, the popularity of soil nails has grown up over other conventional slope stabilization methods due to its economic viability, short construction period and suitability to stabilize slopes at congested places [3, 4]. Primarily, the soil nail is subjected to axial stresses by movement of the unstable soil mass present above the critical slip surface of the slope. Pull-out tests are usually performed to accurately estimate the peak interface shear strength between the soil and grout or soil and nail surface. It helps in simulating actual behavior of soil nail during mobilization of axial force along the soil nail surface. In current design practices, the required reinforcing force is initially computed for the slope, and soil nail layout is determined based on the estimated pullout capacity. The pullout capacity is estimated from analytical and semi empirical methods. However, the pullout behavior of soil nails is complex and is influenced by various parameters viz. overburden pressure, water content, degree of saturation, soil nail roughness, soil properties (type, particle size, soil dilatancy, saturation level, shear strength etc.), testing methods and nail installation procedures [5–15]. Therefore, pullout capacity of nails determined in the laboratory or estimated analytically is usually verified by performing limited number field pullout tests.

9.2 History

The basis of soil nailing goes back to 1960s when the "New Austrian Tunneling Method" was introduced for rock excavation support [16–18]. The inserted steel bars were grouted and subsequently, shotcrete facing was provided for the excavation support. The concept was later expanded for stabilization of soil slopes and excavations. The first research on soil nailing with laboratory model tests and full scale testing of nailed walls was carried out in Germany in 1975 [19]. Further research was initiated in France and US in the 1990s [2].

9.3 Components of Soil Nail System

The basic elements of soil nailed wall are discussed in this section. The cross-section of a typical soil nailed cut slope is shown in Fig. 9.1 [2, 20].

- **Soil nail:** The solid or hollow high yield deformed steel bars known as nails are the key element of a soil nailed structure. It provides primarily tensile resistance in response to the deformation of the unstable mass of the soil nailed wall. The solid bars are installed by driving or inserted into the pre-drilled hole and then grouted. Whereas, hollow bars are installed with a sacrificial drill bit to drill the hole and stay in place as a permanent reinforcement. Now-a-days other materials such as fiber reinforced polymer (CFRP and GFRP) are also used as soil reinforcement.
- **Centralizers and coupler:** Centralizers are used to ensure that soil nail is installed at the center of the drilled hole. Whereas, couplers are employed for joining the sections of soil nails.
- **Grout:** It is generally a mixture of portland cement and water. The grout is injected in the pre drilled hole under gravity or by applying pressure after placement of the soil nail reinforcement. The grout transfers the shear stresses from the unstable soil mass to the soil nail reinforcement and then transfers the

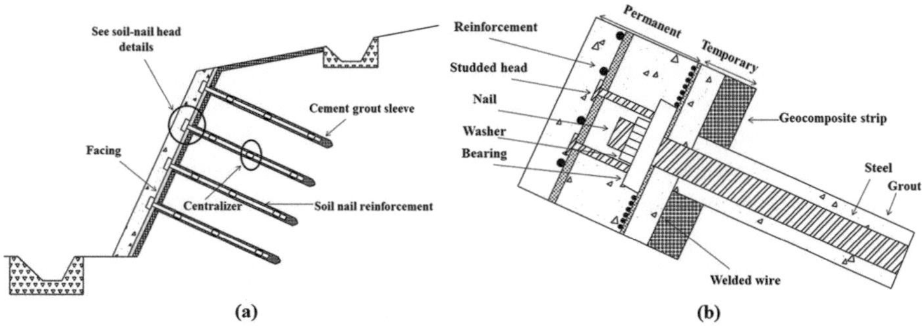

Fig. 9.1 Schematic view of the soil nailed slope [20]: (**a**) Cross sectional view and (**b**) Details of soil nail head

axial stresses from the soil nail to the stable soil mass. It also serves the purpose of providing corrosion protection to the soil nail.

- **Corrosion protection measures:** The additional level of protection against corrosion is provided by applying layers of corrugated synthetic materials such as PVC and heat shrinkable sleeves of HDPE (high density polyethylene).
- **Nail head:** It generally comprises of a square shaped reinforced concrete pad, steel plate, nut and threaded end of the soil nail reinforcement. It enhances the local stability by providing soil nail bearing resistance and transfers bearing loads from the soil mass to the soil nail.
- **Facing:** It comprises of initial and final part. The initial facing usually consists of reinforced shotcrete and is provided on the unsupported excavated face prior to the advancement of excavation lifts. While final facing is provided over the initial facing after the installation of soil reinforcements and generally consists of reinforced shortcrete or precast concrete panels.
- **Connectors:** The soil nail head is connected to the facing using various components such as bearing plates, hex nuts, washers and headed-studs as shown in the Fig. 9.1.
- **Drainage system:** The drainage system serves the purpose of collecting and directing the collected seepage water away from the soil nailed wall. It usually consists of vertical geocomposite strip drains.

9.4 Merits and Limitations

The following section describes the merits and disadvantages of a soil nailing technique with respect to its cost, construction and performance based on the available literature [2, 20].

9.4.1 Advantages

- Soil nailed wall installation is fast and causes less environmental impact.
- Design parameters of soil nailed wall i.e. length, inclination angle and location of soil nail installation can be easily adjusted to the site constraints due to variations in ground conditions when boulders, piles and underground utilities are encountered.
- Cost effective at sites with difficulty to access due to easy mobilization of a small sized construction equipment and can accommodate bends and curves more easily during construction than other techniques.
- Soil nail walls are comparatively flexible and thus can accommodate comparably large settlements, thus performing better under seismic conditions. Also the mode of failure is ductile, therefore it provides early warning signs.
- Cost-effective than conventional concrete gravity walls and ground anchors. Soil nailing results in a cost saving of 10–30% in comparison to tieback walls [21].

9.4.2 *Limitations*

- Not recommended where strict deformation criteria exists as soil nailing technique requires some soil movement to mobilize the interface resistance.
- Unsuitable where high groundwater level exists due to the difficulties in drilling, grouting and excavation.
- Criticized in case of the cohesionless soil as pre drilled hole may collapse during construction.
- Ineffective for deep seated landslides due to difficulty in installation of long soil nail reinforcements.
- The presence of underground utilities such as underground cables, electric wires, drainage systems and buried water pipes may lead to alteration in length, spacing and inclination angles of soil reinforcements.
- A skilled contractor is required for construction of soil nail walls.
- Conventional soil nailing system is prone to corrosion of the steel bars. Therefore, the preventive measure has to be deployed to prevent the corrosion which leads to increase in the overall cost of installation.

9.5 Favorable and Unfavorable Conditions for Soil Nailing

The soil nailing has been substantiated to be economical and technically feasible for the soils which can stand vertically for a height of 1–2 m for 1–2 days and where the drill hole remains stable till grouting is done. In addition, the soil should be free from a large amount of cobbles and boulders so that nails can be easily installed. At the same time, the soil must provide enough bond or frictional resistance on the interface between soil nail interface and the adjacent soil. Soil types that are ideal for soil nailing include dense to very dense granular soils, glacial till with a few boulder and cobbles, residual soils engineered fills, weathered rock (such as compacted completely decomposed granite, stiff to hard fine-grained soils, cemented sands, stiff silts and silts) [2, 22] . Soil nailing is not recommended in dry, poorly graded cohesionless soils, highly corrosive soil or highly corrosive groundwater, plastic clays, clean granular soils with high ground water level, expansive soils and organic silts due to development of inadequate pullout resistance and corrosion of steel bars [2, 23].

9.6 Failure Modes of Soil Nailed Structures

Numerous models and design philosophies of soil nail interactions, nail contribution to stability and overall behavior of the soil-nailed structure have been developed over the years. The design philosophies of gravity retaining wall were adopted to analyze the soil-nailed structures. Stocker et al [19]. presented the design methodology of soil-nailed structure considering the bilinear failure surface and the stability of the nailed-structure was estimated through force equilibrium of the failure wedge. Experimental studies by Gassler and Gudehus [24] showed that the two plane translational failure gives the lowest factor safety than the other failure modes such as deep slip surface, steep circular surface and combined tilting failure. The overall factor of safety was obtained as the ratio of the resisting force including the effect of nails to the driving forces on the failure surface. Gassler [25] proposed a two-step design procedure of the soil-nailed structure. In the first step, various failure modes and different failure surfaces were examined to determine the critical failure surface. In the final step, various partial material safety factors were considered for the material considering the critical failure surface. Stocker and Riedinger [26] extended the design concept of gravity retaining wall to the soil nail design. The two wedge rigid body translational failure mechanism was employed to calculate the external stability of the soil-nailed structures. Internal stability namely the pullout of nail was checked at any level considering the different forces acting on it. Juran et al. [27] presented a kinematical limit analysis approach to determine the stability of the soil-nailed structures. They recommended a four step design methodology for the analyses of soil-nailed structures. The four step design methodology includes determination of global stability, local stability and nail forces at each nail level, and facing stability. They highlighted the importance of local stability of nails which are sometimes critical than the global stability of soil-nailed structures. Most of the design methodologies presently followed for the design of soil-nailed structures are based on the multi criteria approach which satisfies the global as well as local stability of the structures.

FHWA [28] gives the different possible failure mechanism of the soil nailed structure as shown in Fig. 9.2. Soil nailing system is subjected to both internal and

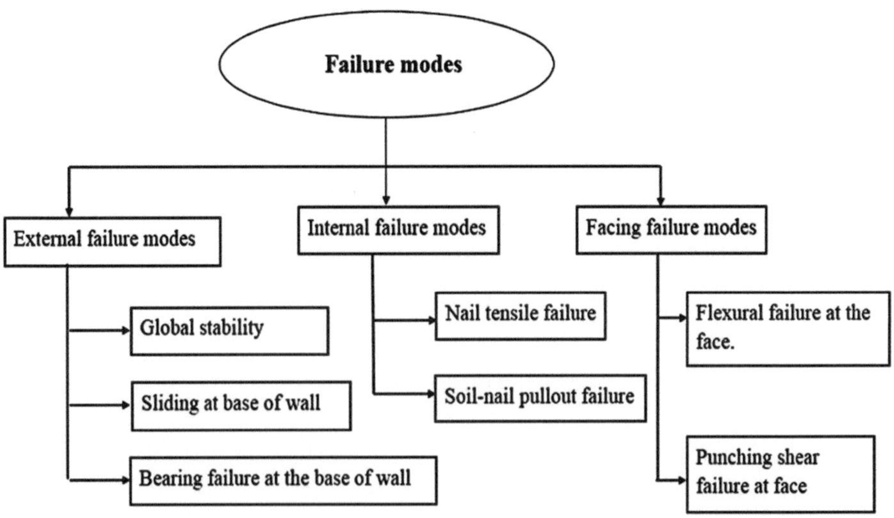

Fig. 9.2 Failure modes of soil nailing system

external failure modes besides the facing failure modes. In case of external failure modes (the failure which occurs behind the failure zone), a soil nailed structure is considered as a rigid body and fails in a similar way like other retaining structures (Fig. 9.2). The external failure modes include overall sliding of the nailed structure, overturning, bearing capacity failure and overall slope failure along a slip surface. Whereas, the internal failure modes (the failure which occurs within the failure zone) include pullout failure of the nail between the nail and soil due to insufficient bond capacity, sliding of soil due to the nail movement and breakage of nails due to tensile failure.

9.7 Applications

Soils nailing system can be used for stabilization of temporary and permanent structures. It has a wide range of applications such as the construction of new slopes, stabilization of existing slopes and renovation of old retaining structures, stabilization of tunnel faces and anchored walls, reinforcement of bridge abutments and failed slopes.

9.8 Design Methods

Generally, different methods based on limit equilibrium analysis are used for the soil nail design. Both external and internal possible failure surfaces throughout the soil mass are examined. Popular methods and the difference in their assumptions regarding assumed failure surface are given in Table 9.1 [29].

9.9 Soil Nail Mechanism

The fundamental mechanism involved in the soil nailing technique is the development of tensile force in the soil nails due to lateral deformation of the structure. The tensile stresses are mobilized within the soil nails mainly as an outcome of the friction interaction between the nail and the surrounding soils and also due to the soil-structure interaction between the nail head and the surrounding soils. However, the tensile forces mobilized at the nail head are very less in the soil nail. The nails are installed in the direction of maximum tensile strain in the soil mass [30]. The maximum tensile force is mobilized at the intersection of soil nail with the potential failure surface. The potential failure surface divides the reinforced soil mass into two zones (active and passive) as shown in Fig. 9.3. The zone in front face of the potential failure surface, which has a tendency to separate itself from the reinforced structure, is termed as the active zone. The frictional shear stresses mobilized on the

Table 9.1 Design methods for soil nail [29]

	Davis method	French method	German method	Kinematical method
Analysis	Limit force equilibrium	Limit momentum equilibrium	Limit force equilibrium	Limit force equilibrium
Input material parameters	Soil parameters (c, φ), limit nail forces, lateral friction	Soil parameters (c, φ), limit nail forces, bending stiffness	Soil parameters (c, φ), lateral friction	Soil parameters (c, φ), limit nail forces, lateral friction
Nail forces	Tension	Tension, shear and bending	Tension	Tension
Failure surface	Parabolic	Circular, any input shape	Bilinear	Log-spiral
Factor of safety (F_c & F_{φ})	1.5	1.5	1	1
Pullout resistance (F_p)	1.5	1.5	1.5–2	2
Tension	Yield stress	Yield stress	Yield stress	Yield stress
Bending		Plastic moment		Plastic moment
Ground water	No	Yes	No	Yes
Loading	Uniform surcharge	Slope, any surcharge	Slope surcharge	Slope surcharge
Structural geometry	Vertical facing	Any input geometry	Inclined facing, vertical facing	Inclined facing, vertical facing

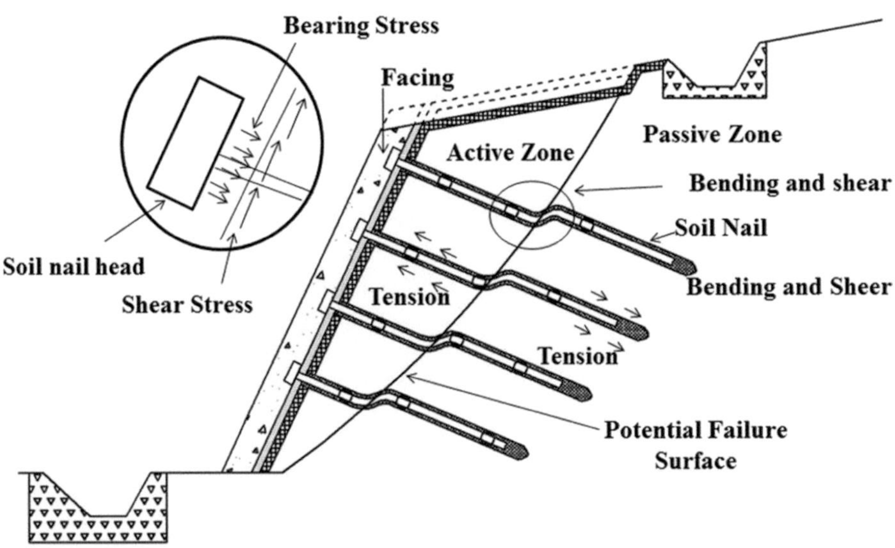

Fig. 9.3 Failure mechanism of soil nail system [20]

surface of the nail are oriented towards the facing and tends to pull out the reinforcements. With the downward deformation of the active zone, the axial displacement along the nails results in mobilization of tensile stresses in the nails till the maximum shear capacity of the nail-soil interface is reached [2, 20, 31, 32].

The zone behind the potential failure surface, where the area is stable and inhibits the failure of the reinforced system is termed as passive zone. The zone tends to restrain the reinforcements from pull-out and thus, the soil nails act to fasten the two zones (active and passive). The soil nail length present in this zone provides the required pullout resistance. The axial force developed varies along the length of soil nail which is different from the anchors where it remains constant along the free length. The soil nail is subjected primarily to tensile forces along with shear forces and bending moments which are generated as the reactions to the slope deformations. However, the contribution of bending and shear resistance in comparison to pullout capacity is very less and therefore can be neglected for the practical purposes.

9.10 Soil Nail Pullout Resistance

The primary interaction between soil nail/soil nail-grout and surrounding soil is the development of interface shear stress between nail/grout and soil surface. To estimate the ultimate shear strength at the interface between soil and grout or soil and soil nail surface, pull-out tests are usually performed. It helps in simulating the actual soil nail behavior during mobilization of tensile forces along nail surface. An extensive theoretical and experimental investigation has been carried out in the last two decades to study the pullout behavior of soil nails [5, 8, 9, 33–37]. Generally, before the installation of permanent soil nails, field pullout tests are carried out to check the adequacy of soil nail bond strength and to verify the applicability of soil nails to the particular location [9]. The pullout load and the corresponding displacement is measured at the nail head. The soil nail pullout behaviour is dependent on large number of uncertain parameters such as overburden pressure, water content, soil nail roughness, soil properties (type, particle size, soil dilatancy, saturation level, shear strength etc.), testing methods and soil nail installation procedures. Therefore, it is difficult to estimate the interface shear strength of soil nails with high accuracy due to the complexity of soil-nail interaction behavior during the field pullout.

9.10.1 Analytical Approaches

Several researchers have developed the various analytical approaches to estimate the ultimate soil nail pullout resistance [35, 38–41]. Table 9.2 provides a summary of the different analytical approaches to estimate ultimate pullout capacity of soil nail. The

Table 9.2 Pullout capacity equations

Authors	Equations
Schlosser and Guilloux [38]	$P = pc' + 2D_{eq}\sigma'_{v}\mu^{*}$
Potyondy [42]	$P = f_c c' + \sigma'_N \tan(f_\varphi \varphi)\Theta$
Wong [43]	$P = \left(\pi Dc' + 2D\sigma'_v \tan\left(\varphi''\right) * l'\right.$
HA68/94 [41]	$P = \lambda l'(c' + \sigma'_N \tan(\varphi')$
Heymann et al. [44]	$P = p(c' + \sigma'_N \tan(\varphi')$
Jewell [40]	$P = p(\sigma'_N f_b \tan(\varphi)$
Mecsi [45]	$P = p(\sigma'_N f_b \tan(\delta)$

ultimate pull-out capacity (P) of soil nail is defined as the summation of shear force mobilized along the entire nail length. Generally, all the equations to estimate the pullout capacity are based on four parameters; Normal stress acting on the soil nail (σ'_n), the coefficient of friction between nail surface and surrounding soil (μ), adhesion between soil and nail (a) and soil nail perimeter (p). Further, Luo et al. [34, 35] proposed an analytical model to predict the pullout behaviour of soil nails embedded in dilative soils. The soil nail is considered as rigid rod whereas soil is assumed as an elastic medium. The theoretical expression to compute the pullout resistance of soil nail indicate that the apparent coefficient of friction decreases with the increment in overburden pressure. Cartier and Gigan [39] proposed coefficient of friction value based on pullout results.

9.10.2 Numericalmodeling

Numerical modeling techniques are widely used to study nail-soil interaction mechanism, distribution of stresses in soil and nails, deformation and serviceability of soil nailing systems. Numerical models can be established after calibrating the test results with laboratory pullout tests on soil nails. Thus, making the numerical investigation more economical and reliable solution for carrying out parametric studies of soil nail pullout behaviour [46–49]. Shafiee [50] carried out one of the first numerical analysis using finite element program. Babu et al. [51] carried out a parametric study to examine the effect of construction procedure, type of facing, the stiffness of facing, and the inclination angle of facing on soil nail behavior. Whereas, Yeo et al. [52] numerically studied the nail-soil interface behavior by simulating the construction procedure (drilling of the hole, soil nail installation, grouting, and then pullout of the nail). The study suggested that initially due to soil nail installation procedure, the radial stress at nail-soil interface was very less. However, it increased with the nail displacement during the pullout. The radial stress developed was strongly influenced by the dilation angle of the soil. The pullout resistance was generated predominantly due to soil shearing, dilatancy and the soil-grout or soil-nail bond strength [53]. Su et al. [49] conducted a parametric study using 3D finite element model for simulating the soil nail pullout behavior. No direct relation was observed between the overburden pressure and pullout capacity. However, pullout force increased with the increment in dilation angle.

9.10.3 Experimental Studies (Lab and Field)

Laboratory pullout testing of soil nails serve the purpose of removing uncertain factors due to parameters resulting from complex ground conditions. Generally, pullout tests are conducted in displacement rate-controlled manner for the measurement of the pullout load displacement response during the peak and post-peak stage. Tei [54] used a 254 mm long, 153 mm wide and 202 mm high pullout box as shown in Fig. 9.4, to conduct soil nail pull-out tests for studying the effect of roughness, stiffness, length and diameter on pullout behavior of soil nails. He concluded that displacement required to mobilize peak pullout force is independent of nail length, diameter and surcharge load for stiff nails with a rough surface. Further, axial stress distribution was found to be linear along the soil nail length.

Milligan and Tei [5] also carried out laboratory pullout tests on stiff and flexible nails. The results suggested that the coefficient of friction between stiff rough nails and the soil is influenced by the friction angle and stiffness of soil, soil dilation and the ratio of a diameter of soil nail to mean particle size of the soil. Further, it was suggested that the use of smooth and extensible nails should be avoided in the field. Franzen [36] developed a 2000 × 4000 × 1500 mm pullout box to study the pullout behavior of different type of soil nails (round steel bar, angle bar, ribbed bar and expansion bolt and) at different stress levels of 25 kPa, 37.5 kPa, 75 kPa and 125 kPa. The study suggested that the pullout capacity in a cohesionless medium is significantly influenced by the roughness of the nail surface, the relative density of

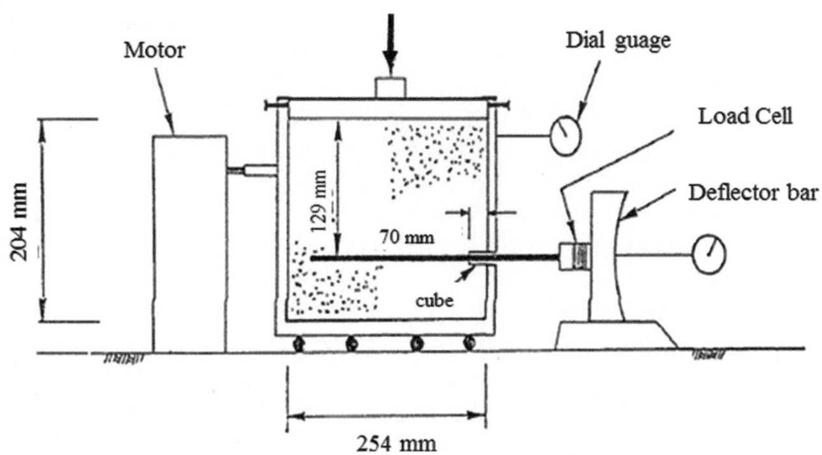

Fig. 9.4 Pullout setup [54]

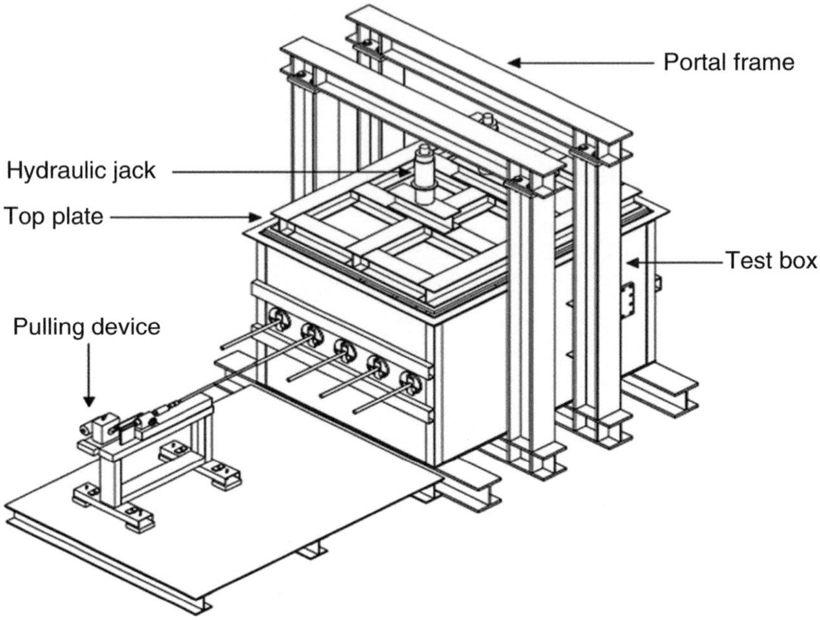

Fig. 9.5 Laboratory pullout test setup [37]

the soil, nail surface area and normal pressure acting on the nail surface. The peak pull-out resistance for driven nails was found to be 50% higher than that for jacked nails. However, the residual pullout resistance was independent of installation method. Junaideen et al. [37] developed a large scale laboratory pullout test apparatus to study the soil nail pullout behavior in loose fills. The pullout box was 2 m long × 1.6 m wide × 1. 4 m high and can take surcharge pressure up to 150 kPa as shown in Fig. 9.5. The pullout tests were conducted on the smooth bar, knurled tube and ribbed bar at surcharge pressure of 12 kPa, 45.5 kPa, 73.5 kPa and 109.5 kPa in a displacement rate of 1.3 mm/min. The study indicated that the pullout behavior of soil nail is significantly influenced by the dilation phenomenon and surface characteristics of the nail. The same test setup was used by Pradhan [55] to examine the pullout behavior of grouted soil nails. Yin and Su [56] developed an instrumented pullout box to conduct pullout tests on soil nails as shown in Fig. 9.6. The setup can be used to study the influence of drilling process, saturation level of soil, overburden pressure and grouting pressure on soil nail pullout capacity. Wu and Zhang [57] carried out laboratory and field pullout tests to examine the pullout behavior of grouted soil nails. The pullout capacity was found to increase linearly with normal stress indicating that it follows the Mohr-Coulomb criteria. In addition, test results suggested that the pullout capacity computed from the laboratory tests correlated very well with the field pullout tests. The similar behavior was concluded by Pradhan, [55].

Fig. 9.6 Instrumented pullout box [56]

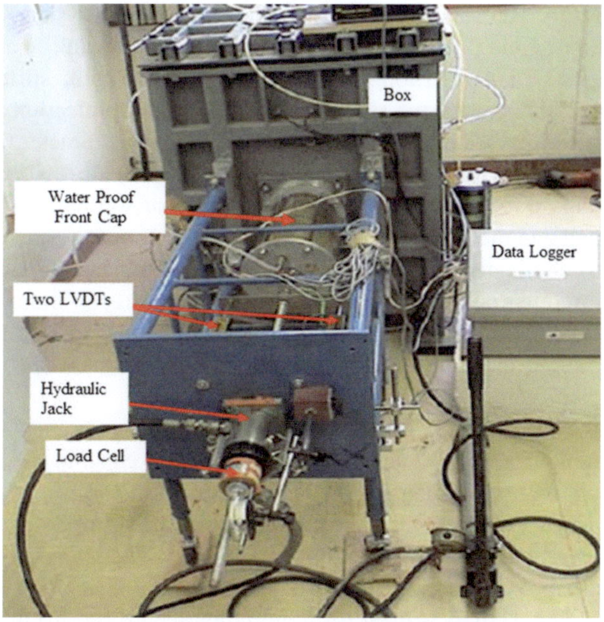

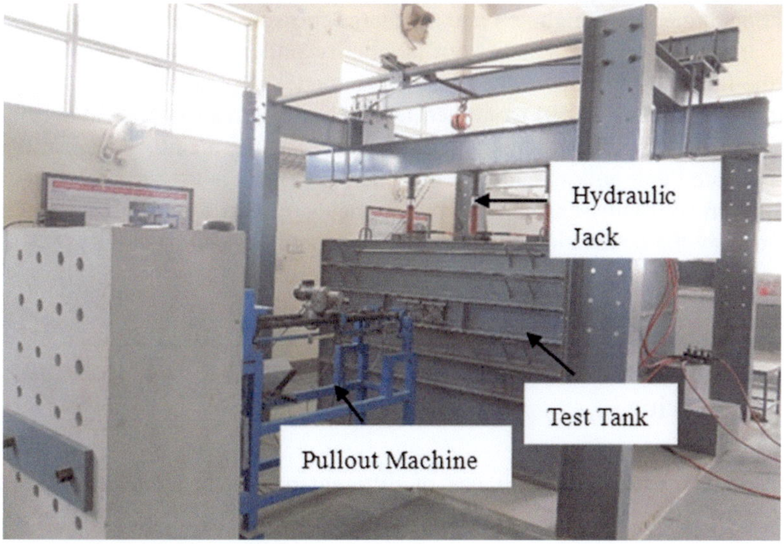

Fig. 9.7 Pullout test facility [58, 59]

An innovative pullout testing facility has been developed by CSIR-CBRI, Roorkee to carry out pullout tests on soil reinforcements (soil nails, anchors and helical soil nails) at the laboratory level. Figure 9.7 shows the overview of the test setup. The setup consists of pullout machine, test tank, surcharge pressure unit and an online data acquisition system. The tank has internal dimensions of $2000 \times 1000 \times 1500$ mm. The sides of the tank stiffened with mild steel angle sections to achieve rigid boundaries under application of different surcharge pressures. A variable circular opening of 125 mm diameter (maximum) at equal distance from sides and at a height of 1020 mm has been provided on the front face of the test tank to facilitate the driving and pullout of soil nails. The distance between the soil nail and the side wall is more than twenty times the nail diameter, which is sufficient to eliminate the boundary effects. The surcharge pressure is applied using a set of three hydraulic jacks of capacity 10 tonnes each mounted on a 25 mm thick rigid steel plate which is placed on top of the finished soil surface of the test tank. The pullout machine conducts pullout tests at a linear speed of (0.7–70 mm/min) in a displacement controlled manner.

Additionally, the developed machine overcomes the limitations of previous pullout arrangements and machines. It is capable of driving and pullout the soil reinforcing elements at different speed horizontally and also at different angles up to 20° with horizontal so as to enable the study of installation effects on the pull out resistance of the soil reinforcing elements. The pullout load and pullout displacement are measured by the load cell and LVDT respectively. A 48 channel data acquisition system (DAS) is used for continuous recording of pullout load and displacements during tests. The facility has been used to study the influence of installation method, inclination angle, nail roughness and surcharge pressure on pullout behavior of soil nails. Samanta et al. [58] carried out laboratory pullout tests on driven soil nails using the facility and proposed a constitutive model to predict the soil nail pullout behavior. Sharma et al. [59] examined the pullout behavior of helical soil nails in dense sand medium by conducting pullout tests using the facility. The results of the study indicated that pullout capacity of helical soil nails is higher than the conventional driven soil nails. Further, the pullout capacity was found to be significantly influenced by the installation angle of helical anchors or soil nails [60].

9.11 Influence of Various Parameters on Soil Nail Pullout Behavior

The section describes the influence of different parameters that have significant effect on the soil-nail pull-out resistance.

9.11.1 Surface Roughness and Properties of Soil and Nail

The surface roughness is a critical parameter for pullout resistance of soil nails. Extensive series of theoretical and experimental investigations have been carried out to study and quantify the effect of surface roughness on soil-nail pullout resistance [42, 61]. The pullout capacity increases with the roughness of soil nails. The smooth soil nails or smooth soil-grout interface have low pullout capacity [62]. Tei [54] and Milligan and Tei [5] conducted series of laboratory pull-out tests on smooth and rough nails in cohesionless soil. The pullout resistance for smooth soil nails was observed to be very less than rough soil nails tested under the same conditions. Further, the nail displacement required to mobilize the maximum pull-out resistance for smooth nails were 50% of those for rough nails. The ratio of size of the nail to the particle size of the surrounding soil also influences the soil nail pull-out resistance. The drill hole roughness has a significant influence on the mobilization of ultimate soil-nail pullout capacity. In the case of a smooth borehole, the pullout capacity will be less due to the low effective normal stress acting on the nails as a result of arching effect. However, irregular surface borehole will lead to mobilization of restrained dilatancy effect and thus increasing the normal stress and consequently the pullout capacity of soil nails [63]. Hong et al. [64] carried out laboratory tests to investigate the effect of drill hole surface roughness on the pullout capacity of model soil nails by creating different surface angles on the plastic rods. The results were verified by the analytical model of Luo et al. [34]. The study suggested that the soil-nail pullout capacity increases linearly with the increase in roughness angle. Similar findings were presented on driven soil nails by Samanta et al. [58].

9.11.2 Soil Dilation

Soil dilation becomes a critical factor in the calculation of pullout resistance of soil nail for dilative soils [31, 54, 65–69]. Schlosser [31] proposed that pullout resistance of soil nail in a granular soil is significantly influenced by soil dilation. In the case of dense granular soils that are composed of larger and more uniform particles with irregular shapes, the particles will rotate and rearrange themselves during pull-out which results in dilation of the soil. Therefore during pullout, shearing takes place in a finite zone around the soil nail. The soil tendency to increase in volume gets constrained by the adjacent soil, which results in an increment in normal stress acting on the nail. The increase in normal stress depends on various factors viz. particle size and shape of soil, the relative density of soil, confining pressure and soil saturation level. The normal stress can increase by two to ten times due to restrained dilation. But it could increase further up to 14 times if dilatancy is fully constrained [70]. The increase in normal stress due to constrained dilation decreases with the increase in overburden pressure. Chai and Hayashi [69] also concluded that in the case of sandy clay, additional normal stress developed because of constrained dilation increases which result in an increase in pullout resistance of soil nail. Wernick [65] conducted pullout tests on cylindrical steel pipes with dimensions of 2500 mm long, 50 to 1000 mm in diameter installed in a sand medium. An increase in normal stress up to eight times of the initial stress was also observed due to soil dilation. Tei [54] stated that soil dilation has an important role in the mobilization of soil-nail pullout resistance. Schlosser [31] suggested to incorporate apparent coefficient of friction (μ^*) to calculate the actual normal stress acting on the reinforcement. The apparent coefficient of friction (μ^*) is defined as τ_{max}/σ_v. Where τ_{max} and σ_v are the mobilized peak shear stress and overburden pressure respectively. Luo et al. [34] developed a theoretical model to estimate the pullout resistance considering dilation behavior of soil and overburden pressure. He further stated that apparent coefficient of friction increases as soil dilation increases but decreases with the overburden pressure. Further, dilation modulus diminishes as diameter of soil nail increases. Therefore, the effect of soil dilation is less significant in estimating pullout resistance of soil nails with large diameters.

The maximum pullout resistance is written as

$$P = A_n q'_{\max} tan\varphi \tag{9.1}$$

Where, $A_n$stands for cross-section area of soil nail, $q'_{\max}$is a parameter which depends on relative density, Poisson's ratio and overburden pressure acting on the soil nail. φ' is effective friction angle of sand.

Apparent friction coefficient for peak and critical state were defined as:

$$\eta_p = \frac{q'_p}{\sigma'_v} \tan\left(\varphi'_{cv} + \psi_{\max}\right) \tag{9.2}$$

$$\eta_{cv} = \frac{q'_{cv}}{\sigma'_v} \tan\left(\phi'_{cv}\right) \tag{9.3}$$

Where,

q'_p is the average normal stress on the soil nail for the peak state.
$q'_{cv}$is the average normal stress on the soil nail for the critical state.
ϕ'_{cv} is the effective friction angle at critical state.
$\psi_{\max}$ is the maximum dilation angle of soil.

Wang and Richwien [71] proposed a formula to calculate the pullout resistance of soil reinforcement by incorporating the effect of reinforcement roughness, friction angle and soil dilation.

$$\tau_f = \frac{f}{1 - \left[\frac{2(1+\upsilon)}{(1-2\upsilon)(1+2k_0)}\right] f \tan\psi} \sigma_m \tag{9.4}$$

Where,

f is the coefficient of friction obtained from direct shear test.
υ is the poisson's ratio.
$\sigma_m$is the overburden pressure.
ψ is the dilation angle of soil.

Juran [72] observed that the normal stress increases during pullout of soil reinforcement in case of loose sand. The similar results were also reported by Junaideen et al. [37]. Yin et al. [73] demonstrated that with the increase in soil dilation, there is an increase in peak shear stress developed at the nail-soil interface.

9.11.3 Overburden Pressure and Grouting Pressure

Many researchers have studied the effect of overburden pressure on soil nail pullout resistance. The actual normal stress acting on the soil nail is dissimilar from the overburden pressure (σ'_v) and is dependent on the nail surface characteristics, installation method, soil type [31]. Various analytical methods as mentioned in Table 9.2 have correlated the normal stress acting on the reinforcement with the overburden pressure. HA 68/94 [41] calculated the effective normal stress taking into consideration lateral earth coefficients in addition to the overburden pressure. Jewel [40] proposed that effective normal stress varies from 0.7 to 1 σ'_v for steep slopes with lightly over consolidated soils. The effective normal stress is close to the overburden pressure for driven soil nails. Whereas, it is very low for grouted nails due to the release of stresses as a result of drilling of holes. Schlosser and Guilloux [38] and Cartier and Gigan [39] suggested that pullout capacity is independent of overburden pressure and the decrease in the apparent friction coefficient was compensated by the increment in effective vertical stress.

The effect of overburden pressure on soil nail pullout resistance still remains unclear due to differing views about its effect. Su [29] and Su et al. [11] carried out laboratory pullout tests in a compacted completely decomposite granite (CDG) soil to investigate the effect of overburden pressure on soil nail pullout capacity. They concluded that the when grouting was done without pressure, the overburden stress has no significant effect on the soil-nail pullout resistance. The drilling process causes stress release in the drill hole nullifying the influence of normal stress on the soil nail. The pullout test data was then verified with the numerical simulation. However, Zhang et al. [14] conducted field pullout tests and concluded that the measured pullout resistance is independent of effective overburden pressure. Further, they further proposed the following equation to estimate the pullout capacity taking into consideration the matric suction of soil nail during field pullout tests

$$P_{ult} = \pi D\left[c' + (u_a - u_w)\tan\varphi^b\right] + \frac{2D\sigma'_V \tan\varphi'}{1 - \frac{2(1+\nu)}{(1-2\nu)(1+2K_0)\tan\varphi' \tan\psi}} \tag{9.5}$$

Where

c' is cohesion value.
u_a is pore air pressure.
u_b is pore water pressure.
φ_b is the contribution of matric suction to shear strength.

D is diameter of soil nail.
ν denotes Poisson's ratio and dilation angle of soil respectively.
K_0 is coefficient of lateral earth pressure.
σ'_v is vertical earth pressure.

The effect of normal stress will be different for grouted nails because of the release of stress taking place around the hole during drilling due to arching effect thus nullifying the influence of overburden pressure. Su [29] and Su et al. [11] investigated the effect of overburden pressure and the process of hole drilling on pullout resistance of soil nail. A significant stress reduction in the soil surrounding the drill hole was observed and pullout resistance was found to be independent of overburden pressure acting on the soil nail in the case of gravity grouting.

Grouted nails generally consist of a steel bar with a diameter of 15–45 mm and 30–80 mm thickness of grout. Generally, nails are cement grouted under gravity and the water/cement ratio is usually kept between 0.4–0.45 [74]. Pullout resistance of the grouted nails generally depends on the level of mechanical interlocking between the cement grout sleeve and the surrounding soil (bond strength). The bond strength is influenced by nail surface condition, soil particle size and the grouting pressure. The coarse gravels with little fines result in grout leakage making it difficult to obtain an intact cement grout sleeve [75]. Low pressure grouting leads to the formation of a smooth soil-grout interface in case of fine grained cohesive soils [76].

Pressure grouted soil nails are inserted by applying high grouting pressure (300 kPa −1000 kPa) [77]. Use of pressure grouted soil nails lead to increase in diameter of soil nails, higher interface shear strength and reduction in the number of reinforcing soil nails [77]. Kim et al. [77] developed a two-dimensional axisymmetric finite element model for simulating the pullout response of pressure grouted soil nails. The model was verified by conducting field pullout tests on gravity grouted and pressure grouted soil nails. The study concluded that application of grouting pressure increases the factor of safety by 50% and 11% as compared to the safety factors for unreinforced and gravity grouted slopes. Various researchers have investigated the interface strength behaviour of grouted soil nails. For grouting pressure less than 350 kPa, pullout resistance increases considerably with the increase in grouting pressure [77–80]. Su [29] stated that for no pressure grouting, peak pullout shear resistance is independent of overburden pressure. However, for constant overburden pressure, peak pullout shear stress increases linearly with increase in grouting pressure. Yeung et al. [81] carried out field pullout tests on GFRP nail in a CDG soil slope. They observed that the pressure grouting increases pullout strength significantly. Grouting pressure increases the pullout resistance of soil nail because of increase in surrounding soil strength due to compaction effect generated by pressure grouting and increase in soil nail roughness because of cement infiltration [12]. The similar finding was also made by Yin and Zhou [13] through the laboratory experiments under a combination of various overburden and grouting pressures. Yin and Zhou, [13] studied the effect of overburden and grouting pressure on soil nail pullout resistance by carrying out laboratory pullout tests in a completely decomposed granite soil. The test results indicated that grouting pressure and overburden stress have an interactional effect on the pullout resistance of soil-nail. The influence of pullout resistance is less significant on the overburden pressure at low grouting pressure but becomes significant with the increase in grouting pressure. The effect of soil dilation is very less for a saturated soil whereas it is significant for unsaturated soil nail. They proposed a relation between interface shear resistance and grouting pressure as shown below.

$$\tau = c_G{}'(p_G) + \sigma_{vi}{}'\mu'_G(p_G) \tag{9.6}$$

Where

τ is the average soil-nail interface shear resistance.
σ_{vi} is the initial vertical effective stress.
p_G is the grouting pressure.
μ'_G is the slope of the fitting line.
c'_G is the interface shear strength for zero initial overburden stress.

The similar behavior was demonstrated by Yin et al. [73]. They proposed an analytical model that is a function of grouting pressure, post installation normal stress and normal stress due to soil dilation for evaluating the maximum shear stress at the nail-soil interface. The analysis showed that there is an interfacial influence of grouting pressure and overburden pressure. The higher the grouted pressure, higher is the effect of overburden pressure and thus mobilizing maximum shear stress at soil nail interface. Pradhan [55] studied the pullout behavior of soil nails installed in loosely compacted soils by conducting laboratory pullout tests. The effect of constrained dilatancy was reported to be insignificant except at very low-stress levels. The pullout capacity was found to increase linearly with the overburden pressure. Further, one-dimensional spring model was used to simulate the peak soil-nail pullout resistance.

9.11.4 Degree of Saturation

Rainwater is the predominant factor for triggering the slope failure. As it will change the weather condition and thus there may be a drop in soil-nail pullout capacity [10]. The initial degree of saturation generates lubrication effect, especially in plastic soils. The shear failure plane at the soil-grout interface and the pullout-displacement characteristics are significantly influenced by the saturation level. Therefore, making the degree of saturation as a critical parameter for the soil nail design. Su et al. [10] suggested that pullout resistance is low for saturated soils in comparison to the partially saturated soils. The shear failure plane shifts from the grout-soil interface into the surrounding soil medium when the degree of saturation increases from 35% to 98%. The significant influence of the degree of saturation was also seen by Pradhan [55] and Su et al. [10] and reported the similar results. Schlosser et al. [66] concluded that the ultimate pullout resistance decreases by 50% when the saturation level increases from optimum moisture content to the full level.

The major parameters that influence the soil-nail pullout resistance have been discussed in detail. The other parameters such as the method of installation also have significant influence on soil nail pullout capacity. Franzen [36] carried out pullout tests on driven and jacked soil nails. The peak pullout resistance for driven nails was 50% greater than that obtained for jacked nails. This may be credited to the increase in the stresses acting around the soil nails due to the displacement of soil volume by the installation of driven nails.

9.12 Advancements in Soil Nailing

Dai et al. [82] proposed an innovative earth retaining structure in which moso bamboo nails were used along with conventional soil nails (Fig. 9.8a). Based on the field tests, the pullout capacity of bamboo nails with branches were found to be 2.5–2.8 times higher than the conventional soil nails. The study concluded that the new system is more economical, easy to construct, reliable and environmental friendly than the conventional soil nailing system. Fiber reinforced polymer

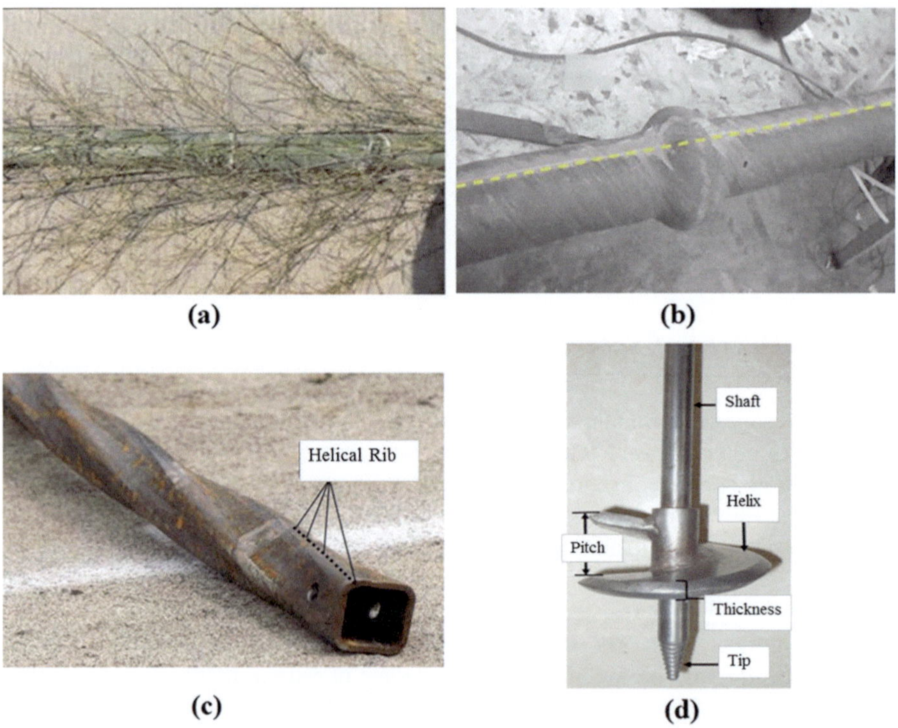

Fig. 9.8 Types of advanced soil nails: (**a**) Moso bamboo soil nails with bamboo branches [82], (**b**) GFRP soil nail with FBG sensors [83], (**c**) Spiral soil nails [86], (**d**) Helical soil nails [59]

materials (CFRP and GFRP) can be employed in place of conventional soil nails to ensure the durability of soil nails due to their technical superiority over steel in terms of corrosion resistance, strength-to-weight ratio, thermal stress and site installation (Fig. 9.8b) [83]. The GFRP soil nail is prepared from glass fibers embedded in a resin matrix to enhance its corrosion resistance and durability characteristics [83]. The pullout behavior of soil nails is different from steel soil nails due to the debonding and slippage at the GFRP-grout interface. Application of GFRP soil nails for slope stabilization purpose was studied by comparing the results of pullout tests on conventional soil nails and GFRP soil nails [83, 84]. The GFRP soil nail is more extensible in comparison to the steel soil nail, and therefore needs additional displacement-based criteria to limit the permissible pullout resistance. Cheng et al. [85] studied the performance of GFRP and CFRP soil nails and concluded that CFRP nail performs better in terms of tensile strength but has low shear strength and there are practical implications while carrying out installation and pullout test. Further, CFRP has a high cost and there is problem in constructing a good nail head for load transfer. Therefore, CFRP soil nails are not suggested to be used for normal applications.

Aziz and Stephens [86] developed a directly driven soil nailing system using spiral nails and demonstrated the use of spiral nails by conducting full scale field tests on instrumented reinforced wall. The spiral nail is a hollow tube, helical in shape with twisted square-section and is made from steel pipe (Fig. 9.8c). Spiral nails offer both frictional and mechanical bonding with the surrounding soil and have immediate loading capability [86]. Generally, wire truss and spider types of fascia are employed with spiral nail system in the field. Helical anchors are usually used to provide resistance against uplift loads, pullout forces and overturning moments. Due to advantages of rapid installation, minimal site disturbance and immediate loading capability, their scope of use has expanded beyond the traditional applications to hydraulic structures, solar panels, retaining walls, tunnels, and tiebacks for retention of unstable slopes, excavations and embankments [87, 88]. Application of helical anchors is seen as an innovative alternative to conventional soil nail applications [59, 60, 89–91]. Helical soil nails are installed by applying sufficient torque and derive their capacity from the helical plates affixed to the nail shaft (Fig. 9.8d). Deardorff et al. [92] described the construction process and results of the instrumented helical soil nail wall. Load cell and inclinometers were installed in the slope to measure the nail pullout resistance and displacement respectively. Results obtained from the load cell and inclinometers are found to be within the range of values provided by Federal Highway Administration design guidelines [2]. Tokhi and Li [91] studied the pullout behavior of helical soil nail in a cohesionless medium. The load-displacement behavior was found to be different from conventional soil nails (Fig. 9.9). In comparison to conventional soil nails, helical soil nails do not show accentuated peak which is followed by a sharp decrease in the residual value [59, 91]. Sharma et al. [59] compared the pullout behavior of helical soil nail with conventional driven soil nail embedded in the dense sand medium. The test results indicated that pullout capacity of helical soil nails is higher than the conventional driven soil nails (Fig. 9.9). Similar to conventional soil nails, the

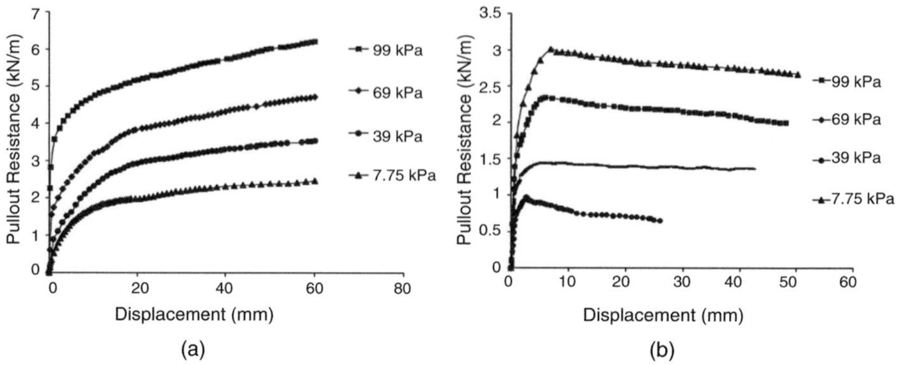

Fig. 9.9 Pullout load – displacement curves for helical and driven soil nails [59]: (**a**) Helical soil nail, (**b**) Driven soil nail

pullout capacity of helical soil nails increases with the normal stress indicating that it follows Mohr-Coulomb failure envelope [59, 91].

9.13 Conclusions

The present chapter discusses the different aspects of the soil nailing technique for slope & excavation protections. Advantages and limitations of this technique, different failure modes, design philosophies and influence of various parameters on pullout capacity of soil nails have been summarized. The following inferences may be drawn from the present study:

(i) Soil nail is an effective slope or excavation protection technique for dense to very dense granular soils, glacial till with a few boulder and cobbles, residual soils engineered fills, weathered rock (such as CDG), stiff to hard fine-grained soils, cemented sands, stiff silts and silts.

(ii) Generally kinematical multi check design methods are followed for soil nailed structure design. External, internal and facing stability are the main criteria of the design of soil nailed structures

(iii) Soil nail pullout response is the governing parameter for the soil nail design. It usually depends on the properties of the soil nail, soil and grout. The roughness of the nail and borehole increases the pullout capacity of soil nail. Soil dilation also leads to increase in the pullout resistance of the nail at low normal stress while its effect diminishes with an increment in the normal stress.

(iv) The method of installation also plays a significant role in development of soil-nail pullout capacity. A higher pullout capacity of the driven nails is observed than the jacked nail. Increasing the grouting pressure leads to an increase in the soil nail pullout capacity. However, few studies also reported the insignificant influence of grouting pressure on the pullout capacity of soil nail.

(v) Studies on CFRP, GFRP, Moso bamboo soil nails, helical and twisted soil nails shows that newly developed soil nails possess several advantages related to installation process, pullout capacity and corrosion resistance over the conventional soil nails.

References

1. Watkins AT, Powell GE (1992) Soil nailing to existing slopes as landslip preventive works. Hong Kong Eng 20:20–27
2. Federal Highway Administration (2015) Geotechnical engineering circular No. 7: soil nail walls – reference manual. FHWA Rep. No. FHWA-NHI-14-007, Washington, DC
3. Yim KP, Watkins AT, Powell GE (1988) Insitu groundreinforcement for slope improvement in Hong Kong. Proc., International Geotechnical Symp. on theory and Practice of Earth Reinforcement, Fukuoka, Japan, pp 363–368
4. Chan RKS (2005) Safe and green slopes—the holistic Hong Kong approach. In Safe and green slopes. Proceedings of the Hong Kong Institution of Engineers, Geotechnical Division annual seminar, vol 4, pp 1–26
5. Milligan GWE, Tei K (1998) The pull-out resistance of model soil nails. Soils Found 38 (2):179–190
6. Burland JB (2002) Reliability of soil nailed slopes in Hong Kong. Geotech Eng Off Hong Kong Gov, Hong Kong
7. Lazarte CA, Baecher GB, Withiam JL (2003) New directions in LRFD for soil nailing design and specifications. In Proceedings of the international workshop on limit state design in geotechnical engineering practice (LSD 2003), Cambridge, MA, vol 26
8. Chu LM, Yin JH (2005) Comparison of interface shear strength of soil nails measured by both direct shear box tests and pullout tests. J Geotech Geoenviron 131(9):1097–1107
9. Pradhan B, Tham LG, Yue ZQ et al (2006) Soil–nail pullout interaction in loose fill materials. Int J of Geomech 6(4):238–247
10. Su LJ, Chan TC, Shiu YK et al (2007) Influence of degree of saturation on soil nail pull-out resistance in compacted completely decomposed granite fill. Can Geotech J 44(11):p1314–p1328
11. Su LJ, Chan TC, Yin JH et al (2008) Influence of overburden pressure on soil–nail pullout resistance in a compacted fill. J geotech geo environ eng 134(9):p1339–p1347
12. Yin JH, Su LJ, Cheung RWM et al (2009) The influence of grouting pressure on the pullout resistance of soil nails in compacted completely decomposed granite fill. Geotechnique 59 (2):103–113
13. Yin JH, Zhou WH (2009) Influence of grouting pressure and overburden stress on the interface resistance of a soil nail. J Geotech Geoenviron 135(9):1198–1208
14. Zhang LL, Zhang LM, Tang WH (2009) Uncertainties of field pullout resistance of soil nails. J Geotech Geoenviron 135(7):966–972
15. Zhou WH, Yuen KV, Tan F (2012) Estimation of maximum pullout shear stress of grouted soil nails using Bayesian probabilistic approach. Int J Geomech 13(5):659–664
16. Rabcewicz L (1964a) The new Austrian tunnelling method. Part I, Water Power 16 (11):453–457
17. Rabcewicz L (1964b) The new Austrian tunnelling method. Part II, Water Power 16 (12):511–515
18. Rabcewicz L (1965) The new Austrian tunnelling method. Part III, Water Power 17(1):19–24
19. Stocker MF, Korber GW, Gassler G, et al (1979) Soil nailing. C.R. Col. Int. Reinforced des. Sols. Paris, pp 469–474

20. GEO (2008) Guide to soil nail design and construction (Geoguide 7). Geotechnical Engineering Office, Civil Engineering and Development Department, Hong Kong, p 97
21. Cornforth DH (2005) Landslides in practice: investigation, analysis, and remedial/preventative options in soils. Wiley, New York
22. Khalil M, Rhodes M, Daly J, Ferris J (1998) Soil nail walls in residual soils. In: Soil improvement for big digs, In geotechnical special publication. American Society of Civil Engineers, Reston, pp 214–225
23. Sheahan TC (2000) A field study of soil nails in clay at the University of Massachusetts—Amhurst national geotechnical experimentation site. In: Benoıt J, Lutenegger AJ (eds) National geotechnical experimentation sites, Geotechnical Special Publication No. 93. ASCE, Reston, pp 250–263
24. Gassler G, Gudehus G (1981) Soil nailing-some aspects of new technique. In: Proceedings of tenth ICSMFE. Balk-ema, Stockholm, pp 943–962
25. Gassler G (1988) Soil-nailing — theoretical basis and practical design. In: Miura N, Ochiai H, Yamanouchi T (eds) Proceedings of the international geotechnical symposium on theory and practice of earth reinforcement, Fukuoka, Kyushu, Japan, October 1988. A.A. Balkema, Rotterdam, pp 283–288
26. Stocker MN, Riedinger G (1990) The bearing behaviour of nailed retaining structures. In: Lambe PC, Hansen LA (eds) Proceedings of design and performance of earth retaining structures, Geotechnical Special Publication, vol 25. ASCE, New York, pp 612–628
27. Juran I, George B, Khalid F, Elias V (1988) Kinematical limit analysis approach for the design of nailed soil retaining structures. In: Proceedings of the geotechnical symposium on theory and practice of earth reinforcement, Japan. AA Balkema, Rotterdam, pp 301–306
28. FHWA (2003) Geotechnical engineering circular No 7—soil nail walls, Report FHWA0-IF-03–017. US Department of Transportation, Federal Highway Administration, Washington, DC
29. Su LJ, (2006) Laboratory pullout testing study on soil nails in compacted completely decomposed granite fill
30. Byrne RJ, Cotton D, Porterfield J et al (1998) Manual for design and construction monitoring of soil nail walls, No. FHWA-SA-96-069R. Federal Highway. Administration, U.S. Department of Transportation, Washington, DC
31. Schlosser F (1982) Behaviour and design of soil nailing. Proc. Symp. on recent developments in ground improvement techniques, Bangkok, 399–413
32. Barley AD, Maddison JD, Jones DB (1997) The use Of soil nails for the stabilization of a new cutting for the realignment ff the 237 A4059 at Lletty Turner Bends. Ground improvement geosystems: densification and reinforcement. Proceedings of the third international conference on ground improvement geosystems, London, 3–5 June 1997, p 459–467
33. Tei K, TAYLOR NR, Milligan GW (1998) Centrifuge model tests of nailed soil slopes. Soils Found 38(2):165–177
34. Luo SQ, Tan SA, Yong KY (2000) Pull-out resistance mechanism of a soil nail reinforcement in dilative soils. Soils Found 40(1):47–56
35. Luo SQ, TANT S, Cheang W et al (2002) Elastoplastic analysis of pull-out resistance of soil. Ground Improvement 6(4):153–161
36. Franzen G (1998) A laboratory and field study of pullout capacity. Doctoral thesis, Chalmers Univ. of Technology, Göteborg, Sweden
37. Junaideen SM, Tham LG, Law KT (2004) Laboratory study of soil-nail interaction in loose, completely decomposed granite. Can Geotech J 41(2):274–286
38. Schlosser F, Guilloux A (1981) Le frottement dens les sols. Revue Francaise de Geotechnique 16:65–77
39. Cartier G, Gigan JP (1983) Experiments and observations on soil nailing structures. In Proceedings of the Eight European Conference on Soil Mechanics and Foundation Engineering (ECSMFE), Helsinki, Vol. 2, p 473–476
40. Jewell RA, Pedley MJ (1990) Soil nailing design: the role of bending stiffness. Ground Eng 23(2)

41. HA 68/94 (1994). Front-face pull-out in the absence of facing elements or wrap-round reinforcement. Geotechniques and Drainage, Section 1 Earthworks, Vol.4, Part 4
42. Potyondy JG (1961) Skin friction between various soils and construction materials. Geotechnique 11(4):339–353
43. Wong HY (1995) Soil nails design manual for slopes (with worked example). Architectural Services Department, Hong Kong
44. Heymann G, Rohde AW, Schwartz K et al (1992) Soil nail pullout resistance in residual soils. In: Proceedings of the international symposium on earth reinforcement practice, Kyushu, Japan, vol 1, pp 487–492
45. Mecsi J (1997) Some practical and theoretical aspects of grouted soil anchors. In Ground anchorages and anchored structures: Proceedings of the international conference organized by the Institution of Civil Engineers and held in London, UK, on 20–21 March 1997, Thomas Telford Publishing, p 119–130
46. Smith IM, Su N (1997) Three-dimensional FE analysis of a nailed soil wall curved in plan. Int J Numer Anal Methods Geomech 21(9):583–597
47. Benhamida B, Unterreiner P, Schlosser F (1997) Numerical analysis of a full scale experimental soil nailed wall. Proc. 3rd int. conf. on ground improvement geosystems, London, p 452–8
48. Kim JS, Kim JY, Lee SR (1997) Analysis of soil nailed earth slope by discrete element method. Comput Geotech 20(1):1–14
49. Su LJ, Yin JH, Zhou WH (2010) Influences of overburden pressure and soil dilation on soil nail pull-out resistance. Comput Geotech 37(4):555–564
50. Shafiee S (1986) Numerical simulation of the behaviour of soil nailing. Interaction of soil nail and behaviour of the structure. PhD Thesis, Paris
51. Babu GLS, BRS M, Srinivas A (2002) Analysis of construction factors influencing the behavior of soil nailed earth retaining walls. Ground Improv 6(3):137–143
52. Yeo KC, Lo SR, Yin JH (2007) Installation method and overburden pressure on soil nail pullout test. New Horizons in Earth Reinforcement. In: Otani J, Miyata Y, Mukunoki T (eds) Proc., 5th Int. Symp. on Earth Reinforcement. Taylor and Francis, Kyushu, pp 321–327
53. Li KS, Lo SR (2007) Discussion of comparison of interface shear strength of soil nails measured by both direct shear box tests and pullout tests by Lok-Man Chu and Jian-hua Yin. J Geotech Geoenviron Eng 133(3):344–346
54. Tei K (1993) A Study of soil nailing in sand. PhD thesis, University of Oxford, London
55. Pradhan B (2003) Study of pullout behaviour of soil nails in completely decomposed granite fill. M.Phil thesis, The University of Hong Kong
56. Yin JH, Su LJ (2006) An innovative laboratory box for testing nail pull-out resistance in soil. ASTM Geotech Test J 29:451
57. Wu JY, Zhang ZM (2009) Evaluations of pullout resistance of grouted soil nails. In slope stability, retaining walls, and foundations: Selected papers from the 2009 Geo Hunan International Conference, p 108–114
58. Samanta M, Sharma M, Punetha P et al (2017) Pullout capacity of soil nails in cohesionless soil and its constitutive modelling, Conference on Numerical Modeling in Geomechanics. IIT Roorkee, India
59. Sharma M, Samanta M, Sarkar S (2016) A study on comparison of pullout behavior of helical and conventional driven soil nails. In. proceeding of International Geotechnical Engineering Conference on Sustainability in Geotechnical Engineering Practices and Related Urban Issues, Mumbai, India, (Submission number: 133)
60. Sharma M, Samanta M, Sarkar S et al (2017) A laboratory study on inclined pullout capacity of helical anchors in sand medium. Proceeding of Sixth Indian Young Geotechnical Engineers Conference (6IYGEC2017) NIT Trichy, India, 10–11th March, 2017
61. Pedley MJ (1990) The performance of soil reinforcement in bending and shear (Doctoral dissertation, University of Oxford)
62. Clouterre (1991) Recommendations Clouterre 1991. US Department of Transportation, Federal Highway Administration

63. Plumelle BC, Schlosser F, Delage P et al (1990) French national research project on soil nailing Clouterre. In: Proceedings of a conference on design and performance of earth retaining structures, Geotechnical Special Publication No. 25, Ithaca, USA, pp 660–675
64. Hong CY, Zhang YF, Guo JW et al (2015) Experimental study on the influence of drillhole roughness on the pullout resistance of model soil nails. Int J Geomech 16(2):04015047
65. Wernick E (1978) Stress and strain on the surface of anchors. Revue Fracaise de geotechnical, n°3, Janvier 3:113–119
66. Schlosser F, Jacobsen HM, Juran I (1983) General report—soil reinforcement, specialty session 5. In: Proceedings of the 8th European conference on soil mechanics and Foundation Engineering, vol 3, Espoo, Finland, pp 1159–1180
67. Palmeira EM, Milligan GWE (1989) Scale and other factors affecting the results of pull-out tests of grids buried in sand. Geotechnique 39(3):511–542
67. Farrag K, Acar YB, Juran I (1993) Pull-out resistance of geogrids reinforcements. Geotext Geomembr 12(2):133–159
69. Chai XJ, Hayashi S (2005) Effect of constrained dilatancy on pull-out resistance of nails in sandy clay. Ground Improv 9(3):127–135
70. Guilloux A, Schlosser F, Long NT (1979) Etude du frottement sable-armature en laboratoire. Proc. of Int. Conf. of Soil Reinforcement, Paris, France, pp 35–40
71. Wang Z, Richwien W (2002) A study of soil-reinforcement interface friction. J Geotech Geoenviron 128(1):92–94
72. Juran I (1985) Reinforced soil systems-application in retaining structures. Geotech Eng 16 (1):39–82
73. Yin JH, Hong CY, Zhou WH (2011) Simplified analytical method for calculating the maximum shear stress of nail-soil interface. Int J Geomech 12(3):309–317
74. Schlosser F, Guilloux A (1981) Le frottement dans le renforcement des sols. Rev Fr Géotech 16:65
75. CIRIA (2005) Soil nailing – best practice guidance, Report No C637, p286. Construction Industry Research & Information Association, London
76. Juran I, Elias V (1991) Ground anchors and soil nails in retaining structures. In Foundation engineering handbook. Springer, US, pp 868–905
77. Kim Y, Lee S, Jeong S et al (2013) The effect of pressure-grouted soil nails on the stability of weathered soil slopes. Comput Geotech 49:253–263
78. Elias V, Juran I (1989) Soil nailing for stabilization of highway slopes and excavations, Report FHWA-RD-89-198. Federal Highway Administration, U.S. Department of Transportation, Washington, DC
79. Elias V, Juran I (1991) Soil nailing for stabilization of highway slopes and excavations, Technical report FHWA-RD-89- 198. Federal Highway Administration, U.S. Department of Transportation, Washington, DC
80. Patra CR, Basudhar PK (2005) Optimum design of nailed soil slopes. Geotech Geol Eng 23 (3):273–296
81. Yeung AT, Cheng YM, Lau CK et al (2005) An innovative Korean system of pressure grouted soil nailing as a slope stabilization measure. The HKIE Geotechnical Division 25th annual seminar, Hong Kong, vol 1, pp 43–49
82. Dai ZH, Guo WD, Zheng GX et al (2016) Moso bamboo soil-nailed wall and its 3D nonlinear numerical analysis. Int J Geomech 16(5):04016012
83. Zhu HH, Yin JH, Yeung AT (2010) Field pullout testing and performance evaluation of GFRP soil nails. J Geotech Geoenviron 137(7):633–642
84. Yeung AT, Cheng YM, Tham LG (2007) Field evaluation of a glass-fiber soil reinforcement system. J Perform Constr Facil ASCE 21(1):26–34
85. Cheng YM, Au SK, Yeung AT (2015) Laboratory and field evaluation of several types of soil nails for different geological conditions. Can Geotech J 53(4):634–645

86. Aziz ES, Stephens TJ (2013) Cost and schedule savings from directly-driven soil nail and snnovative fascia systems. In: Geo-Congress. Stability and Performance of Slopes and Embankments III, pp 1704–1718
87. Spagnoli G, Gavin K (2015) Helical piles as a novel foundation system for offshore piled facilities. Proceedings of Abu Dhabi international petroleum exhibition and conference, Society of Petroleum Engineers, Abu Dhabi, UAE 2015, November 9–12. https://doi.org/10.2118/177604-MS
88. Spagnoli G (2016) A CPT-based model to predict the installation torque of helical piles in sand. Mar Georesour Geotechnol 35(4):1–8
89. Perko HA (2009) Helical piles: a practical guide to design and installation. Wiley, Hoboken
90. FSI (2014) Technical manual: helical piles and anchors, hydraulically driven push piers, polyurethane injection & supplemental support systems, 2nd edn. Foundation Support Works, Omaha, pp 33–39
91. Tokhi H, Ren G, Li J (2016) Laboratory study of a new screw nail and its interaction in sand. Comput Geotech 78:144–154
92. Deardorff D, Moeller M, Walt E (2010) Results of an instrumented helical soil nail wall. In Earth Retent Conf 3:262–269

Chapter 10
Bioengineering as an Effective and Ecofriendly Soil Slope Stabilization Method: A Review

Piyush Punetha, Manojit Samanta, and Shantanu Sarkar

Abstract Soil-bioengineering is a cost-effective and eco-friendly alternative to the conventional methods of soil slope stabilization and erosion control. Numerous techniques such as fascines, bush layering, vegetated gabions etc. have been developed to enhance the soil slope stability, arrest soil erosion and improve the aesthetic aspect of a project, using plants as well as inert materials. Nevertheless, a limited control on the properties of the plants and the complex interaction of plant roots with the soil and other materials poses a challenge for the accurate design of soil-bioengineering techniques. The design of bioengineering techniques involves accurate evaluation of the root and root-soil properties. Different methods have been developed for the analysis of root and soil-root system that can aid in a better understanding of the complex phenomenon. The Present study provides a review on different aspects of bioengineering techniques for soil slope stabilization measures, especially, the existing techniques of physical modeling, laboratory scale testing and numerical techniques for evaluating the effect of root system on the strength properties of soil-root matrix. The different failure modes of the soil-root system i.e. adhesion failure, tension failure and progressive failure are briefly discussed. The present review will be useful for the design of bioengineering measures for soil slope stabilization or erosion control.

Keywords Bioengineering · Slope stability · Root system · Remedial measures · Root strength · Factor of safety

---

P. Punetha
Academy of Scientific and Innovative Research, CSIR-Central Building Research Institute, Roorkee, Uttarakhand, India

M. Samanta (✉) · S. Sarkar
Geotechnical Engineering Group, CSIR-Central Building Research Institute, Roorkee, Uttarakhand, India

S. P. Pradhan et al. (eds.), *Landslides: Theory, Practice and Modelling*, Advances in Natural and Technological Hazards Research 50,
https://doi.org/10.1007/978-3-319-77377-3_10

10.1 Introduction

Soil bioengineering deals with the use of plants or plant parts for engineering applications involving the control of soil erosion, shallow mass movements, stream-bank protection etc. [1–4]. The use of vegetation for soil slope protection and erosion control is not a new technique. This technique has been used in different parts of the world, especially Asia and Europe, from a very long time. However, there was no rational or scientific basis for the proper investigation and design of vegetation for the engineering applications [5]. Consequently, the last few decades have been devoted to understand the governing soil-plant interaction mechanism (to enhance soil stability) and develop a scientific rationale for the proper design of vegetation for engineering applications.

Soil bioengineering is a subset of biotechnical engineering. Both these terms are often used synonymously but biotechnical engineering or stabilization involves the use of plant, or plant parts, either alone or in conjugation with inert materials such as steel, concrete, rocks etc. to enhance the soil stability [6, 7]. The biotechnical engineering techniques can be classified into various categories depending on the site conditions and intended use. These include surface protection or erosion control, stabilization using plants and techniques involving both plants and inert materials [8]. An example of the biotechnical engineering technique is the use of trees in conjugation with small retaining walls to support unstable soil slopes. In this technique, the trees are used to provide additional stability as the roots of trees can anchor a loose mass of soil to a hard stratum or reinforce a layer of loose soil and reduce the deformation, provide drainage for surface runoff, prevent debris movement, reduce pore water pressure by absorbing moisture etc. [9, 10].

The soil bioengineering technique is basically an amalgam of biological sciences and classical engineering disciplines. It is a multidisciplinary subject which involves the contribution from specialists involved in different fields such as geotechnical engineering, botany, landscape architecture, hydrology etc. [11]. This technique is different from the traditional methods of slope stabilization due to limited control on the properties of the stabilizing material.

Soil bioengineering is most commonly used for erosion control, however, several studies have shown that it is capable of preventing shallow earth movements [12]. Despite several challenges in the utilization of vegetation, different softwares and tools are now available which can assist in the selection of a suitable bioengineering technique as an alternative to traditional methods of slope protection and control of soil erosion [13–15]. Before adopting a suitable bioengineering technique, it is essential to understand the existing conditions (type of material viz. soil, debris or rock, climate etc.) as well as its effect on the environment. For e.g. bioengineering can be used to stabilize the soil slopes as well as debris slopes in which the roots of plants will hold the soil/debris and prevent their movement, however, on the contrary, it could generate discontinuities in the rock mass and reduce the stability of rock slopes.

This paper presents a brief review on the plant-soil interaction mechanism, the effect of different parameters on the interaction mechanism, physical and numerical modeling, advantages, disadvantages and challenges in the field of soil-bioengineering. It must be noted that the present paper only provides a brief review of the aspects and therefore, the readers may refer to the publications in the reference section for further information.

10.2 Advantages and Limitations of Soil-Bioengineering

The advantages of soil bioengineering technique are as follows:

- ***Enhanced soil slope stability:*** The roots of the plants reinforce the soil and prevent shallow slope movements [1, 11]. The increment in soil stability can be in the form of enhanced cohesion or friction angle, pore water pressure reduction, reduction in the surface runoff etc.
- ***Arrest soil erosion:*** The fine roots of the plants hold the soil and prevent erosion. Moreover, they arrest the movement of small sediments and reduce the velocity of surface runoff [1, 11, 12].
- ***Ecology:*** It serves as a habitat for several wild animals, and regulates the temperature and moisture at the soil near the vegetation which promotes the growth of different type of organisms. Moreover, it improves the aesthetic aspect of the projects by fitting in with the existing landscape and preventing damage to the nearby forest or any agricultural land [8, 9]. Soil bioengineering techniques such as hydro-seeding are also used to improve the appearance of the bare rock slopes (improve the aesthetics) [16, 17].
- ***Economy:*** It requires a very little initial cost. However, at the later stages, the effectiveness of the soil bioengineering technique depends on proper maintenance and care [1, 8, 10, 11].
- ***Hydrology:*** The Plants play a crucial role in modifying the hydrology of a particular site. The leaves of the plants intercept the rainfall and cause evapotranspiration, roots of the plant absorb moisture from the soil and hence reduce the pore-water pressure (which is a triggering factor for the soil movement in a slope) [11, 12].
- ***Improved stability with time:*** The bioengineering techniques involve the use of living materials, therefore, they take some time to resist the loads or soil movements. However, once established, the roots propagate over a large depth with time and hence, improve the stability of the soil slope with time [1, 5]. Unlike the abiotic components, the effectiveness of plants increases with time.

The limitations of soil-bioengineering include:

- ***Limited influence zone of roots:*** The roots of the plants can extend only upto a depth of few meters (2–3 m), therefore, soil-bioengineering can't be used to prevent deep-seated soil movements [11].
- ***Effect of wind:*** The wind plays a crucial role in the stability of soil slope, especially when the slope is covered with trees. The trees attract wind loading and can induce a lateral load on the slope due to the tree-wind interaction. However, the magnitude of loading due to wind is very small as compared to the other disturbing/driving forces. Moreover, the falling of trees after a storm could disrupt the normal functioning of the transportation operations [15].
- ***Effect on nearby structures:*** The roots of trees often damage the nearby structures such as footings, retaining walls etc. In some cases, it can even lead to failure of these structures [15].

- ***Limited installation time:*** It is essential to install the vegetation at the right time (usually during the dormant period) to ensure a good survival rate [9, 18]. Therefore, the success of a soil bioengineering project depends on meticulous planning.
- ***Design difficulty:*** It is very difficult to design a bioengineering project due to the difficulty in quantification of the interaction between the soil and plants [10]. Consequently, some researchers even adopted a factor of safety of 8 during the design of bio-engineered slopes [19].
- ***Increase in percolation:*** The plants make the soil surface rough and enhance the percolation of water. This enhanced percolation can reduce the stability of a soil slope [1, 11].
- ***Quality control:*** The success of a bioengineering project depends on the quality of work during the installation and maintenance. In the case of inert materials, a high factor of safety can be used to compensate for the poor quality of work, however, the same is not applicable to bioengineering [9].
- ***Slow rate of stabilization:*** The vegetation does not stabilize the soil immediately after the installation, but it takes some time for the roots to develop and strengthen the soil [8].
- ***Environmental considerations:*** The properties of the soil, namely pH, nutrient and metal concentrations should be within the acceptable limits. The properties of the soil and the environmental conditions may restrict the application of certain plant species and techniques in a particular slope. Fertilizers or various treatment techniques are also required for the soil to promote favorable growing conditions [20]. Soil texture and slope direction also play an important role in the selection of a suitable bioengineering technique.
- ***Site-specific nature:*** The soil bioengineering techniques are highly site-specific. This is due to the difference in the soil properties, climate etc. in the different sites. A detailed investigation of all the parameters that could affect the soil slope stability before and after the installation of vegetation must be conducted before designing a bioengineering project [9].
- ***Adverse effect on rock slopes:*** The presence of plants can have detrimental effects in the rock slopes and can even induce slope failure by reducing the existing strength of the rock mass. The roots of the plants may penetrate through the existing discontinuities present in the rock mass and may widen them or even generate secondary discontinuities [21].

10.3 Challenges in Soil-Bioengineering

Numerous challenges exist with the bioengineering techniques. Simon and Steinemann [22] coined out five major challenges: (a) lack of adequate data for the design of protection or restoration work; (b) vulnerability of bioengineering projects immediately after the installation due to plant mortality or other factors; (c) previously applied ineffective protection measures; (d) accessibility of sites; and (e) need of participation from local residents. The limited availability of plants that can adapt the site conditions, limited time for installation, availability of skilled labor, high maintenance requirement are also some of the challenges to the implementation of the soil-bioengineering projects. The selection of an appropriate species of plant and generation technique (cutting or seedling) also plays a very crucial role. The selection depends on the site conditions, role type (reinforcement or erosion control), adaptability, root-soil volume ratio of the plant, resistance to diseases, pests, weeds etc. [5, 8, 23]. A poor selection of the species or the generation technique can lead to the failure of the bio-engineering project [18].

10.4 Application Techniques of Bioengineering/Biotechnical Engineering to Improve Stability of a Soil Slope

Numerous bioengineering/biotechnical engineering techniques have been developed to enhance the soil slope stability and control soil erosion [1, 2, 8, 9, 12, 24–27]. Some of these techniques are briefly discussed below (Figs. 10.1, 10.2, and 10.3):

- ***Fascines:*** Initially trenches are dug and fascines of live cuttings of branches are laid over the trenches. Wooden poles or steel pegs are used to hold the fascines. After fixing the fascines in position, the trench is backfilled with soil (Fig. 10.1a).
- ***Groove structures:*** This technique is similar to fascines but the only difference is the addition of one or two small rooted trees alongside the fascines of live material (Fig. 10.1b).

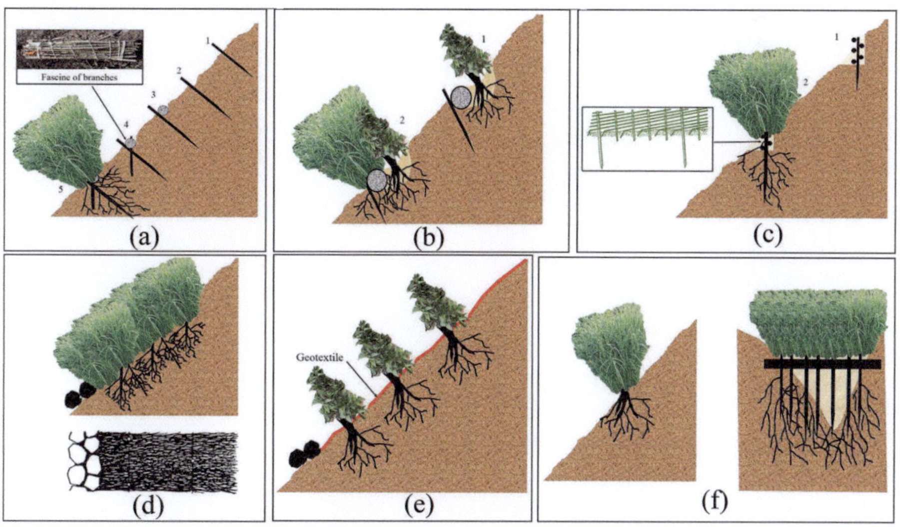

Fig. 10.1 Different biotechnical engineering techniques: (**a**) Fascines, (**b**) Groove structures, (**c**) Wattle fences, (**d**) Bushmattress construction, (**e**) Biotechnical technique involving geotextile, (**f**) Vegetated palisades [8]

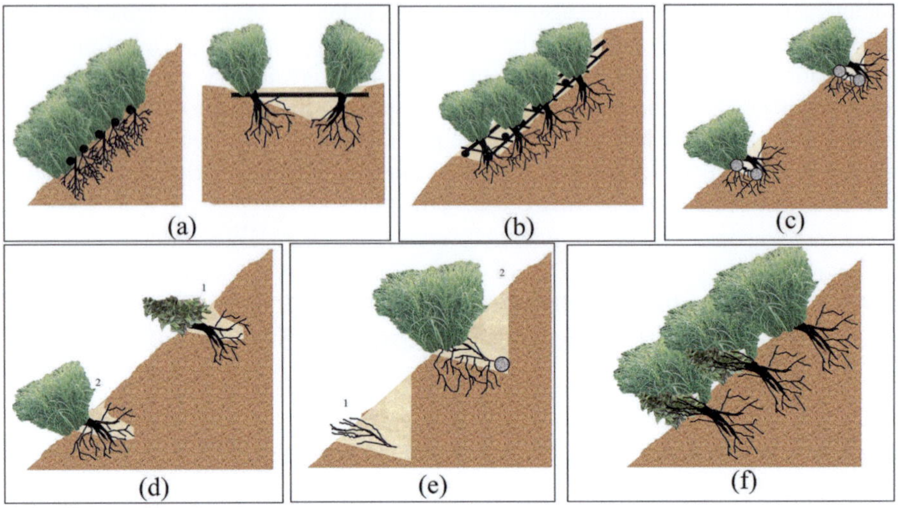

Fig. 10.2 A few biotechnical engineering techniques: (**a**) Branch layer in gullies, (**b**) Live slope gratings, (**c**) Cordon construction, (**d**) Hedge layer construction, (**e**) Brush layer construction, (**f**) Hedge brush layer construction [8]

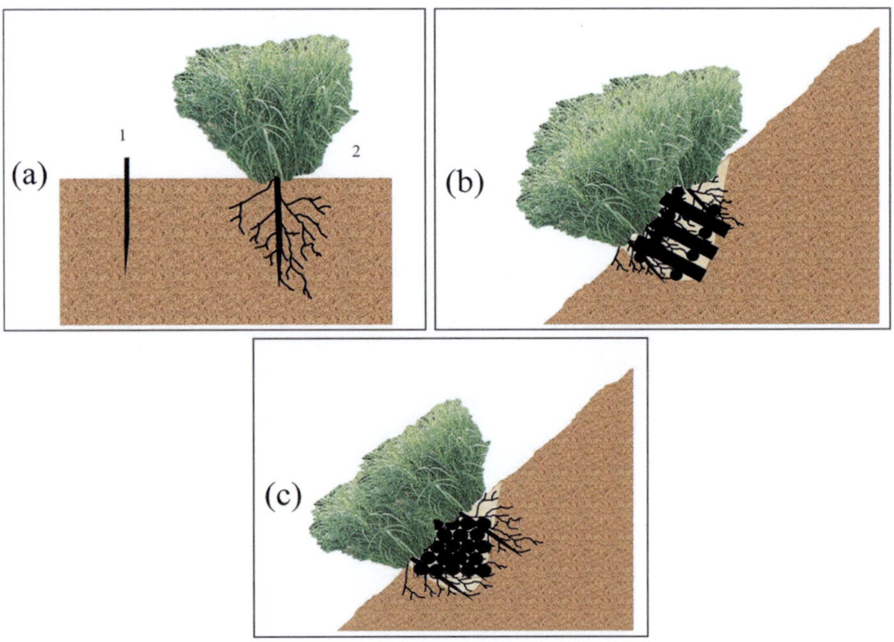

Fig. 10.3 Other biotechnical engineering techniques: (**a**) Placing of cutting, (**b**) Vegetated crib wall, (**c**) Vegetated gabion [8]

- ***Wattle fences:*** In this method, a fence is produced using wooden vertical poles and live cuttings of branches which are woven around the poles. The ends of these cuttings are inserted into the soil to allow germination (Fig. 10.1c).
- ***Bushmattress construction:*** In this technique, live cuttings of branches or stem of a plant (that can germinate through cuttings) are spread throughout the slope and secured at the ends by using wooden pegs, poles or fascines. A layer of soil is used to cover the branches to allow their germination (Fig. 10.1d).
- ***Biotechnical technique involving geotextile:*** This technique involves the use of geotextiles along with the vegetation. The geotextile is laid over the slope and the seeds are sown or the cuttings are planted over the slope. The geotextile initially protects the small plant (seedlings) and later on decomposes (and provides nutrients to the soil which further aids in the plant growth) once the plant has fully grown (Fig. 10.1e).
- ***Vegetated palisades:*** This technique is especially used in the case of a V-shaped gully. A palisade is formed by placing live cuttings side by side along the gully and the top of the cuttings is tied to a horizontal bar (either living or dead) which connect the two sides of the gully (Fig. 10.1f).
- ***Branch layer in gullies:*** This method involves placement of live cuttings in a fish bone pattern along a gully. This mattress of plants is supported at regular intervals using horizontal poles or bars which connects the two sides of a gully (Fig. 10.2a).
- ***Live slope gratings:*** In this method, the live cuttings of plants are placed alongside a framework of parallel or crossed poles of wood, concrete, metal or plastic. In addition to this, seeding is conducted to enhance the stability of the slope (Fig. 10.2b).
- ***Cordon construction:*** In this technique, terraces are constructed and covered with poles of dead material (placed diagonally) and branches of conifers. This surface is then covered with soil and live cuttings are planted over the terrace (Fig. 10.2c).
- ***Hedge layer construction:*** In this method, terraces are constructed and live root plants are placed over the terraces side by side. The plants are then covered with soil to about two third of their length (Fig. 10.2d).
- ***Brush layer construction:*** In this method, inclined terraces are constructed and covered with live cuttings in a crisscross pattern such that only one fourth part of the cutting reaches beyond the slope surface. This terrace is then backfilled with soil. The immediate reinforcement in this method is provided by the installation of cuttings while the adventitious rooting stabilizes the slope in a long run (Fig. 10.2e).
- ***Hedge brush layer construction:*** This method is similar to brush layer construction. The only difference is the addition of live rooted healthy plants alongside the cuttings (Fig. 10.2f).
- ***Placing of cuttings, wall joint plantings, vegetated stone walls and rock piles:*** This method involves placement of live cuttings of plants vertically into the soil either alone or in conjugation with stone walls such that one-fourth part of the cutting protrudes beyond the surface (Fig. 10.3a).

- ***Crib wall construction with branch layering:*** In this method, a crib wall is constructed using wood, concrete, steel or plastic material. During the backfilling of soil inside the wall, layers of live cuttings of plants are inserted at an angle of 10°. The length of the cuttings should be sufficient so that some part of the cutting is stuck to the existing soil while about 0.25 m of cutting protrudes outside the wall (Fig. 10.3b).
- ***Vegetated gabions:*** In this method, live cuttings are inserted within and between the gabion boxes, during their construction. The live cuttings develop roots after some time and enhance the stability (Fig. 10.3c).

  It must be noted that a combination of different species is better than a single species of plants for increasing the soil slope stability [28].

10.5 Mechanism of Soil Stabilization

The plants enhance the stability of a soil slope both mechanically as well as hydrologically. The presence of vegetation in a soil slope contributes to the slope stability by increasing the resisting force in the form of root cohesion or friction angle, reducing the pore water pressure, reducing the weight of the soil mass by absorbing the moisture, reducing the surface runoff, intercepting the rainfall etc. [1, 8, 10, 11, 15, 19, 28–44]. Further, the roots crossing the failure surface act as tensile reinforcement, which further improves the stability of a soil slope. Numerous case studies have shown that the slopes with vegetation are statistically less prone to mass movements as compared to slopes without vegetation [29, 30]. The falling of trees reduces the stability of a slope and makes it vulnerable to mass movement. However, this phenomenon doesn't occur immediately after falling of trees, as it is a slow process. The mass movement occurs once the roots decay and lose their strength [15, 45]. Moreover, the cutting of trees generates suction in the clayey soil and it readily absorbs moisture and becomes soft, thereby promoting the mass movement [15]. Figure 10.4 shows a typical cross section of the soil-root system along with the different components of forces acting on it. Table 10.1 shows the different mechanisms of soil-plant interaction along with the beneficial and adverse effects of plants on the stability of soil slope.

The increase in resisting force in the soil slope due to plants is usually expressed in terms of root cohesion or the apparent friction angle. The root cohesion is derived from the suction generated by the roots and the ability of the roots to hold the soil particles together [1, 11]. It is more pronounced for grasses and shrubs in cohesionless soils as compared to the trees. It (root cohesion) is often localized at a shallow depth because the fine root distribution occurs at a shallow depth [2, 15]. The presence of tap roots enhances the apparent friction angle of the soil [28]. It must also be noted that with an increase in biomass content, the shear strength of soil increases linearly [8]. Furthermore, the roots increase the ductility of the soil and hence, can provide an indication before the actual soil slope failure [38].

The tensile reinforcement provided by the roots of the plants has a significant contribution towards the soil slope stability [19, 28, 37]. This reinforcement depends on

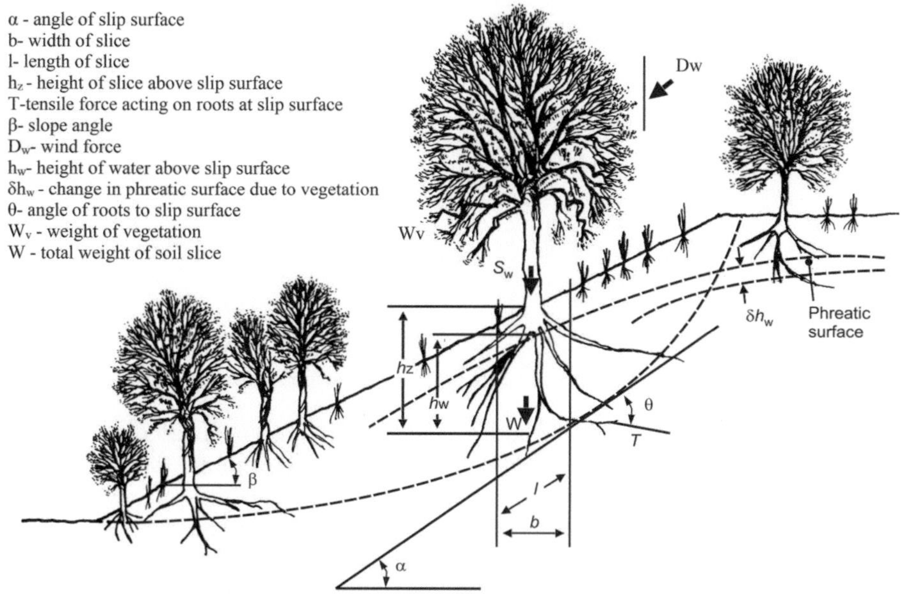

Fig. 10.4 Typical section of a slope along with different components of soil-root system [1]

Table 10.1 Effect of vegetation on slope stability [5, 54]

S. no.	Different mechanisms involved	Influence
Hydrological mechanisms		
1	Interception of rainfall by foliage, causing evaporative and absorptive losses which reduce the rainfall available for infiltration	Beneficial
2	Roots extract moisture from the soil which is lost to the atmosphere via transpiration, leading to a reduction in pore water pressure	Beneficial
3	Stems and roots increase the roughness of the ground surface and the permeability of soil, leading to an increased infiltration capacity	Adverse
4	Depletion of soil moisture may lead to desiccation cracking in the soil, resulting in a higher infiltration capacity	Adverse
Mechanical mechanisms		
5	Roots reinforce the soil and increase its shear strength	Beneficial
6	Tree roots may anchor into firm strata and provide support to the upslope soil mantle through buttressing and arching	Beneficial
7	Weight of trees surcharges the slope and increases the normal and downhill force component	Adverse/ Beneficial
8	Vegetation exposed to wind transmits dynamic forces to the slope	Adverse
9	Roots hold the soil particles at the ground and reduce their susceptibility to erosion	Beneficial

several factors including the mechanical properties of roots (tensile strength, modulus of elasticity), soil (shear strength, modulus of subgrade reaction) and the root architecture (number, size, shape, orientation of roots, root area ratio) [28, 38, 46–

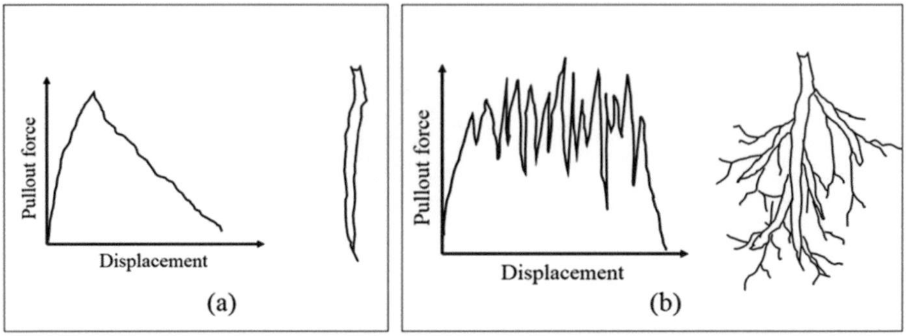

Fig. 10.5 Pullout force displacement curve for root: (**a**) Without branches, (**b**) With branches [52]

48]. These parameters are liable to both spatial as well as temporal variations [28, 38, 49–51]. The amount of reinforcement provided by the roots to the soil depends on the parametrs such as the pullout resistance of roots and the root architecture [15, 19]. The pullout resistance of the vegetation is slightly less than or equal to the tensile strength of the roots [15, 19, 37]. The pullout resistance of the roots in soil depends on several factors including the type of soil, type of root etc. A typical pullout force-displacement curve for roots with and without branching is shown in Fig. 10.5. It can be observed that the presence of branched roots influences the pullout behavior to a large extent. Without branching, the roots fail by simple pullout and a distinct peak is observed in the pullout force-displacement curve (Fig. 10.5a) whereas in case of branching, a progressive failure occurs with the breakage of branches and multiple peaks (Fig. 10.5b) are observed in the pullout force-displacement curve [12, 18, 19, 30, 37, 52]. Thus, during a pullout test, the failure could occur due to the tensile failure in the main root, the tensile failure in the branch roots leading to a progressive failure, the failure in the soil-root interface or a combination of these [27]. Moreover, the tensile strength of roots depends on the lignin and cellulose content which may vary depending on the temperature, availability of moisture, height etc. [18, 53].

The root distribution plays a significant role in the soil slope stability. However, it is very difficult to quantify the effect of root distribution because it depends on a large number of parameters such as the soil conditions and the environment [12, 20, 54]. Moreover, it also depends on the type of the plant. For instance, the roots of herbaceous plants are classified into extensive, intensive or mixed rooters depending on the vertical and lateral extent of roots. Similarly, the roots of trees or woody plants can be classified into tap, heart and flat rooters depending on the orientation and distribution of different root parts [8].

The effect of mass of the vegetation can have a beneficial or adverse effect on the soil slope stability depending on the location of the vegetation within the slope. The presence of heavy vegetation at the toe portion increases the stability while the presence at some height can increase the driving force [19, 37].

The plants (specifically the roots) can also reduce the pore water pressure and the weight of soil mass by absorbing the moisture. However, the contribution of plants in reducing the pore water pressure is very small as compared to other mechanisms.

It is often quite difficult to monitor the pore water pressure due to large seasonal fluctuations in the water level [15]. Moreover, the suction generated within the soil is highly dependent on the root density. With an increase in the root density, the micro-pores present in the soil get pre-occupied with the roots and this hinders the generation of matric suction [28]. Some researchers even argued that the hydrological effects should be ignored while assessing the slope stability, as these mechanisms are unavailable under extreme conditions such as heavy rainfall [8]. Whereas, some of them included the hydrological components in the slope stability models, despite their spatial and temporal variability [55, 56]. The pore water pressure in the soil is also reduced through evapotranspiration. However, in some cases, evapotranspiration can even lead to swelling and shrinking of clayey the soils. The influence of evapotranspiration is also very limited as it depends on the seasonal variations, for e.g. during the autumn season, the rate of evapotranspiration decreases as the deciduous plants shed their leaves [38].

It is clear that both the mechanical and hydrological mechanisms depend on several parameters. Moreover, it has been found that the mechanical and hydrological components of root reinforcement sometimes counteract each other. Therefore, it is very crucial to understand each mechanism clearly before incorporating them in the design of bio-engineered projects.

10.6 Modeling the Behavior of Plant Roots in Soil

10.6.1 Field or In-Situ Testing

The mechanical effects of vegetation on the soil slope stability are assessed using different field testing procedures such as in-situ direct shear test, vertical and horizontal pullout tests [18, 30, 44, 48, 57–59]. The field tests are especially useful to study the effect of different parameters on the root-soil interaction mechanism. Apart from mechanical tests, the estimation of parameters related to root architecture must be given due consideration. Different techniques have been developed to study these parameters including the core break sampling, root counting using profile trench, direct extraction of plant from soil without damaging the roots and image analysis [18, 60–64]. Usually, the root density decreases with an increase in depth as well as an increase in horizontal distance from the stem of a plant [65].

10.6.2 Laboratory Testing

Owing to the difficulty in conducted the field tests, several researchers have conducted laboratory tests on plants, especially the roots, to evaluate their mechanical behavior. Three types of test are most commonly conducted: element tests (tension test of roots, direct shear test or tri-axial test for soil), root-soil interaction tests (pullout test, direct shear test or tri-axial test on vegetated soil) and reduced scale model tests on slopes containing vegetation using centrifuge [10, 12, 20, 38, 54, 66–71]. However, scaling can have a significant effect on the pullout resistance of the roots and therefore, the results of the reduced scale model tests must be interpreted with caution [72].

The laboratory testing can be conducted on live roots or their analogues (fibers) [73–76]. The advantage of using analogue roots is the control on the mechanical properties and the geometry, which helps in assessing the effect of different parameters on the soil-root behavior [38]. Furthermore, in order to reduce the heterogeneity of soil, some researchers have used idealized soil [67]. However, the accuracy of the results obtained from these experiments is objectionable because the natural interface of root and soil is different from the fiber (analogues roots)-soil interface [10].

The element tests on roots are usually conducted using universal testing machines. The tensile strength of roots varies between 5 and 60 MPa and the modulus of elasticity usually varies between 200 and 600 MPa [15, 27]. Moreover, it has been found that the tensile strength of roots depends on the diameter of the roots (especially on the cellulose content) and therefore, it (tensile strength) decreases with an increase in the root diameter [15, 27, 38, 53, 80]. The roots with a smaller diameter are also stiffer as compared to the roots with large diameter [10]. The direct shear tests and triaxial tests on soils with vegetation indicate that the increase in shear strength of the soil is due to an increase in cohesion (direct shear) and apparent frictional angle (triaxial) [68, 71, 77–79]. During the shearing, the fibers may either stretch or bend, depending on the particle size of the soil [71]. The surface roughness of the roots plays an essential role in the soil-root interaction [71, 81]. The increase in surface roughness of the roots increases the shear strength of the root-soil interface. Other properties of interest are the normal stress at the root-soil interface, soil properties and the contact stress. Another important aspect of laboratory testing is the evaluation of the root characteristics. It is a usual practice to evaluate the root characteristics on the shear/failure plane after the tests. One such characteristic is the root area ratio which is the ratio of total area of roots crossing the shear plane to the total area of the plane [10]. Root area ratio significantly affects the shear strength of the vegetated soil. For e.g. a change in root area ratio can compensate for the effect of suction and confining stress [28].

10.6.3 Numerical Modeling of Roots

The numerical modeling of roots is highly complex owing to a large number of parameters that could not be evaluated properly. Nevertheless, a large number of researchers have tried to model the root-soil interaction and evaluated the improvement in the stability of a soil slope due to root reinforcement. The roots of plants basically increase the shear strength of the soil. Most commonly, a term known as root cohesion is incorporated in the classical Mohr's Coulomb equation for the soil, to account for the increase in shear strength.

$$\tau = c' + c_r + \sigma' tan\varphi' \quad (10.1)$$

Where, τ is the shear strength; c' is the cohesion intercept; c_r is the root cohesion; σ' is the effective stress; φ' is the angle of shearing resistance. The in-situ tensile (pullout) resistance of plants, T_f is given by:

$$T_f = \sigma_{rf} A_{root} \quad (10.2)$$

Where, σ_{rf} is equal to or less than the tensile strength of roots (depending on the governing mode of failure); A_{root} is the area of the root. The failure during a pullout test may occur either due to the tensile failure of the roots or due to the mobilization of adhesion between soil and roots [27]. The tensile (pullout) resistance due to the mobilization of the adhesion between soil and roots is given by:

$$T_f = a\pi dL \quad (10.3)$$

Where, d and L are the diameter and length of roots; 'a' equals to the cohesion intercept for cohesive soils; for cohesionless soils, 'a' is given by:

$$a = K_{xy} tan\delta \quad (10.4)$$

Where, K_{xy} is the lateral earth pressure and tan δ is half of tan φ. As explained in the preceding section, the pullout load-displacement behavior depends on the root architecture. The plant with single main root and fewer branch roots show a single peak value in the pullout load-displacement behavior while multiple peaks are observed for the plant with multiple branch roots. The ultimate load, in this case corresponds to the load at which the final branch root fails [27].

Wu et al. [30], represented the root cohesion in Eq. 10.1 in terms of tensile strength of root and the root area ratio (Eq. 10.5). This equation is based on the force equilibrium and holds good for roots with small dimension and spacing. This approach of modeling the soil-vegetation interaction is relatively simple and requires the in-situ tensile strength as the input.

$$c_r = 1.2\, \sigma_t A_r \quad (10.5)$$

Where, σ_t is the tensile strength of the root and A_r is the root area ratio.

This can be derived by considering a root inclined with the horizontal at an angle 'θ' as shown in Fig. 10.6a.

The shear strength increment due to roots is given by [15, 27]:

$$S_r = T_x + T_z tan\varphi = T(cos\theta + sin\theta\ tan\varphi) \quad (10.6)$$

Where, T is the tensile force on the reinforcement; φ is the frictional angle. In terms of stresses, Eq. 10.6 can be represented as:

$$s_r = \frac{\sigma_r A_{root}(cos\theta + sin\theta\ tan\varphi)}{A} = A_r \sigma_r (cos\theta + sin\theta\ tan\varphi) \quad (10.7)$$

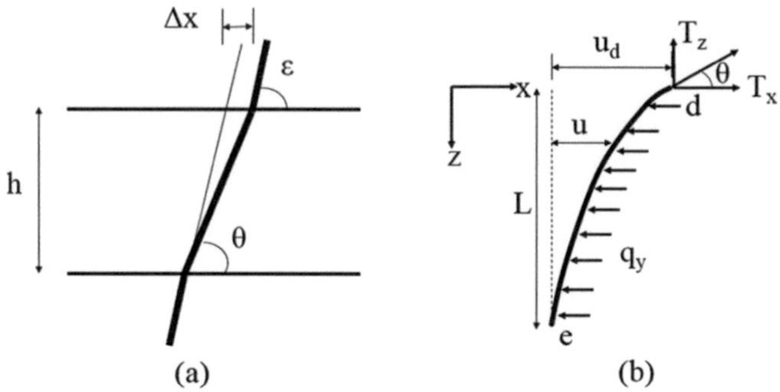

Fig. 10.6 (**a**) Deformation in shear zone, (**b**) Deformed bar [27, 75]

Where, σ_r is the tensile stress in the root; A_{root} is the area of root; A is the area of the shear plane.

$$\theta = \tan^{-1}\left(\frac{1}{\frac{\Delta x}{z} + \tan^{-1}\varepsilon}\right) \tag{10.8}$$

In case of multiple roots the root area ratio is represented as:

$$A_r = \left(\sum A_i\right)/A \tag{10.9}$$

Where, A_i is the area of root i.

At failure, σ_r becomes σ_{rf} (in situ shear strength of reinforcement) and the value of θ ranges between 48 and 72°. Putting $\varphi = 35°$, s_r in Eq. 10.7 becomes numerically equal to c_r in Eq. 10.5. This root cohesion ranges from 1 to 25 kPa depending on the type of vegetation, soil and the environmental conditions [15]. It must be noted that the factor 1.2 in Eq. 10.5, as suggested by [82, 83], overestimates the root cohesion values [10, 18, 84–87]. Therefore, the equation must be used with due care and 1.2 should not be used as a standard value. As the increase in shear strength of the soil due to root-reinforcement depends on the depth, the Eq. 10.5 can be modified as:

$$c_r(z) = K\sum_{j=1}^{n} T_{rj}\,(A_r(z)) \tag{10.10}$$

Where, n is the number of diameter classes at depth z and T_{rj} is the tensile strength of jth diameter class.

The above method is based on the breakage of roots at the shear plane during the pullout test but in actual, the pullout could also occur due to the mobilization of adhesion between the root and soil. The increment in soil cohesion due to the mobilization of adhesion can be evaluated using the following set of equations [10, 83]:

$$c_r = K\alpha A_r(cos\theta + sin\theta\ tan\varphi) \tag{10.11}$$

$$K = \left(\frac{4z\tau_b E}{d}\right)^{0.5} \tag{10.12}$$

$$\alpha = (sec\theta - 1)^{0.5} \tag{10.13}$$

$$\tau_b = z\gamma(1 - sin\varphi)ftan\varphi \tag{10.14}$$

Where, τ_b is the root-soil bond stress; z is the depth below the soil surface; f is the friction ratio between root-soil interface and intact soil; E is the elastic modulus of root; d is the diameter of root; γ is the unit weight of the soil.

The contribution of roots in enhancing the soil slope stability can also be evaluated by considering the root as a structural element embedded inside soil as shown in Fig. 10.6b. This approach is used for large roots with non-uniform geometry or configuration. In this method, the flexural rigidity of the root and the soil reaction are taken into account. The analysis becomes similar to a pile embedded in soil. This approach requires the root deformation data as an input. The interaction between root and the soil is given by:

$$EI\left(\frac{d^4u}{dz^4}\right) - T_z\left(\frac{d^2u}{dz^2}\right) = qd \tag{10.15}$$

Where, EI is the flexural rigidity of the roots; u is displacement; q is the soil reaction; d is the average diameter of the root. At limiting point (yielding of soil), q becomes equal to the bearing pressure at the yield of soil (q_y). Passive failure occurs in the soil once the yield pressure is reached. With an increase in pullout displacement, the passive failure zone increases until failure occurs due to tension or pullout. The Eq. 10.15 can be solved by considering the root as a flexible cable or a beam/ pile. The root is considered as a flexible cable when the component of tension in upward direction is greater than the flexural rigidity (i.e. $T_z > EI$) whereas, it is considered as a beam or pile when T_z is smaller than EI. Thus the equation can be solved using the flexible cable solution or the pile solution [27].

The above-described method can also be used to evaluate the behavior of live poles installed in the slope. The failure of the live poles occurs in a manner similar to the piles such as in flow mode, short pile mode or long pile mode as shown in Fig. 10.7 [88–91]. Figure 10.7a shows the flow mode of failure in which the soil reaction in soil above the shear plane at a depth 'z' becomes equal to the bearing pressure at yield. This causes the soil to flow around the pole. Figure 10.7b shows the long pile mode of failure which takes place due to bending. Figure 10.7c shows the short pile failure mode, which occurs in the soil below the failure plane. The failure can also occur due to the shear failure of pole at the failure plane [92]. The development of soil pressure (q) and bending moment (M) along the length of the idealized root are also shown in the figure.

Another approach is the fiber bundle method which is based on the progressive mode of failure of roots [84]. In this method, equal stress is applied on all the roots such that the roots with large diameter share a high proportion of load. If the load on

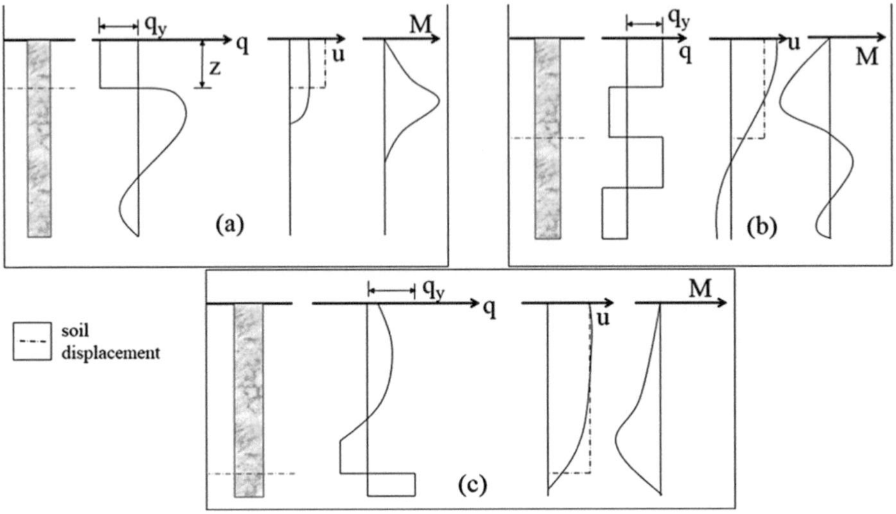

Fig. 10.7 Failure modes of live poles: (**a**) Flow mode, (**b**) Long-pile mode, (**c**) Short-pile mode [27]

a particular root exceeds its strength (leading to breaking of root), the calculation is continued for unbroken roots while the broken root is removed from the analysis. Mickovski et al. [10] used similar method in which they evaluated the minimum tensile stress required to break the weakest root. This stress was then multiplied to the area of all the unbroken roots available at the failure plane. This process was repeated until all the roots break [93]. The maximum value of the stress at failure gives the value of peak-root reinforcement, which can further be used for the slope stability analysis. However, the results obtained using this method contradicted the experimental results as the actual failure occurred due to the stretching of large diameter roots (rather than breaking) but the model predicted the breaking of large diameter roots (as they possess lesser strength as compared to the small diameter fibers). Therefore the stress can be replaced by the stiffness of roots as the governing parameter to simulate the actual failure mechanism.

The behavior of roots in the soil can also be studied using finite element analysis (FEM). The fundamental equation used in FEM is:

$$\bar{\bar{K}}\,\bar{\delta} = \bar{R} - \overline{F_p} - \overline{F_0} \tag{10.16}$$

Where $\bar{\bar{K}}$ is the stiffness matrix; $\bar{\delta}$ is displacement vector; $\bar{R}$ is nodal force vector; $\overline{F_p}$ is boundary load vector; $\overline{F_0}$ is initial load vector. Different elements can be used to represent the roots (beam element) and the root-soil interface (slide-line contact element) [94, 95]. The behavior of root can be taken as elastic upto the yield and the soil can be modeled using Drucker–Prager model [96] or the Cap model [97]. These models can only be used after validation with the published literature. The factor of safety in FEM analysis can be computed using a soil strength reduction approach which involves a continuous reduction in soil shear strength till failure [44, 98–100]. This method is the most accurate method available, however, the only limitation is that it requires a large amount of input data (material properties and root architecture). Nevertheless, yet another approach to model the soil-root interaction is based on the composite material behavior. In this approach, the roots are considered as fibers incorporated in a matrix of soil. A few authors tried to model the suction generated by the roots of the plants. They incorporated the complex soil-plant-atmosphere interactions to quantify the pore pressure dissipation and matric suction produced by the plants in soil [101].

Root architecture plays a very important role in the root-soil interaction mechanism [102–104]. The role of root architecture in soil-root interaction can be modeled using FEM. Different idealized root structures are shown in Fig. 10.8. Previously, 2-D analyses were conducted assuming a homogenous distribution of biomass throughout the slope [105]. However, in the actual scenario, the biomass distribution is heterogeneous and a 3-D analysis is required to incorporate the heterogeneity [44]. Through FEM analysis, it was found out that the pattern of plants (inter-plant spacing along the slope direction etc.) within a slope and root morphology affect the stability of soil slope and the volume of mobilized soil to a large extent [44]. Instead of pre-determining the root architecture, different softwares are available nowadays to simulate the propagation of roots at different environmental conditions [46, 104, 106]. These softwares simulate the branching as well as growth of roots using special functions derived from the experimental data [104].

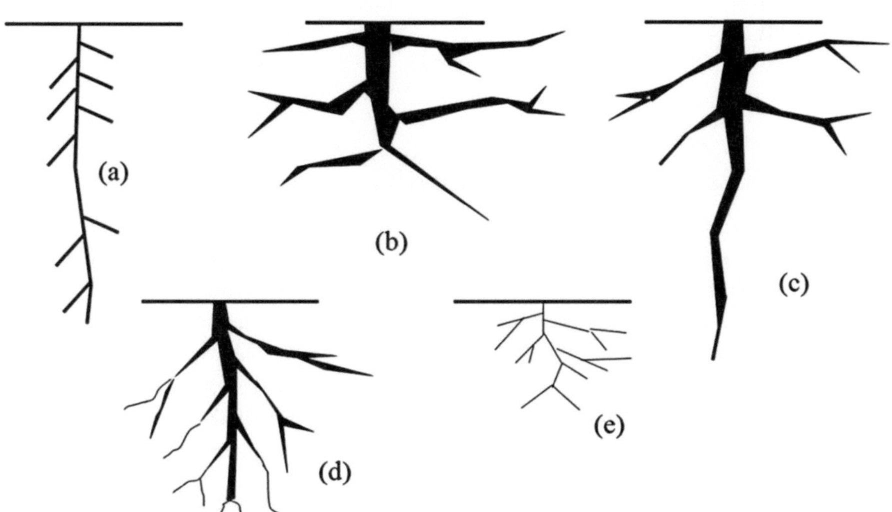

Fig. 10.8 Different idealized root structures of the plant: (**a**) V-type, (**b**) H-type, (**c**) VH-type, (**d**) R type, (e) M-type [4]

10.7 Stability Analysis with Root System

10.7.1 Infinite Slope Stability Analysis

To include the effect of soil reinforcement by the roots (of vegetation) on the stability of soil slope, a simple method proposed by Barker [107] may be used. Barker [107] presented the stability of an infinite slope considering the stabilizing and destabilizing effects of the plant roots on the slope. This method can be applied to slopes where the slip surfaces are parallel to the ground surface and the depth to length ratio of the sliding mass is very small. Infinite slope stability analysis determines the stability in terms of factor of safety, which is defined as the ratio of the stabilizing to the destabilizing forces acting on the element (computed using the force equilibrium approach). The factor of safety (FS) for an infinite slope with no vertical surcharge pressure, can be determined by the following set of equations:

$$FS = A\frac{\tan\varphi}{\tan\beta} + B\frac{(C + S_r)}{\gamma H} \tag{10.17}$$

$$A = \frac{(1 - R_u)}{cos^2\beta} \tag{10.18}$$

$$B = \frac{1}{\cos\beta\sin\beta} \tag{10.19}$$

$$R_u = \frac{\gamma_w}{(\gamma + \gamma\tan\beta\tan\alpha')} \tag{10.20}$$

Where, β = slope angle of natural ground; α' = seepage angle; φ = angle of internal friction; C = soil cohesion; S_r = root adhesion determined from Eqs. 10.7, 10.10–10.11; ⍰ = unit weight of soil; $⍰_w$ = unit weight of water; H = thickness of sliding mass.

10.7.2 Finite Slope Stability Analysis

The stability of a finite slope (cut, fill, embankment, excavation etc.) reinforced with plants can be assessed with a simple limit equilibrium approach using the method of slices. However, the method proposed by Greenwood [108] can be used to determine the stability of a finite slope reinforced with plant roots. This method takes into account both the mechanical and hydrological components of the plant root reinforcement to evaluate the factor of safety (FS). The factor of safety (FS) using the method of slices is given by:

$$FS = \frac{\sum [C'L + (W\cos\alpha - UL - (U_2 - U_1)\sin\alpha)\tan\varphi]}{\sum W\sin\alpha} \tag{10.20}$$

Greenwood [108] modified the Eq. 10.20 by incorporating the mechanical and hydrological components of root reinforcement. The factor of safety using this modified method is given by:

$$FS = \frac{\left(\begin{array}{c}\sum (C' + C_r)L + \{(W + W_v)\cos\alpha - (U + \Delta U_v)L- \\ ((U_2 + \Delta U_2) - (U_1 + \Delta U_1))\sin\alpha - D_w \sin(\alpha - \beta) + T\sin\theta\}tan\varphi\end{array}\right)}{\sum[(W + \Delta W_v)\sin\alpha + D_w\cos(\alpha - \beta) - T\cos\theta]} \tag{10.21}$$

Where, C' is the effective cohesion at the base of slice; L is the length of the base of slice; W is the weight of soil; α is the inclination of base of slice to horizontal; β is the slope angle; φ is effective angle of friction at the base of slice; U is the water pressure on the base of slice; U_1 and U_2 are interslice water forces; C_r is the enhanced cohesion due to the roots; W_v is the weight of vegetation; ΔU_v is the change in water pressure due to vegetation; ΔU_1 and ΔU_2 are changes in interslice water forces due to vegetation; D_w is the wind force; T is the tensile force on roots and θ is the angle of roots to slip surface.

For more realistic analysis, a computer-based finite element program can be used which can model the complex geometry and the behavior of soil and root system and evaluate the stability of the soil slope.

10.8 Conclusions

The accurate evaluation of the strength parameters of the soil-root matrix is essential to design the bioengineering methods effectively for the soil slope stabilization. These strength parameters depend on numerous factors, therefore physical and numerical models have been developed by several researchers to evaluate them as accurately as possible and incorporate them in the slope stability analysis. The accurate evaluation of these parameters increases the accuracy of the slope stability analysis and hence, optimizes the design of the bioengineering method. The present study provides a brief review of the application of bioengineering techniques for soil slope stabilization. The various advantages, limitations, challenges and different methods of application of bioengineering for soil slope stabilization have been discussed. Different physical and numerical models for evaluating the effect of root reinforcement effect on the soil strength have been described briefly. The different mechanisms of root reinforcement along with the factors affecting them have also been discussed. The following conclusions may be drawn from the present study

- Bioengineering can be used effectively for shallow soil slope stabilization or erosion control.
- The Design of bioengineering measures for a particular project is highly site specific and depends on the environmental conditions, properties of slope, nature of project and plant species. Sometimes, the combination of bioengineering with

other slope stabilization measures can be more effective in controlling the slope instability problems than bioengineering alone.

- The Accurate evaluation of the composite strength parameters for a soil-root system is the most important step for a safe and satisfactory design of bioengineering project. However, it is a tedious and challenging task as the composite strength parameters of the soil-root system depend on a number of factors including the soil strength, root geometry and branching, type of root, root area ratio, strength and failure mechanism.
- Numerous techniques have been developed to determine the composite strength parameters of the soil-root system. These include field pullout tests, laboratory pullout tests, triaxial and direct shear tests. Along with the physical model tests, several numerical models have been developed to investigate the role of plants in enhancing the soil slope stability. Moreover, advanced computer-based finite element programs can be used to realistically model the soil and complex geometry of the root system.
- The Factor of safety for the root reinforced finite and infinite slope can be determined by taking into consideration the extra stabilizing and destabilizing forces due to the vegetation or plantation on the slope.

References

1. Coppin NJ, Richards IG (2007) Use of vegetation in civil engineering. Construction Industry Research and Information Association, London
2. Gray DH, Sotir RB (1995) Biotechnical stabilization of steepened slopes, Transportation research record. Transportation Research Board, National Research Council, Washington, DC, pp 23–29
3. Schiechtl HM, Stern R (1996) Ground bioengineering techniques for slope protection and erosion control. Blackwell Science Ltd., London
4. Morgan RP, Rickson RJ (2003) Slope stabilization and erosion control: a bioengineering approach. Taylor & Francis, Madras
5. Phillips C, Marden M (2006) Use of plants for ground bioengineering and erosion sediment control in New Zealand. In: Proceedings of soil water. Too good to lose. Joint annual conference NSW Stormwater Industry Association and the International Erosion Control Association, 27 June, 2006
6. Schiechtl HM (1980) Bioengineering for land reclamation and conservation. University of Alberta Press, Edmonton
7. Gray DH, Leiser AT (1982) Biotechnical slope protection and erosion control. Van Nostrand Reinhold Company, New York
8. Donat M (1995) Bioengineering techniques for stream-bank restoration. A review of central European practices. Watershed restoration program. Ministry of Environment, Lands and Parks, and Ministry of Forests, Vancouver
9. Howell JH, Sandhu SC, Vyas N et al (2006) Introducing bio-engineering to the road network of Himachal Pradesh. J Indian Roads Congr 67(3):84. W07025 (1–11)
10. Mickovski SB, Hallett PD, Bransby MF et al (2009) Mechanical reinforcement of soil by willow roots: impacts of root properties and root failure mechanism. Soil Sci Soc Am J 73 (4):1276–1285
11. Freer R (1991) Bio-engineering: the use of vegetation in civil engineering. Constr Build Mater 5(1):23–26

12. Mafian S, Huat BBK, Ghiasi V (2009) Evaluation on root theories and root strength properties in slope stability. Eur J Sci Res 30(4):594–607
13. Stokes A, Mickovski SB, Thomas BR (2004) Eco-engineering for the long-term protection of unstable slopes in Europe: developing management strategies for use in legislation. Landslides: Eval Stabilization 2:1685–1690
14. Mickovski SB, Van Beek LPH (2006) A decision support system for the evaluation of eco-engineering strategies for slope protection. Geotech Geol Eng 24(3):483–498
15. Norris JE, Greenwood JR (2006) Assessing the role of vegetation on soil slopes in urban areas. In: Proceedings of the 10th IAEG International Congress, Nottingham, September, 2006
16. Albaladejo-Montoro J, Alvarez RJ, Querejeta J et al (2000) Three hydro-seeding revegetation techniques for soil erosion control on anthropic steep slopes. Land Degrad Dev 11(4):315–325
17. Gao GJ, Yuan JG, Han RH et al (2007) Characteristics of the optimum combination of synthetic soils by plant and soil properties used for rock slope restoration. Ecol Eng 30 (4):303–311
18. Preti F, Giadrossich F (2009) Root reinforcement and slope bioengineering stabilization by Spanish Broom (Spartium junceum L.) Hydrol Earth Syst Sci 13(9):1713
19. Greenwood JR, Norris JE, Wint J (2004) Assessing the contribution of vegetation to slope stability. Proc ICE-Geotech Eng 157(4):199–207
20. Gray DH, Sotir RB (1996) Biotechnical and soil bioengineering slope stabilization: a practical guide for erosion control. Wiley, New York
21. Price DG (2008) Engineering geology: principles and practice. Springer, New York
22. Simon K, Steinemann A (2000) Soil bioengineering: challenges for planning and engineering. J Urban Plan Dev 126(2):89–102
23. Van Kraayenoord CWS, Hathaway RL (1986) Plant materials handbook for soil conservation. Volume 1: principles and practices. Water and soil miscellaneous publication (93)
24. Schiechtl HM, Stern R (1997) Water bioengineering techniques: for watercourse, bank and shoreline protection. Blackwell Science, Oxford
25. Georgi NJ, Stathakopoulos I (2006) Bioengineering techniques for soil erosion protection and slope stabilization. Neapolis University Pafos, Cyprus
26. Choudhury PK, Das A, Goswami DN et al (2008) Bio-engineering approach with jute geotextile for slope stabilization. In: Geosynthetics in civil and environmental engineering. Springer, Berlin, pp 863–867
27. Wu TH (2013) Root reinforcement of soil: review of analytical models, test results, and applications to design. Can Geotech J 50(3):259–274
28. Veylon G, Ghestem M, Stokes A et al (2015) Quantification of mechanical and hydric components of soil reinforcement by plant roots. Can Geotech J 52(11):1839–1849
29. Burroughs ER, Thomas BR (1977) Declining root strength in Douglas-fir after felling as a factor in slope stability. USDA For. Serv Res INT-190
30. Wu TH, McKinnell WP, Swanston DN (1979) Strength of tree roots and landslides on Prince of Wales Island, Alaska. Can Geotech J 16(1):19–33
31. Wu TH (1984) Effect of vegetation on slope stability, Transportation Research Record (965). Transportation Research Board, Washington, DC, p 3764
32. Wu TH (1984) Soil movements on permafrost slopes near Fairbanks, Alaska. Can Geotech J 21(4):699–709
33. Sidle RC, Pearce AJ, O'Loughlin CL (1985) Hillslope stability and land use. American Geophysical Union, Washington, DC
34. Sidle RC (1992) A theoretical model of the effects of timber harvesting on slope stability. Water Resour Res 28(7):1897–1910
35. Wu W, Sidle RC (1995) A distributed slope stability model for steep forested basins. Water Resour Res 31(8):2097–2110
36. Abernethy B, Rutherfurd ID (2001) The distribution and strength of riparian tree roots in relation to riverbank reinforcement. Hydrol Process 15(1):63–79
37. Greenwood JR, Morgan RP, Coppin NJ et al (2001) Bioengineering: a field trial at Longham Wood Cutting. CIRIA, London

38. Bransby MF, Davies MCR, Mickovski SB et al (2006) Stabilisation of slopes by vegetation reinforcement. In: Physical modelling in geotechnics, two volume set. Proceedings of the sixth international conference on physical modelling in geotechnics. 6th ICPMG'06, 4–6 August 2006. CRC Press, Hong Kong
39. Stokes A, Ghani MA, Salin F et al (2007) Root morphology and strain distribution during tree failure on mountain slopes. In: Eco-and ground bio-engineering: the use of vegetation to improve slope stability. Springer, Dordrecht, pp 165–173
40. Wu TH (2007) Root reinforcement: analyses and experiments. In: Eco-and ground bio-engineering: the use of vegetation to improve slope stability. Springer, Dordrecht, pp 21–30
41. Mickovski SB et al (2008) Slope stability and erosion control: ecotechnological solutions. Springer, Dordrecht
42. Tang J, Pilesjö P, Miller PA et al (2014) Incorporating topographic indices into dynamic ecosystem modelling using LPJ-GUESS. Ecohydrology 7(4):1147–1162
43. Tang J, Miller PA, Crill PM et al (2015) Investigating the influence of two different flow routing algorithms on soil–water–vegetation interactions using the dynamic ecosystem model LPJ-GUESS. Ecohydrology 8(4):570–583
44. Temgoua AGT, Kokutse NK, Kavazović Z (2016) Influence of forest stands and root morphologies on hillslope stability. Ecol Eng 95:622–634
45. Hoskins CG, Rice P (1992) Vegetation and embankment dams. In: Proceedings BDS conference, Stirling, pp 329–338
46. Dupuy L, Fourcaud T, Stokes A (2005) A numerical investigation into factors affecting the anchorage of roots in tension. Eur J Soil Sci 56(3):319–327
47. Reubens B, Poesen J, Danjon F et al (2007) The role of fine and coarse roots in shallow slope stability and soil erosion control with a focus on root system architecture: a review. Trees 21 (4):385–402
48. Genet M, Kokutse N, Stokes A et al (2008) Root reinforcement in plantations of Cryptomeria japonica D. Don: effect of tree age and stand structure on slope stability. For Ecol Manag 256 (8):1517–1526
49. Pollen N (2007) Temporal and spatial variability in root reinforcement of streambanks: accounting for soil shear strength and moisture. Catena 69(3):197–205
50. Mao Z, Saint-André L, Genet M et al (2012) Engineering ecological protection against landslides in diverse mountain forests: choosing cohesion models. Ecol Eng 45:55–69
51. Mao Z, Jourdan C, Bonis ML et al (2013) Modelling root demography in heterogeneous mountain forests and applications for slope stability analysis. Plant Soil 363(1–2):357–382
52. Riestenberg MM (1987) Anchoring of thin colluvium on hillslopes by roots of sugar maple and white ash, dissertation, University of Cincinnati
53. Genet M, Stokes A, Salin F et al (2005) The influence of cellulose content on tensile strength in tree roots. Plant Soil 278(1):1–9
54. Greenway DR (1987) Vegetation and slope stability. In: Anderson MG, Richards KS (eds) Slope stability. Wiley, Chichester, pp 187–230
55. Simon A, Collison AJ (2002) Quantifying the mechanical and hydrologic effects of riparian vegetation on streambank stability. Earth Surf Process Landf 27(5):527–546
56. Pollen-Bankhead N, Simon A (2010) Hydrologic and hydraulic effects of riparian root networks on streambank stability: is mechanical root-reinforcement the whole story? Geomorphology 116(3):353–362
57. Nilaweera NS (1994) Influence of hardwood roots on soil shear strength and slope stability in Southern Thailand. Ph. D. Dissertation, Asian Institute of Technology, Bangkok
58. Norris JE, Greenwood JR (2000) In situ shear and pull out testing to demonstrate the enhanced shear strength of root reinforced soil. In: Proceedings of the 8th International Symposium on Landslides, Cardiff, pp/ 1123–1128
59. Norris JE, Greenwood JR (2003) In-situ shear box and root pull-out apparatus for measuring the reinforcing effects of vegetation. In: Myrvoll F (ed) Field measurements in geomechanics, Oslo. Swets and Zeitlinger, Lisse, pp 593–597

60. Vogt KA, Persson H (1991) Root methods. In: Techniques and approaches in forest tree ecophysiology. 477–502
61. Schmid I, Kazda M (2002) Root distribution of Norway spruce in monospecific and mixed stands on different soils. For Ecol Manag 159(1):37–47
62. Bischetti GB, Chiaradia EA, Simonato T et al (2005) Root strength and root area ratio of forest species in Lombardy (Northern Italy). Plant Soil 278(1):11–22
63. Tosi M (2007) Root tensile strength relationships and their slope stability implications of three shrub species in the Northern Apennines (Italy). Geomorphology 87(4):268–283
64. Böhm W (2012) Methods of studying root systems, vol 33. Springer, Berlin
65. Laclau JP, Arnaud M, Bouillet JP et al (2001) Spatial distribution of Eucalyptus roots in a deep sandy soil in the Congo: relationships with the ability of the stand to take up water and nutrients. Tree Physiol 21(2):129–136
66. Waldron LJ, Dakessian S (1982) Effect of grass, legume, and tree roots on soil shearing resistance. Soil Sci Soc Am J 46:894–899
67. Abe K, Ziemer RR (1991) Effect of tree roots on a shear zone: modeling reinforced shear stress. Can J For Res 21:1012–1019
68. Operstein V, Frydman S (2000) The influence of vegetation on soil strength. Proc Inst Civ Eng-Ground Improv 4(2):81–89
69. Huat BBK, Ali FH, Maail S (2005) The effect of natural fiber on the shear strength of soil. Am J Appl Sci 9–13
70. Ali FH, Osman N (2007) Soil-roots composite: correlation between shear strength and some plant properties. Electron J Geotech Eng 12
71. Graf F, Frei M, Böll A (2009) Effects of vegetation on the angle of internal friction of a moraine. For Snow Landsc Res 82(1):61–77
72. Mickovski SB, Bransby MF, Bengough AG et al (2010) Resistance of simple plant root systems to uplift loads. Can Geotech J 47(1):78–95
73. Gray DH, Ohashi H (1983) Mechanics of fiber reinforcement in sand. J Geotech Eng 109:335–353
74. Wu TH, McOmber RM, Erb RT et al (1988) Study of soil-root interaction. J Geotech Eng 114:1351–1375
75. Shewbridge SE, Sitar N (1989) Deformation characteristics of reinforced soil in direct shear. J Geotech Eng 115:1134–1147
76. Shewbridge SE, Sitar N (1996) Formation of shear zones in reinforced sand. J Geotech Eng 122:873–885
77. Stauffer SD, Holtz RD (1995) Stress-strain and strength behavior of staple fiber and continuous filament-reinforced sand. Transp Res Rec 1474:82–95
78. Consoli NC, Montardo JP, Prietto PDM et al (2002) Engineering behavior of a sand reinforced with plastic waste. J Geotech Geoenviron 128(6):462–472
79. Yetimoglu T, Salbas O (2003) A study on shear strength of sands reinforced with randomly distributed discrete fibers. Geotext Geomembr 21(2):103–110
80. Commandeur PR, Pyles MR (1991) Modulus of elasticity and tensile strength of Douglas-fir roots. Can J For Res 21(1):48–52
81. Tang C, Shi B, Gao W et al (2007) Strength and mechanical behavoir of short polypropylene fiber reinforced and cement stabilized clayey soil. Geotext Geomembr 25:194–202
82. Wu TH (1976) Investigation of landslides on prince of Wales Island. Ohio State University, Alaska
83. Waldron LJ (1977) The shear resistance of root-permeated homogeneous and stratified soil. Soil Sci Soc Am J 41(5):843–849
84. Pollen N, Simon A (2005) Estimating the mechanical effects of riparian vegetation on stream bank stability using a fiber bundle model. Water Resour Res 41(7):W07025. 1–11
85. De Baets S, Poesen J, Reubens B et al (2008) Root tensile strength and root distribution of typical Mediterranean plant species and their contribution to soil shear strength. Plant Soil 305 (1–2):207–226

86. Preti F, Dani A, Laio F (2010) Root profile assessment by means of hydrological, pedological and above-ground vegetation information for bio-engineering purposes. Ecol Eng 36 (3):305–316
87. Schwarz M, Preti F, Giadrossich F et al (2010) Quantifying the role of vegetation in slope stability: a case study in Tuscany (Italy). Ecol Eng 36(3):285–291
88. Broms BB (1964) Lateral resistance of piles in cohesive soils. J Soil Mech Found Div 90 (2):27–64
89. Broms BB (1964) Lateral resistance of piles in cohesionless soils. J Soil Mech Found Div 90 (3):123–158
90. Viggiani C (1981) Ultimate lateral load on piles used to stabilize landslides. In: Proceedings of the 10th international conference on soil mechanics and foundation engineering, vol 3, pp 555–560
91. Poulos HG (1995) Design of reinforcing piles to increase slope stability. Can Geotech J 32 (5):808–818
92. Steele DP, MacNeil D, Barker D et al (2004) The use of live willow poles for stabilising highway slopes, TRL REPORT TRL 619. TRL Limited, Crowthorne
93. Hidalgo RC, Kun F, Herrmann HJ (2001) Bursts in a fiber bundle model with continuous damage. Phys Rev E 64(6):066122
94. El-Khouly MA (1995) Analysis of soil-root interaction. Ph.D. dissertation, Ohio State University, Columbus
95. Frydman S, Operstein V (2001) Numerical simulation of direct shear of root reinforced soil. Proc Inst Civ Eng-Ground Improv 5(1):41–48
96. Drucker DC, Gibson RE, Henkel DJ (1957) Soil mechanics and work hardening theories of plasticity, Transactions of the American Society of Civil Engineers 122. ASCE, New York
97. DiMaggio FL, Sandler IS (1971) Material model for granular soils. J Eng Mech 97:935–950
98. Brinkgreve RBJ, Vermeer PA (1999) Plaxis: finite element code for soil and rock analyses: version 7: [user's guide]. Balkema, Rotterdam
99. Dawson EM, Roth WH, Drescher A (1999) Slope stability analysis by strength reduction. Geotechnique 49(6):835–840
100. Krahn J (2003) The 2001 RM Hardy lecture: the limits of limit equilibrium analyses. Can Geotech J 40(3):643–660
101. Indraratna B, Salim W, Rujikiatkamjorn C (2011) Advanced rail geotechnology: ballasted track. CRC Press, London
102. Ennos AR (1989) The mechanics of anchorage in seedlings of sunflower, Helianthus annuus L. New Phytol 113(2):185–192
103. Niklas KJ, Molina-Freaner F, Tinoco-Ojanguren C et al (2002) The biomechanics of Pachycereus pringlei root systems. Am J Bot 89(1):12–21
104. Dupuy L, Fourcaud T, Stokes A (2005) A numerical investigation into the influence of soil type and root architecture on tree anchorage. Plant Soil 278(1):119–134
105. Chok Y, Kaggwa G, Jaksa M et al (2004) Modelling the effects of vegetation on stability of slopes. In: Proceedings of 9th Australia New Zealand conference on geomechanics, Auckland, pp 391–397
106. Dunbabin VM, Postma JA, Schnepf A et al (2013) Modelling root–soil interactions using three–dimensional models of root growth, architecture and function. Plant Soil 372 (1–2):93–124
107. Barker DH (1986) Enhancement of slope stability by vegetation. Ground Eng 19(3):11–15
108. Greenwood JR (2006) SLIP4EX–A program for routine slope stability analysis to include the effects of vegetation, reinforcement and hydrological changes. Geotech Geol Eng 24 (3):449–465

Part IV
Landslide Numerical Modelling Techniques

Chapter 11
Optimization Techniques in Slope Stability Analysis Methods

Koushik Pandit, Shantanu Sarkar, and Mahesh Sharma

Abstract The estimation of factor of safety (FoS) or design reliability of slopes is a pre-requisite for an efficient and safe application of landslide mitigation measures for ensuring long-term slope stability. The evaluation of stability of slopes, especially in a hilly region with wide variations in its geological formation is an upfront challenging task for geologists as well as for geotechnical engineers and till date, has been tackled using several optimization algorithms and slope stability analysis methods. The purpose of this book chapter is to present an up-to-date along with an overall review of the slope stability analysis methods which have used different optimization algorithms for deterministic and probabilistic or stochastic evaluation of FoS or reliability index, respectively, including some case studies from published literatures. This review shows that the FoS or reliability of slopes obtained by applying the commonly established analysis methods coupled with optimization algorithms, using both the deterministic and the probabilistic approaches, may vary in their values as well as in their computational effort and errors encountered.

Keywords Slope stability · Factor of safety · Reliability index · Optimization techniques · Stochastic analysis · Heuristic algorithms

11.1 Introduction

The slope stability problem poses critical questions in the event of a landslide and various civil engineering applications, such as a road-cut or a building construction in a hilly terrain or an embankment or an earth dam or any other excavation on slopes. The stability crisis comes up whenever the old or new slope profile becomes unable to retain its natural or engineered shape under various geological, geotechnical and loading conditions. Gravity plays the most prominent loading condition in

K. Pandit (✉) · S. Sarkar · M. Sharma
Geotechnical Engineering Group, CSIR-Central Building Research Institute, Roorkee, Uttarakhand, India

S. P. Pradhan et al. (eds.), *Landslides: Theory, Practice and Modelling*, Advances in Natural and Technological Hazards Research 50,
https://doi.org/10.1007/978-3-319-77377-3_11

case of a slope, apart from other forces like seismic (in the event of an earthquake) and hydrostatic (in the event of massive rainfall due to water seepage through soil pores or cracks in a rock mass) ones. These controlling factors for a slope's overall stability may cause catastrophic repercussions in a landslide incidence in the form of considerable loss of human and animal lives, and, damage to communication routes, other man-made constructions, human settlements and natural resources like agricultural fields and forest lands. A world-wide scenario of landslide disaster is presented in Table 11.1. During the decade (2006–2015), the Asian countries have globally accounted for (i) 40% of the total number of disasters, but for 67% of landslides; (ii) 51% of the total number of people killed by disasters, but for 77% of deaths from landslides; (iii) the highest (79%) proportion of people in the world affected by disasters, but for 95% of people affected by landslides. All these data have been extracted from [1].

In India, the Himalayan region (North-West and North-East States) and the Western Ghats are the two main regions of high vulnerability in terms of landslide proneness (Fig. 11.1, [2]). According to [3], approximately 30% of the world's landslides occur in the Himalayan ranges (~3400 km stretch), the youngest and most active mountain system in the world. There are recorded occurrences of two landslides (on an average) per sq. km in North Sikkim and Garhwal regions in the Himalayas; whereas, the mean rate of land loss is of almost 120 m per km per year and annual soil loss is about 2500 tons per km^2 [3, 4]. Each year, India faces a monetary loss of roughly 100 crores (1 billion) to Rs. 150 crores (1.5 billion) at the current prices [4].

The above discussion highlights the need and importance of developing an efficient assessment methodology for slope stability analysis. Depending upon the working principles, slope stability analysis methods can be grouped as deterministic and probabilistic approaches. The basic differences in these two approaches are mentioned in Table 11.2 below.

Optimization algorithms are useful and powerful techniques in both deterministic and probabilistic (or reliability or stochastic) slope stability analysis. Locating the critical slip surface of a slope with the minimum factor of safety for deterministic approach, or with the minimum reliability index for probabilistic approach can be characterized as a difficult N-P (Non-deterministic Polynomial-time) type optimization problem [5]. With the advancement of computing power and human understanding, diversified optimization techniques (algorithms) have been developed by researchers and put into solving difficult problems in engineering fields. Some of these algorithms like the linear programming (LP), non-linear programming (NLP), dynamic programming (DP), sequential quadratic programming (SQP), interior point method (IPM) and intelligent heuristic algorithms like the genetic algorithm (GA), simulated annealing (SA), swarm intelligence (SI) and harmonic search optimization (HSO) have garnered increased attention of researchers in solving slope stability analysis problems due to their inherent advantages. Optimization problems may be deterministic or stochastic depending upon the approach (Table 11.2) of slope stability analysis adopted by the user. Some quantifiable differences in between deterministic and stochastic optimization techniques are tabulated below (Table 11.3).

There are three main sections in this present chapter. While, Sect. 11.2 gives a brief outline of the mathematical background of various optimization algorithms,

Table 11.1 The world-wide scenario of landslide disaster during the years 2006–2015

Description	2006	2007	2008	2009	2010	2011	2012	2013	2014	2015	Total
Reported landslides	20	10	12	28	32	17	13	11	15	19	177
People killed	1638	271	504	649	3427	309	501	235	943	1000	9477
People affected (in thousands)	432	9	5	44	2460	7	4	1	19	50	3031
Estimated damage (in millions of US $, 2015 prices)	49	ND	ND	175	1430	ND	ND	ND	273	8	1935

Data extracted from World Disasters Report 2016 [1]
ND No Data Available

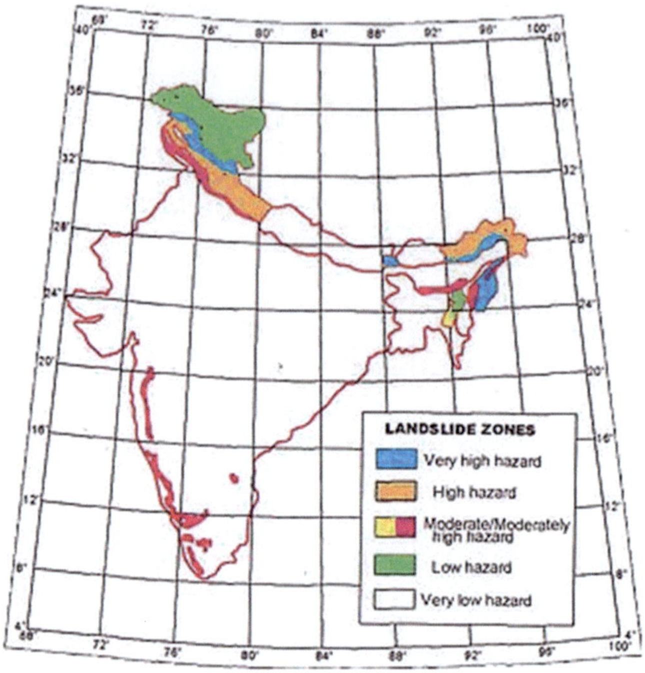

Fig. 11.1 A general landslide hazard map of India [2]

Table 11.2 Differences in working philosophy of deterministic and probabilistic approaches in slope stability analysis

Observations	Deterministic approach	Probabilistic approach
Pre-requisites for applying the approach in slope stability analysis	Entry and exit co-ordinates of all the probable slip surfaces in a slope have to be predefined and given as an input in a user written code or commercially available software based on limit equilibrium method	The user-written code searches for a critical slip surface location based on theory of probability. Hence the probability distribution functions for all the variables have to be pre-known
Optimization function	The expression for factor of safety (FoS) of a slope	The expression for reliability index (β) of a slope
Optimization goal	Factor of safety (FoS) of the slope along all the identified slip surfaces are minimized	The reliability index (β) of the slope is minimized
	The critical slip surface will have the minimum FoS value	The critical probabilistic slip surface will have the minimum reliability index (β_{min}) or maximum probability of failure
Computational effort	A general shape critical slip surface is presumed and geotechnical parameters (or variables) involved in the calculations are represented with certainty by a single value (statistical mean value of each variable)	The variability of the factors governing slope stability is taken into account in the design process by their statistical mean and standard deviation values, and the concept of probability of failure is employed

Table 11.3 Key differences in deterministic and stochastic optimization techniques for slope stability analysis

Observations	Deterministic optimization technique	Probabilistic optimization technique
Optimization search techniques	Always arrive at the same final solution through the same sequence of solutions, although they may depend on the initial solution	The solutions are considered stochastic and their orders are different depending on random variables
Working philosophy	These methods are based on gradient information of the objective function and constraints. However, the acquisition of gradient information can be costly or even altogether impossible to obtain	These methods are derivative-free, and are applicable to any optimization problem regardless of the linearity or non-linearity of its objective function and constraints
Error in final solution	These methods follow deterministic transition rules, and hence are entitled to be trapped by local minima of an attained solution in any iteration. This leads to an erroneous final solution	These methods use stochastic techniques and include randomness in moving from one solution to the next. This phenomenon gives an added advantage to these methods in avoiding local minima which helps in global search for a critical slip surface

Sect. 11.3 deals with application of those optimization algorithms which have found popularity among the research fraternity for solving slope stability analysis problems in practice. Finally, a synopsis of all the popular optimization techniques along with their advantages and limitations in application point of view in slope stability analysis problems has been presented in Sect. 11.4.

11.2 Optimization Techniques: Mathematical Background for Slope Stability Analysis

The main objective of any optimization algorithm is to find the conditions that give the maximum or minimum value of a function [6]. Hence, the optimization techniques can be applied in slope stability analysis to find the critical failure surface which gives the lowest factor of safety (for deterministic analysis) or the minimum reliability index (for probabilistic analysis). In the following sub-sections, mathematical background behind various popular optimization algorithms for slope stability analysis has been discussed.

11.2.1 Linear Programming (LP)

In a linear programming (LP) optimization problem, a linear objective function subjected to linear constraints is optimized [7]. The general LP problem can be expressed in a standard matrix form as below:

$$\text{Maximize}: f(\boldsymbol{X}) = \boldsymbol{c}^T\boldsymbol{X}, \text{ subject to the constraints}: \boldsymbol{aX} = \boldsymbol{b} \text{ and}, \boldsymbol{X} \geq 0 \tag{11.1}$$

Where,

$$\boldsymbol{X} = \begin{Bmatrix} x_1 \\ x_2 \\ \vdots \\ x_n \end{Bmatrix}, \ \boldsymbol{b} = \begin{Bmatrix} b_1 \\ b_2 \\ \vdots \\ b_m \end{Bmatrix}, \boldsymbol{c} = \begin{Bmatrix} c_1 \\ c_2 \\ \vdots \\ c_n \end{Bmatrix} \text{and}, \boldsymbol{a} = \begin{bmatrix} a_{11} & a_{12} & \dots & a_{1n} \\ a_{21} & a_{22} & \dots & a_{2n} \\ \vdots & \vdots & \vdots & \vdots \\ a_{m1} & a_{m2} & \dots & a_{mn} \end{bmatrix} \tag{11.2}$$

In (11.1) and (11.2), $\boldsymbol{b}_i$, $\boldsymbol{c}_j$ and $\boldsymbol{a}_{ij}(i = 1, 2, \ldots, m; j = 1, 2, \ldots, n)$ are known constants, and $\boldsymbol{x}_j$ are the decision variables.

It is to be noted here that the objective function is to be maximized, not minimized for a LP problem in its standard form. If the problem to be addressed is to minimize the objective function: $f(\boldsymbol{X})$, then it has to be simply converted to maximize the negative of the same objective function: $-f(\boldsymbol{X})$.

For LP problems, involving more than two variables or several constraints, it is suitable to use a solution method which can be written in a computer coding language. One such method for the solution of the LP problems is the Simplex method [8, 9] which is carried out by performing elementary row operations like transformations on a matrix called the Simplex tableau. This tableau consists of the augmented matrix corresponding to the constraint equations along with the coefficients of the objective function. Among several solution methods, the Simplex method remains the most efficient and popular method for solving general LP problems [6].

A major disadvantage of using the original Simplex method is that it requires a large amount of computer storage and time. This limitation can be surpassed by using some other technique like the revised Simplex method [10] which allows for greater computational efficiency by enabling sparse matrix operations (out of scope of this chapter for discussion).

11.2.2 Non-linear Programming (NLP)

In a non-linear programming (NLP) optimization problem, either the objective function and/or some of the constraints are non-linear [11]. For the solution of a constrained NLP optimization problem, one has to do the following:

Find $\boldsymbol{X}$ which minimizes: $f(\boldsymbol{X})$, subject to the constraints:

$$\begin{aligned} g_j(\boldsymbol{X}) \leq 0, j = 1, 2, \ \ldots, m \\ h_k(\boldsymbol{X}) = 0, k = 1, 2, \ \ldots, p \end{aligned} \tag{11.3}$$

For simple optimization problems, the constraints may not have any effect on the optimum (minimum) point, and the constrained minimum solution becomes exactly the same as the unconstrained minimum solution. In those cases, for a point X^* becoming the point at which the objective function $f(X)$ will have relative minimum value, the following necessary (Eq. 11.4) and sufficient (Eq. 11.5) conditions are to be satisfied:

$$\frac{\partial f}{\partial x_i}(X = X^*) = 0, i = 1, 2, \ldots, n \tag{11.4}$$

And, the Hessian matrix,

$$J_{X^*} = [J]_{X^*} = \left[\frac{\partial^2 f}{\partial x_i \partial x_j}(X^*)\right] = positive\ definite \tag{11.5}$$

In practical problems, the constraints will have some effect on the location of the optimum point, X^* and the optimum solution will occur on a constraint boundary. Therefore, in all such cases, the Kuhn – Tucker necessary conditions [12] are imposed.

There are many numerical techniques to get the solution of a constrained NLP problem. Broadly, these methods are either (i) the direct methods, like the random search methods [13], complex method [14], sequential linear programming (SLP) method (also known as the cutting plane method) [15, 16], sequential quadratic programming (SQP) method (also known as the projected Lagrangian method) [17], methods of feasible directions [18], Zoutendijk's method [19], Rosen's gradient projection method [20, 21], generalized reduced gradient method [22] etc., or (ii) the indirect ones, like the transformation of variables technique [23], the interior penalty function methods (also known as barrier methods since the constraint boundaries act as barriers) [24], the exterior penalty function methods [25], the augmented Lagrange multiplier (ALM) method (combines the Lagrange multiplier and the penalty function methods) [26, 27] etc. The basic difference in between the direct and indirect methods is that the constraints are treated in an explicit manner in the direct methods, whereas, the constrained problem is solved as a sequence of unconstrained minimization problems in most of the indirect methods. Among the direct methods, the sequential quadratic programming (SQP) approach is one of the most effective methods for solving the NLP optimization problems associated with slope stability analysis using the upper bound theorem [28].

Numerical methods used for solving NLPs have limited information about the problem. Hence, they can recognize a local maximum or minimum, but fail to easily determine location of the global optimum [29]. Also, it is extremely difficult for even a simple optimization problem to determine the influence of the constraints on the minimum point, in advance. Again, if the objective function is not differentiable, then using Eqs. (11.4) and (11.5), the optimum point cannot be identified.

11.2.3 Dynamic Programming (DP)

Sometimes, in practical optimization problems, the decisions are to be made sequentially at a number of stages. For example, a ground-radar-controlled missile chasing a moving target, or minimizing the construction cost of a water tank, which can be seen as components of a pile or mat foundation system, a set of steel or RCC columns, and an overhead tank, in a series of construction process. These kinds of problems are called sequential decision problems or multistage decision problems. Dynamic programming (DP) technique was originally introduced by Bellman [30] for optimization of such multistage decision problems.

A multistage decision process can be represented schematically as shown in Fig. 11.2. The stages n, n − 1, . . ., i, . . ., 2, 1 are labelled in decreasing order, generally.

For the ith stage, the input state vector is denoted by s_{i+1} and the output state vector as s_i. Now, the output from stage (i + 1) must be fed as the input to the next stage, i. The state transformation (or design equations) and return functions, respectively, can be expressed as:

$$s_i = t_i(s_{i+1}, x_i) \tag{11.6}$$

$$R_i = r_i(s_{i+1}, x_i) \tag{11.7}$$

Here, x_i denotes the vector of decision variables at stage i. The objective of a multistage decision problem is to find $(x_1, x_2, \ldots, x_n)$ so as to optimize some function of the individual stage returns, say, $f(R_1, R_2, \ldots, R_n)$ and satisfy Eqs. (11.6) and (11.7). A multistage decision problem can be solved by dynamic programming (DP) algorithms if the objective function can be represented as the composition of the individual stage returns (Eq. 11.8).

$$f = \sum_{i=1}^{n} R_i = \sum_{i=1}^{n} R_i(x_i, s_{i+1}) \tag{11.8}$$

The DP algorithm optimizes each of these objective functions at every stage, stores the solution at each stage and uses this solution in the next stage for finding the optimal solution. The overall optimal solution is found by combining the optimal

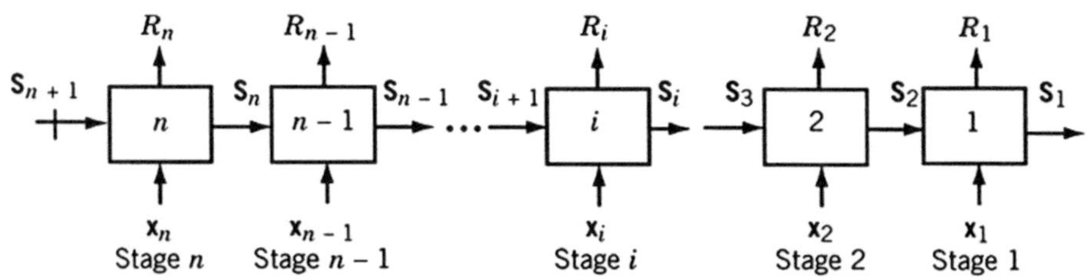

Fig. 11.2 Representation of a multistage decision problem (initial value problem) [6]

solutions for each stage. A linear programming (LP) problem can be formulated as a DP problem by converting the LP problem into a series of multistage decision problem.

When implemented in slope stability analysis problems, DP techniques have two added advantages over conventional approaches like the limit equilibrium method (LEM): (i) DP methods allow consideration of more complex stress-strain relation by always satisfying all the equilibrium conditions, and (ii) DP algorithms can determine arbitrarily shaped (multi-modal) slip surfaces in two or three dimensions, contrary to the LEM which is limited to certain slip surface geometries [31]. DP solutions usually give a lower factor of safety (FoS) values for slopes than those obtained in the LEM outcomes [32].

11.2.4 Interior Point Method (IPM)

Optimization algorithms based on the interior point method (initially suggested by Neumann in 1948 [33], and later improved by Karmakar in 1984 [34]) can solve both linear and nonlinear convex optimization problems. The interior point method (IPM) got its name so since it finds improved search directions for getting the optimum solution point, specifically inside the interior of the feasible region [6]. The Simplex method [8, 9] searches along the boundary of the feasible region, and hence, requires huge computing time for large LP problems. Karmakar's method [34] is fast and can get solutions of LP problems which are beyond the capabilities of the Simplex method.

Karmarkar's method [34] requires the LP problem to be represented in the following form:

$$\text{Minimize}: f(\boldsymbol{X}) = \boldsymbol{c}^T\boldsymbol{X}, \text{subject to the constraints}: \boldsymbol{aX} = \boldsymbol{0}, x_1 + x_2 + \ldots + x_n = 1 \text{ and}, \boldsymbol{X} \geq \boldsymbol{0} \tag{11.9}$$

Where, $\boldsymbol{X} = \{x_1, x_2, \ldots, x_n\}^T$, $\boldsymbol{c} = \{c_1, c_2, \ldots, c_n\}^T$, and $\boldsymbol{a}$ is an m x n matrix. In addition, the optimum value of f must be zero for the problem and usually, an interior feasible starting solution to the Eq. 11.9 for the first iteration is taken as:

$$\boldsymbol{X}^{(1)} = \left\{\frac{1}{n}, \frac{1}{n}, \ldots, \frac{1}{n}\right\}^T \tag{11.10}$$

11.2.5 Heuristic Algorithms

Heuristic algorithms proceed to an optimum solution by trial and error or by loosely defined rules. They are often applied when exact solutions are inevitably computationally expensive, but approximate solutions are sufficient under acceptable error limits [35]. These modern optimization methods, developed in recent times, have materialized as popular choice of methods for solving complex problems in a much lesser time than the classical methods [6]. Usually, heuristic methods require only the function values, and not their derivatives. Following are some of the heuristic methods which have gained huge popularity among researchers in recent times for solving slope stability problems.

11.2.5.1 Genetic Algorithm (GA)

Genetic algorithm (GA) [36, 37] is a stochastic (random) global search and optimization method which is based on the Darwin's theory of "Survival of the fittest". GA employs a genetic search procedure which uses the basic elements of natural genetics (i.e., reproduction, crossover, and mutation) and natural selection. GA is naturally suitable for solving unconstrained maximization problems since it tries to maximize a usually non-negative fitness function, $F(\mathbf{X})$, which is equivalent to the objective function, $f(\mathbf{X})$ of an unconstrained maximization problem. The following transformation equation converts an unconstrained minimization problem into an equivalent maximization problem by the fitness function:

$$F(\mathbf{X}) = \frac{1}{1 + f(\mathbf{X})} \tag{11.11}$$

The steps involved in a GA for obtaining optimization solutions to a problem can be represented through a flowchart (Fig. 11.3).

For avoiding the false local minima as the final solution and finding the critical factor of safety of slopes, many researchers [39, 40] have ascertained that the Simple Genetic Algorithm (SGA) is a superior method than simple optimization routines, 'brute force' approaches and the 'Monte Carlo' approach.

11.2.5.2 Simulated Annealing (SA)

Simulated annealing (SA) method was initially proposed by Kirkpatrick in 1984 [41]. The slow cooling phenomenon of the molten metal is called annealing. To achieve the minimum function value in an optimization problem, annealing is simulated by introducing a temperature-like parameter, and using the concept of Boltzmann's probability distribution [42]. This concept in SA is implied according to the relation:

$$P(\boldsymbol{E}) = e^{-\frac{\boldsymbol{E}}{\boldsymbol{kT}}} \tag{11.12}$$

Where, P(**E**) denotes the probability of achieving the energy level **E** of a thermodynamic system in thermal equilibrium at temperature **T**, and **k** is called the Boltzmann's constant.

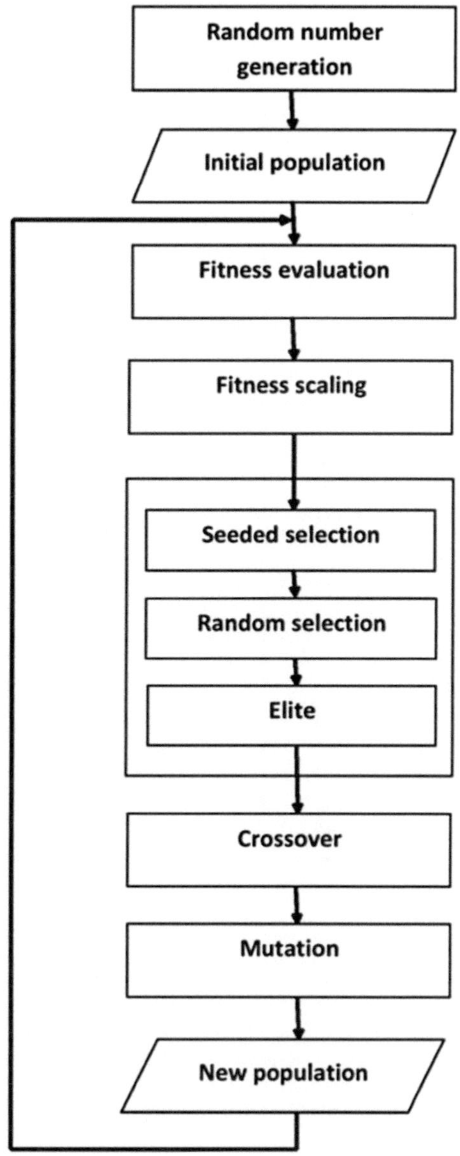

Fig. 11.3 Flowchart of a simple genetic algorithm [38]

Convergence of the SA algorithm depends on (i) the initial temperature T_0, (ii) the number of iterations before reducing the temperature, and (iii) the temperature reduction factor. Following figure (Fig. 11.4) shows a simple flowchart depicting various steps involved in a simulated annealing optimization.

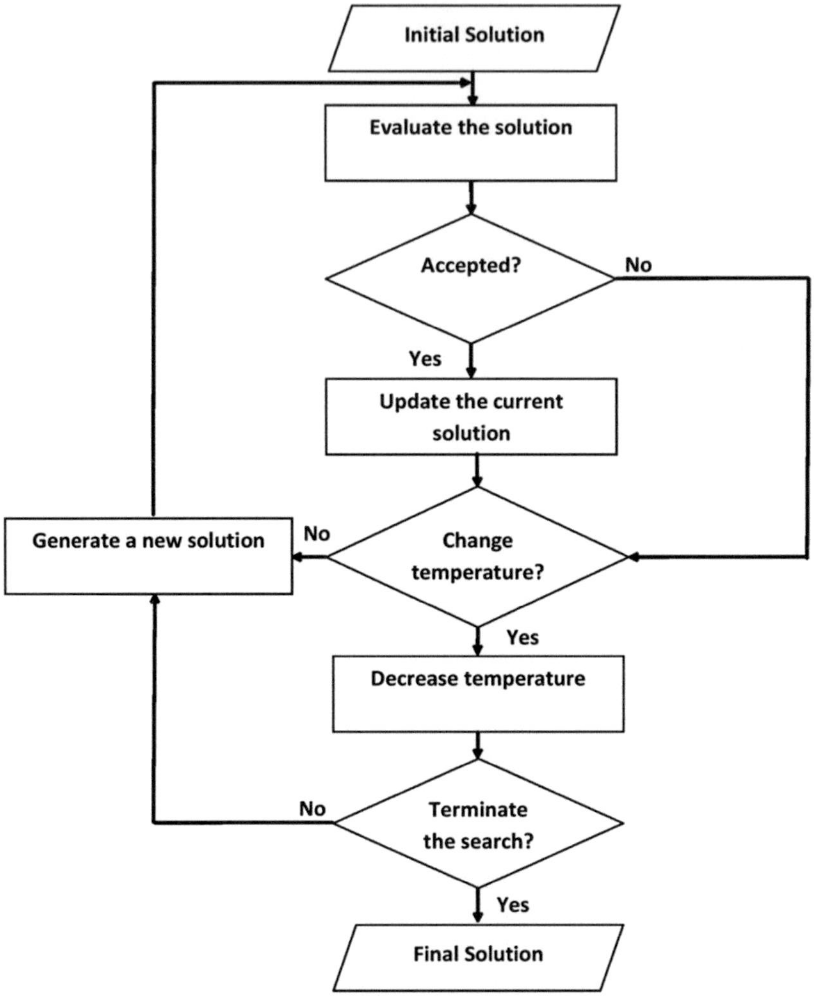

Fig. 11.4 A simple flowchart for Simulated Annealing (SA)

11.2.5.3 Swarm Intelligence (SI)

The working principle [43] of optimization techniques inspired by swarm intelligence (SI) is to mimic the social behaviour of (i) flocks of birds or schools of fish (Particle Swarm Optimization Method), (ii) a colony of ants (Ant Colony Optimization Method), (iii) a swarm of bees (Artificial Bee Colony Optimization Method), etc. These SI approaches are advantageous over traditional optimization techniques with respect to their inherent robustness and flexibility in engineering applications [44]. Following are the two booming SI approaches of optimization techniques getting popular in slope stability analysis problems:

Particle Swarm Optimization (PSO)

The particle swarm optimization (PSO) method was developed by Kennedy and Eberhart in 1995 [45]. The PSO algorithm searches in the problem space for an optimal solution by adjusting the motion of each particle of a swarm by collecting intel from its neighbouring particles. The updated velocity of any individual particle (i) in terms of velocity (v_i^t) and position or solution vector (x_i^t) at time t can be expressed as:

$$v_i^{t+1} = x_i^t + v_i^t = v_i^t + \alpha\varepsilon_1(x_i - g^*) + \beta\epsilon_2(x_i - x_i{}^*), x_i^{t+1} \tag{11.13}$$

Where, ε_1 and ϵ_2 are two random vectors, while α and β are constants (often called the learning parameters).

From Eq. 11.13, it is evident that the velocity of each particle is modified iteratively according to its current velocity and by the difference between its personal best position, P_{best} (i.e., the best solution found by the particle so far), and the best position found by particles in its neighbourhood, L_{best}. As the number of iterations goes up, the swarm focuses increasingly on a search space containing high-quality solutions, g_{best} [46]. This PSO concept is mainly suitable to use for continuous optimization problems [47].

Ant Colony Optimization (ACO)

Ant colony optimization (ACO) was developed and improved by Dorigo et al. [48–50]. Blum and Li [43] explained the ant colony behaviour as: (i) at first, ants look for food randomly in the surroundings of their nest; (ii) on finding a food source, the ant evaluates its quantity and quality; (iii) the ant then leaves a trail of pheromone in its return trip to the nest; (iv) this pheromone trail guides other ants to the food source. The ants generally choose the paths marked by strong pheromone concentrations [44], thus, leading them to the shortest path to the food source from their nest [6]. This concept is mainly suitable to use for combinatorial optimization [51].

11.2.5.4 Harmonic Search Optimization (HSO)

Harmony search optimization (HSO) was developed by Geem et al. in 2001 [52] and is inspired by the effort to find the harmony in music, which is analogous to find the optimal solution in an optimization problem [53]. The diversification of a local solution in HSO is basically controlled by (i) the randomization process, which explores the search space more widely and efficiently, and (ii) the pitch adjustment, which ensures that the newly generated solution is not too far from existing good solutions. The relatively simple structure of the HSO algorithm makes it straightforward to combine it with other heuristic algorithms [54], such as the PSO.

11.3 Optimization Techniques: Applications in Slope Stability Analysis

Optimization techniques employed for slope stability analysis are generally of two types: deterministic and stochastic (or probabilistic).

In deterministic slope stability analysis approach, the single mean value of each geotechnical parameter is used to calculate the factor of safety (FoS) of the slope. The FoS, defined as the ratio between resisting and disturbing forces involved in the slope stability problem, can be computed using methods of limit equilibrium, limit analysis, rigid finite element, discrete element, finite difference and finite element. Among these, the limit equilibrium method (LEM) is the most popular one for slope stability analysis. Most of the LEMs are based on the method of slices, such as the methods formulated by Fellenius in 1936 [55]; Taylor in 1948 [56]; and Bishop in 1955 [57], all of which employ circular slip surfaces; and the methods formulated by Morgenstern and Price in 1965 [58]; Spencer in 1967 [59]; Janbu in 1973 [60] and Sarma in 1973 [61], all of which employ non-circular slip surfaces.

In the method of slices, the potential sliding mass is sub-divided into a number of slices, so that the trial slip surface gets discretized into several nodal points (designated as $A_1, A_2, \ldots, A_6$) as shown in Fig. 11.5. These nodes are connected with each other by either smooth curves or straight lines depending upon the shape (circular or non-circular) assumed by the programmer. Then the expression for FoS is obtained by any of the LEMs discussed above. Finally, an optimization algorithm is written to search for the critical slip surface (designated as $B_1, B_2, \ldots, B_6$ in Fig. 11.5), for which the FoS value of the slope will get minimized for a particular combination of the variables present in the expression for FoS.

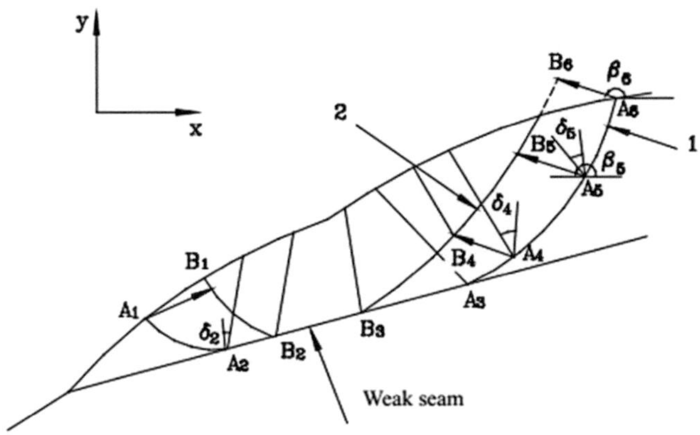

Fig. 11.5 The optimization process for locating the critical failure mode [62]

Commonly, the task of locating the critical deterministic slip surface linked with the lowest FoS value can be converted into a constraint optimization problem as below:

$$\text{Minimize the objective function} : FoS = \; f(X, Y, P), \quad \text{subjected to some kinematic constraints.} \tag{11.14}$$

In the above equation, the factor of safety is expressed as a function of (i) a set of co-ordinates, (X, Y), defining the geometric shape and location of the trial slip surfaces, and (ii) a set of various geotechnical parameters, P, with their mean values provided as inputs to the optimization algorithm.

In stochastic slope stability analysis approach, some or all engineering properties of slope materials are considered to be varying spatially and quantitatively, while being assigned with a probability distribution function (PDF). In this approach, random objective functions or random constraints are involved [63] and instead of computing FoS of a slope, the reliability index (β) associated with the concept of reliability or probability of failure (P_f) is employed. The relationship between the relative likelihood of occurrence of an event and the numerical value associated with the event is depicted by a probability distribution function (PDF) [64]. Typical pattern of the PDFs for some important variables affecting the slope stability are shown in Fig. 11.6.

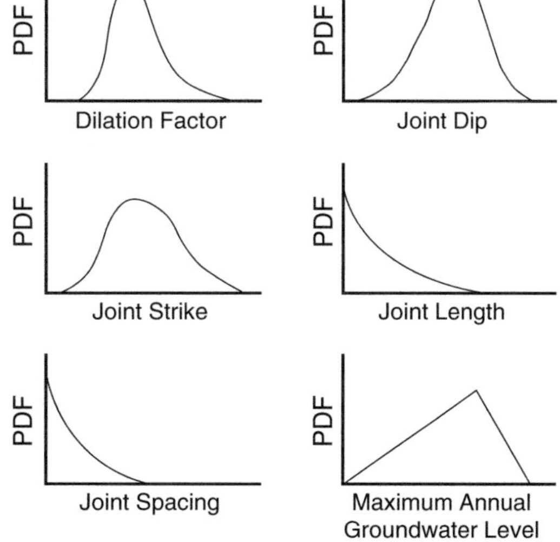

Fig. 11.6 Typical pattern of the PDFs for some important variables (cohesion, friction angle, dip, dilation, joint dip, strike, length and spacing, groundwater level) affecting slope stability [65]

The stochastic optimization approach seeks to search for a critical probabilistic slip surface with the minimum value of reliability index (β_{min}) or the maximum probability of failure ($P_{f,\ max}$). This can be expressed as an optimization problem of the form:

$$\text{Minimize the objective function} : \beta = f(X, Y, P), \quad \text{subjected to some kinematic constraints.} \tag{11.15}$$

In the above equation, (X, Y) is a set of co-ordinates defining the geometric shape and location of the probable slip surfaces, and P is a set of variable geotechnical parameters with their statistical properties provided as inputs to the optimization algorithm.

The reliability index (β) can be computed by probabilistic methods like the First-Order/Second-Order, Second-Moment (FOSM [66]/SOSM [67]) reliability method, the Monte-Carlo Simulation (MCS) method [68] etc. After a probabilistic analysis, if the distribution of safety factors found to be normally distributed, then Eq. (11.16) is used to calculate the reliability index (β):

$$\beta = \frac{\mu - 1}{\sigma} \tag{11.16}$$

Where, μ is the mean and σ is the standard deviation of factor of safety values, respectively.

Practically, the distribution of FoS values is often best fitted by a Log-normal rather than a Normal distribution [69]. For a Log-normal distribution of safety factors, the Eq. (11.17) is used to calculate the Log-normal reliability index (β_{LN}):

$$\beta_{LN} = \frac{\ln\left[\frac{\mu}{\sqrt{1+V^2}}\right]}{\sqrt{\ln\left(1 + V^2\right)}} \tag{11.17}$$

Where, μ is the mean and σ is the standard deviation of factor of safety values, respectively. Also, V stands for co-efficient of variation of FoS, defined as, $V = \frac{\sigma}{\mu}$.

To have a cautious assurance of a safe slope design, the minimum reliability index (β_{min}) should be at least 3 or more, as a thumb rule [69].

In the following sub-sections, a comprehensive review of various optimization methods in deterministic and stochastic slope stability analysis has been presented.

11.3.1 Reviewing Optimization Techniques for Deterministic Slope Stability Analysis

From the review of the literatures involving deterministic stability analysis of slopes with soil and/or rock materials by using optimization techniques, it is evident that researchers have chosen various optimization algorithms to maximize or minimize different objective functions, while a range of analytical approaches like limit equilibrium methods, limit analysis, finite difference method, and rigid finite element or finite element analysis have been tried to find the most critical slip surface associated with the least value of factor of safety for the slope. In the limit analysis of slope stability problems, a linear Mohr-Coulomb (MC) failure criterion is commonly used [70], probably because the MC failure surface is a linear one, and this along with the slip surface can be easily programmed in a LP optimization code as small segments of circles.

Table 11.4 summarizes some influential and pioneering research studies that have applied different optimization techniques in the domain of deterministic slope stability analysis.

11.3.2 Reviewing Optimization Techniques for Stochastic Slope Stability Analysis

Extensive research has been performed on the application of different optimization algorithms for the deterministic analysis of slopes in the past, while stochastic slope stability studies based on the theory of probability and reliability have gained immense popularity in the recent times.

A summary of some significant and leading research studies that have made use of different optimization techniques in the purview of stochastic slope stability analysis has been presented in Table 11.5.

11.4 Concluding Remarks

Slope stability analysis is of utmost importance in geotechnical engineering and design of mitigation measures. In general, finding the critical slip surface having the lowest factor of safety in a deterministic analysis or the least reliability index in a stochastic analysis can be formulated as an optimization problem. Optimization techniques are basically numerical methods, and search for an approximate solution by iterations, starting from an initial solution. Different optimization methods for slope stability analysis have been developed in the past. Each of them has some unique advantages, along with certain inherent limitations. In this book chapter, an extensive literature review has been carried out on deterministic and stochastic slope stability analysis methods based on various optimization algorithms. The major observations from the review of literatures have been pointed out below:

- An extensive research has been performed till date in locating the critical slip surface and finding the minimum factor of safety in the deterministic approach, while the probabilistic approach is becoming the research hotspot in recent times.

Table 11.4 Summary of research works carried out using optimization techniques for deterministic slope stability analysis in chronological order of the year of publication

Year of publication	Authors	Analytical method used to derive the safety factor (FoS) eqn./value	Optimization algorithm used to minimize the safety factor (FoS) eqn.	Contributions/key findings from the study
Limit equilibrium method (LEM)				
1980	Baker [71]	Spencer's method of slices [59]	Dynamic programming algorithm [30]	Applicable to slopes of any geometry, layering, pore pressure and external loading; the algorithm finds lower safety factor than other methods
1988	Chen and Shao [72]	Generalized method of slices by Chen and Morgenstern [73]	Nelder and Mead's simplex method [74], the method of steepest descent [75], and Davidon-Fletcher-Powell (DFP) method [76, 77] and its modification [72]	Exercised different optimization algorithms for slope stability analysis to investigate their practicability; modified the DFP method for minimization procedure
1989	Bardet and Kapuskar [78]	Generalized piece-wise linear slope failure mechanism	Downhill simplex algorithm (Nelder–Mead method) [74]	Verified the code with some numerical cases with circular and non-circular slip surfaces; firmly established the use of Simplex method in slope stability analysis as a multipurpose and easily implementable program
1999	Goh [79]	Multiple-wedge slope stability method [80]	Genetic algorithm (GA) methodology [37]	Demonstrated performance of a Fortran program code "GAWEDGE" for three non-homogeneous slope examples; ascertained that the code is suitable for layered soils with weak zones; found the code and pattern search method of same efficiency and accuracy

(continued)

Table 11.4 (continued)

Year of publication	Authors	Analytical method used to derive the safety factor (FoS) eqn./value	Optimization algorithm used to minimize the safety factor (FoS) eqn.	Contributions/key findings from the study
2002	McCombie and Wilkinson [39]	Bishop's simplified method [57]	Simple genetic algorithm (SGA) [37]	Concluded that performance of the SGA is superior to a brute force or a Monte Carlo approach to find the critical circular slip surface having the lowest safety factor
2003	Cheng [81]	Generalized method of slices by Chen and Morgenstern [73]	Original simulated annealing algorithm [41]	Addressed complicated cases with circular or non-circular failure surface; precision of the global minimum can be specified by the user which is a unique feature; the code can incorporate a very thin soft band in slope profile; also proposed a new method (double QR factorization method) for determination of FoS, which is useful for deep seated non-circular slope failures in achieving fast convergence
2005	Zolfaghari et al. [40]	Morgenstern–Price method [58]	Simple genetic algorithm (SGA) [37]	Examined SGA for critical non-circular failure surface in finite or infinite natural slopes for earthquake and surcharge loading; made no assumptions for shape of the failure surface; implied no restrictions on positions of initiation and termination points of the failure surface

(continued)

Table 11.4 (continued)

Year of publication	Authors	Analytical method used to derive the safety factor (FoS) eqn./value	Optimization algorithm used to minimize the safety factor (FoS) eqn.	Contributions/key findings from the study
2005	Karaulov [82]	Method of sections	Simplex algorithm [8, 9]	Concluded that for the collapse diagram adopted for the state of the slope, Simplex method is sufficiently reliable to determine the system of interaction forces, which is most unsuitable with respect to the slope's stability
2007	Cheng et al. [83]	Arbitrary initial trial failure surface (proposed)	Modified particle swarm optimization (MPSO) method	Modification to PSO code eliminated the need of a pre-determined number of trials; added a termination criterion; FoS values obtained and critical slip surfaces derived by using MPSO were very close to those by PSO; number of evaluations required by MPSO was much lower than that by PSO code
2008	Sun et al. [84]	Spencer's method of slices [59]	Combination of spline curves and genetic algorithm	Observed superior results by using spline curves for a certain nodal points of slip surface as compared to the general LE methods with line segments of slip surfaces
2008	Cheng et al. [85]	Different methods for generating trial slip surfaces	Improvised algorithm over Harmonic search optimization (HSO) algorithm [52]	Applied five types of procedures for generating trial slip surfaces; found the improved HSO algorithm efficient and effective for minimizing factor of safety

(continued)

Table 11.4 (continued)

Year of publication	Authors	Analytical method used to derive the safety factor (FoS) eqn./value	Optimization algorithm used to minimize the safety factor (FoS) eqn.	Contributions/key findings from the study
2009	Li et al. [86]	Method of sections	A mixed search procedure based on PSO [45] and HSO [52]	An initial solution obtained by HSO algorithm is used in the implementation of PSO algorithm in locating critical slip surfaces of soil slopes; demonstrated efficiency of the proposed mixed search algorithm through case studies
2009	Kahatadeniya et al. [87]	Morgenstern–Price method [58]	Ant colony optimization (ACO) algorithm [48–50]	Treated initiation point and shape of the slip surface as search variables; presented four examples with varying complexity; concluded that ACO technique is one of the best global search algorithms for slope stability analysis
2011	Sabhahit and Rao [88]	Generalized slip surface method by Janbu [60]	Genetic algorithm (GA) [37]	Located critical slip surface of non-homogeneous slopes; concluded that the accuracy of GA depends on its population size, generation number, probability of both crossover and mutation
2015	Gao [89]	Spencer's method of slices [59]	Meeting ACO algorithm	Meeting ACO algorithm overcomes limitations of ACO and performs faster than ACO in locating the critical slip surface; verified the effectiveness of the new algorithm by two typical examples

(continued)

Table 11.4 (continued)

Year of publication	Authors	Analytical method used to derive the safety factor (FoS) eqn./value	Optimization algorithm used to minimize the safety factor (FoS) eqn.	Contributions/key findings from the study
2015	Kalatehjari et al. [90]	The method of columns [91], and definition of directional FoS (DFoS) [92]	Particle swarm optimization (PSO) algorithm [45]	Demonstrated the efficiency and effectiveness of their presented code in determining the 3D shape of the failure surface in soil slopes by validating against two example problems from the literature and comparing with laboratory test results
2016	Regmi and Jung [93]	Spencer's method of slices [59]	Dynamic programming [30] algorithm	Determined the location and shape of the failure surface as well as were able to predict the time of a slope failure due to a rainfall event
Limit analysis method (LAM)				
1997	Donald and Chen [94]	Upper bound theorem of classical plasticity	Various methods of optimization provided into a computer code, EMU (Energy Method Upper-bound)	Easily formulated than conventional methods; reaches the least upper bound solution; can address practical problems involving complicated slope profiles and heterogeneous materials
1999 and 2002	Kim et al. [95, 96]	Lower-bound and upper-bound limit analysis	Linear programming method [7]	Presented theoretical basis and procedure for limit analysis of slopes with complex geometries, soil profiles, and groundwater patterns; for the same slope, Bishop's method [57] provided solutions above the upper-bound solutions from [95, 96] for high water table and low friction angle of soil

(continued)

Table 11.4 (continued)

Year of publication	Authors	Analytical method used to derive the safety factor (FoS) eqn./value	Optimization algorithm used to minimize the safety factor (FoS) eqn.	Contributions/key findings from the study
2003	Pastor et al. [97]	Lower and upper bounds using both the static and kinematic approaches	Interior point optimization [33, 34]	Gave precise bound solutions to the well-known problem of the height limit of a Tresca or von Mises vertical slope under gravity; proved that by using their approach, linearization of slope stability problems remain efficient, and both rigorous and global
2011	Liu et al. [98]	Upper bound limit analysis of slope stability	Quadratic Programming (QP) -free algorithm based on penalty function and active-set strategy	SQP-type algorithms are time consuming and it is hard to search the optimum solution; QP-free algorithm has global convergence with an arbitrary starting point and in this, it is optional to maintain feasibility of iteration; concluded that QP-free algorithm performed better than SQP-type algorithms in solving problems of slope stability
2015	Jia et al. [99]	Mohr-Coulomb failure criterion to formulate the objective function and safety factor determined from optimization	Discontinuity topology optimization (DTO) as a linear programming (LP) method	Made advancement in limit analysis of slope stability problems by using adaptive refinement scheme for DTO problems to add or remove inter-node connections; DTO can treat problems involving pore water pressures; solved some benchmark examples in 2D; proposed method can be extended to

(continued)

Table 11.4 (continued)

Year of publication	Authors	Analytical method used to derive the safety factor (FoS) eqn./value	Optimization algorithm used to minimize the safety factor (FoS) eqn.	Contributions/key findings from the study
				3D problems and rock slope failure models having fracture or joint
2015	Rongfu and Gaopeng [100]	Upper bound theorem of classical plasticity based on inclined slices technique and translational velocities	Multi-variable nonlinear sequential quadratic programming (SQP)	Derived the general equation of safety factor with linear and nonlinear M-C failure criterion; illustrated the effectiveness of the inclined slices technique; introduced a multi-tangent method to consider the non-homogeneity of the normal stress in a rock mass
Rigid finite element method (RFEM)				
1995	Lingxi and Xiong [101]	Combined limit analysis and rigid finite element method (RFEM)	Quadratic programming (QP)	Proposed the elastic body-seams model (EBSM) based on RFEM; both the combined RFEM and finite element stress analysis method and EBSM are effective for computing deformation and damage in a discontinuous media like rock masses
2003	Chen et al. [28]	Upper bound limit method based on RFEM	Sequential quadratic programming (SQP) [17]	Developed a code for addressing 2D and 3D slope stability problems; proposed method is simpler than a similar method employing linear finite elements used by [102, 103] and is superior to the

(continued)

Table 11.4 (continued)

Year of publication	Authors	Analytical method used to derive the safety factor (FoS) eqn./value	Optimization algorithm used to minimize the safety factor (FoS) eqn.	Contributions/key findings from the study
				method developed by [94] in modelling non-homogenous soil and complex boundary conditions
2005	Chen et al. [104]	Upper bound approach of limit analysis based on RFEM	Feasible SQP algorithm (FSQP)	In FSQP, replaced the objective function by an exact penalty function; presented three numerical examples to illustrate the usefulness of the FSQP algorithm in the slope stability analysis
2013	Liu and Zhao [105]	Upper and lower bound limit analysis using the RFEM	Primal-dual interior-point method	Proposed a novel approach of RFEM based limit analysis of slope stability to consider of both sliding and rotational failure mechanisms in general; validated the proposed methodology by solving three classical soil or rock slope stability problems; proposed method provided an efficient and safe way for practical slope design
Finite element method (FEM)				
1988	Yamagami and Ueta [106]	Finite element method (FEM) stress analysis	Dynamic programming (DP) optimization method	Enhanced Baker's approach (1980) [71] by combining the DP method with finite element stress analysis to evaluate the stresses acting along the critical slip surface and FoS of the slope

(continued)

Table 11.4 (continued)

Year of publication	Authors	Analytical method used to derive the safety factor (FoS) eqn./value	Optimization algorithm used to minimize the safety factor (FoS) eqn.	Contributions/key findings from the study
1997	Yamagami and Jiang [91]	3D limit equilibrium method and FEM	Dynamic programming and random number generation	Determined location and shape of critical slip surface in 3D and its associated FoS for a slope of arbitrary geometry, with layered soils, and/or a water level
1997	Kim and Lee [107]	FEM	Optimization strategy based on the Broyden–Fletcher–Goldfarb–Shanno (BFGS) method [108] and the feasible direction method [18]	Proposed search method is applicable to both the FEM and the LEM for finding critical slip surfaces of homogenous slopes; predicts location of critical slip surface in harmony with the actual failure surface of a studied embankment slope
2002	Lyamin and Sloan [109, 110]	Lower bound and upper bound limit analysis for one-, two- and three-dimensional continua	Non-linear programming	No need for linearization of the yield surface in the proposed method; modelled 3D geometries with no difficulties; found the new scheme faster than an equivalent LP routine with linear yield surface, for 2D cases
2003	Pham and Fredlund [111]	Finite element stress analysis	Dynamic programming (DP) method	Developed a computer program DYNPROG; eliminates the need of making any assumptions regarding the LEMs of slices; in general, the proposed method finds the more critical

(continued)

Table 11.4 (continued)

Year of publication	Authors	Analytical method used to derive the safety factor (FoS) eqn./value	Optimization algorithm used to minimize the safety factor (FoS) eqn.	Contributions/key findings from the study
				slip surface as it computes FoS on a lower side than that estimated by other conventional LEMs
2010	Kremen and Tsompanakis [31]	Stress analysis performed by FEM	Dynamic programming (DP) optimization method	Applied a different approach, avoiding numerous drawbacks of conventional methods; performed sensitivity analysis to evaluate the effect of various waste strength parameters on factor of safety of the landfill
2016	Yang et al. [112]	FEM	Developed optimization algorithms for finding spherical and ellipsoidal slip surfaces in 3D	Proposed method can dodge some limitations of the existing 3D limit equilibrium methods; reduces the search range during calculation process; validated proposed method against a classical 3D benchmark soil slope problem

Table 11.5 Summary of research works carried out using optimization techniques for stochastic slope stability analysis in chronological order with respect to the year of publication

Year of publication	Authors	Analytical method used to derive the performance function eqn.	Optimization algorithm used to minimize the performance function eqn.	Contributions/key findings from the study
1997	Low and Tang [113]	Generalized slip surface method by Janbu [60]	Generalized reduced gradient algorithm [22]	Computed the Hasofer-Lind reliability index [66] as the first order reliability method (FORM); proposed method included an automated search algorithm for finding the critical slip surface
2003	Bhattacharya et al. [114]	Spencer's method of slices [59]	Monte Carlo technique for slope stability analysis implemented by [115]	Proposed algorithm needs no priori assumptions about shape of the slip surface; facilitates a direct search for the critical probabilistic surface; verified the process by applying into several cases
2004	Griffiths and Fenton [116]	Random finite element method (RFEM) based on elasto-plasticity and random field theory	Monte Carlo simulation [68]	RFEM method (i) makes no a priori assumptions about the shape or location of the critical failure mechanism, (ii) enables slope failure to develop naturally by seeking out the most critical mechanism; thus making it advantageous over traditional probabilistic slope stability techniques
2007	Xue and Gavin [117]	Bishop's simplified method [57]	Genetic algorithm (GA) [37]	Determined the probability of failure from the reliability index of the slope by considering soil properties as random

(continued)

Table 11.5 (continued)

Year of publication	Authors	Analytical method used to derive the performance function eqn.	Optimization algorithm used to minimize the performance function eqn.	Contributions/key findings from the study
				variables; transformed the random variables into polar co-ordinates to avoid complexities regarding definition of the limit state function
2007	Cho [118]	Spencer's method of slices [59]	Monte Carlo simulation [68] and feasible direction method [18]	Adopted FORM to determine the critical failure surface and to perform sensitivity analyses; studied effects of uncertainty due to spatial variability of soil properties pertaining to probabilistic assessment of layered slopes
2008	Hong and Roh [119]	Generalized method of slices by Chen and Morgenstern [73]	Sequential quadratic programming (SQP) method [120]	Considered FORM for estimating the reliability index of earth slopes; found reliability of a slope to be sensitive to the probability distribution function adopted for the input variables; noticed that probability of failure decreases as spatial correlation of soil properties decreases
2009	Tan and Wang [121]	Non-linear finite element reliability analysis (FERA) and slip surface stress analysis (SSA) technique	Accelerating convergence algorithm	Adopted the limited step length iteration method (LSLIM) to calculate the reliability index; demonstrated competence and strength of the proposed

(continued)

Table 11.5 (continued)

Year of publication	Authors	Analytical method used to derive the performance function eqn.	Optimization algorithm used to minimize the performance function eqn.	Contributions/key findings from the study
				method by solving some numerical examples; LSLIM and the proposed algorithm reduced the iteration number significantly
2010	Khajehzadeh et al. [122, 123]	Spencer's method of slices [59]	Harmony search optimization (HSO) algorithm [52] and particle swarm optimization (PSO) [45]	Computed the Hasofer-Lind reliability index [66]; demonstrated the effectiveness and robustness of the proposed HSO algorithm by comparing the results with published cases in earth slopes; results indicated that for layered soils, the searched critical probabilistic surfaces are not the same as the critical deterministic surface
2010	Zhang and Zhao [124]	Inclined slices technique and upper bound theorem in plasticity theory	Winner's polynomial chaos expansion for stochastic optimization	Presented a probabilistic numerical approach for stability analysis of soil and rock slopes; treated cohesion, friction angle and the pore pressure ratio as random variables; proposed probabilistic approach comparing to conventional deterministic LEM gave accurate estimates of safety factor and probability of failure of the slopes

(continued)

Table 11.5 (continued)

Year of publication	Authors	Analytical method used to derive the performance function eqn.	Optimization algorithm used to minimize the performance function eqn.	Contributions/key findings from the study
2011	Farah et al. [125]	Stochastic finite element method (SFEM), LEMs: Spencer's method of slices [59] and Bishop's Method [57]	Optimization technique for finding the critical probabilistic slip surface developed by [126]	Considered spatial variability of soil properties; used the proposed SFEM and two LEMs to locate the position of critical surface; proposed approaches showed a small difference amid the reliability index values and position of the critical probabilistic slip surface
2015	Zeng et al. [127]	Spencer's method of slices [59]	Custom-made genetic algorithm (GA)	Proposed a new approach to spot the fully specified representative slip surfaces (RSSs) of layered soil slopes; used FORM to determine the probability of failure; re-examined three typical benchmark-slopes with layered soils to demonstrate the efficiency, accuracy and robustness of the suggested method

- In the traditional approach of slope stability analysis, the use of limit equilibrium methods is the most popular choice and hence, a lot of research papers can be found in this area where different optimization techniques have been applied.
- Among slope stability analysis methods, the well-accepted choice of modern optimization techniques (also called heuristic methods) include genetic algorithm, simulated annealing, particle swarm intelligence, ant colony optimization, and harmonic search optimization.

- The heuristic methods are capable of finding superior results than conventional simplex or gradient based methods as the latter methods can avoid local minima and reach a global solution during the optimization process.
- For slope stability problems, the modern optimization methods work fine for simple slope geometry and soil conditions. For complicated problems, like a slope with a soft band of weak soil, the factor of safety becomes very sensitive to the exact location of the critical solution and the inconsistency in final outcomes from different global optimization methods become quite large.
- The way in which the slip surface is defined has an important effect on the speed and eventual success of the heuristic algorithm.
- Due to the lack of an unambiguous termination criterion for the iteration process in any modern optimization technique, a pre-determined number of trials have to be specified by the programmer, and the minimum value of the objective function from different trials is taken as the global minimum solution.
- There is no existing guideline for selecting the optimum number of trials for the minimization analysis; hence, the users usually specify a large number of trials. This in turn, makes the computation time-taking and almost infeasible for practical applications for a quick decision making in case of a complicated case study.

To conclude, no particular optimization method may perhaps surpass all the other methods in all cases of slope stability analysis in computational efficiency and superiority of the final outcome. Still, application of heuristic optimization methods in this area is highly recommended as they are easy to program, simple to comprehend, adaptable to various boundary and loading conditions, and can avoid local minima, hence inherently superior to conventional optimisation procedures. Among the heuristic methods, the simple genetic algorithm (SGA) performs satisfactorily even if several 'depressions' (local minima) are present in the optimization process. In simulated annealing (SA), the quality of the final solution remains unaffected by the primary guesses, but computational effort increases with poorer starting guesses. Again, particle swarm optimization (PSO) shares many similarities with GA, except the fact that the PSO has no evolution operators (crossover and mutation) like the GA. For normal cases, the PSO appears to be effective and efficient over various conditions, and becomes the natural choice of optimization technique to be used in general cases. For special cases, such as when the objective function is highly discontinuous, the simulated annealing (SA) method performs well to provide a more stable solution. The coupling of the PSO and harmonic search (HS) is a new approach in the use of global optimization for slope stability problems. This hybrid method is less efficient (still effective) for simple problems, hence, not recommended for normal cases, although it can be used in complicated slope stability problems as it provides a more stable solution and is less attracted by the local minima. Finally, after analyzing the recent developments in the heuristic global optimization methods, and availability of powerful and easy-to-learn software packages based on the finite element method, the authors suggest using a hybrid slope stability analysis method, like the PSO and FEM or the SA and FEM, in general.

References

1. World Disasters Report (2016) Resilience: saving lives today, investing for tomorrow. Available from: http://media.ifrc.org/ifrc/publications/world-disasters-report-2016/. Last accessed on 28.03.2017
2. National Disaster Management Guidelines – Management of Landslides and Snow Avalanches (2009, June) A publication of the National Disaster Management Authority, Government of India, New Delhi
3. National Disaster Management Plan (2016, May. A publication of the National Disaster Management Authority, Government of India, New Delhi. Available from: http://www.ndmindia.nic.in. Last accessed on 28.03.2017
4. Disaster Management in India (2011) A publication of: Ministry of Home Affairs, Government of India, New Delhi. Available online from: www.undp.org/content/dam/india/docs/disaster_management_in_india.pdf. Last accessed on 28.03.2017
5. Taha MR, Khajehzadeh M (2010) Slope stability assessment using optimization techniques: an overview. Electron J Geotech Eng (EJGE) 15:1901–1915
6. Rao Singiresu S (2009) Engineering optimization: theory and practice, 4th edn. Wiley, Hoboken
7. Ferguson TS (n.d.) LP – Linear programming: a concise introduction. UCLA Department of Mathematics. Available from: www.math.ucla.edu/~tom/LP.pdf. Last accessed on 29.03.2017
8. Dantzig GB (1963) Linear programming and extensions. Princeton University Press, Princeton
9. Dantzig GB, Thapa MN (1997) Linear programming 1: introduction. Springer, New York. LLC. ISBN 978-0-387-22633-0
10. Morgan SS (1997) A comparison of simplex method algorithms. MSc. thesis. University of Florida. Available from: https://web.archive.org/web/20110807134509/http://www.cise.ufl.edu/research/sparse/Morgan/index.htm. Last accessed on 29.03.2017
11. Bertsekas DP (1999) Nonlinear programming, 2nd edn. Athena Scientific, Cambridge, MA. ISBN 1-886529-00-0
12. Kuhn HW, Tucker A (1951) Nonlinear programming. In: Proceedings of the 2nd Berkeley symposium on mathematical statistics and probability. University of California Press, Berkeley
13. Fox RL (1971) Optimization methods for engineering design. Addison-Wesley, Reading
14. Box MJ (1965) A new method of constrained optimization and a comparison with other methods. Comput J 8(1):42–52
15. Cheney EW, Goldstein AA (1959) Newton's method of convex programming and Tchebycheff approximation. Numer Math 1:253–268
16. Kelly JE (1960) The cutting plane method for solving convex programs. J SIAM VIII (4):703–712
17. Powell MJD (1978) A fast algorithm for nonlinearity constrained optimization calculations. In: Watson GA et al (eds) Lecture notes in mathematics. Springer-Verlag, Berlin
18. Zoutendijk G (1960) Methods of feasible directions. Elsevier, Amsterdam
19. Zoutendijk G (1966) Nonlinear programming: a numerical survey. SIAM J Control Theory Appl 4(1):194–210
20. Rosen JB (1960) The gradient projection method of nonlinear programming, part I: linear constraints. SIAM J 8:181–217
21. Rosen JB (1961) The gradient projection method of nonlinear programming, part II: linear constraints. SIAM J 9:414–432
22. Gabriele GA, Ragsdell KM (1977) The generalized reduced gradient method: a reliable tool for optimal design. ASME J Eng Ind 99:384–400
23. Box MJ (1966) A comparison of several current optimization methods and the use of transformations in constrained problems. Comput J 9:67–77
24. Carroll CW (1961) The created response surface technique for optimizing nonlinear restrained systems. Oper Res 9:169–184

25. Zangwill WI (1967) Nonlinear programming via penalty functions. Manag Sci 13(5):344–358
26. Hestenes MR (1969) Multiplier and gradient methods. J Optim Theory Appl 4:303–320
27. Rockafellar RT (1973) The multiplier method of Hestenes and Powell applied to convex programming. J Optim Theory Appl 12(6):555–562
28. Chen J, Yin JH, Lee CF (2003) Upper bound limit analysis of slope stability using rigid finite elements and nonlinear programming. Can Geotech J 40:742–752. https://doi.org/10.1139/T03-032
29. Chinneck JW (2015) Practical optimization: a gentle introduction. Carleton University, Ottawa. Available online at: www.sce.carleton.ca/faculty/chinneck/po.html Last accessed on 30.03.2017
30. Bellman R (1957) Dynamic programming. Princeton University Press, Princeton
31. Kremen A, Tsompanakis Y (2010) Application of dynamic programming to evaluate the slope stability of a vertical extension to a balefill. Waste Manag Res 28:373–382. https://doi.org/10.1177/0734242X09354767
32. Kremen A (2014) Improve your slope stability analyses by using dynamic programming. In: A blog post. Available online from: http://www.cornerstoneeg.com/2014/10/29/improve-slope-stability-analyses-dynamic-programming/. Last accessed on 30.03.2017
33. Dantzig GB, Thapa MN (2003) Linear programming 2: theory and extensions. Springer, New York
34. Karmarkar N (1984) A new polynomial-time algorithm for linear programming. Combinatorica 4:373–395
35. Cook SA (1983) An overview of computational complexity. Commun ACM 26(6):401–408
36. Rechenberg I (1965) Cybernetic solution path of an experimental problem. Library Translation 1122, Royal Aircraft Establishment, Farnborough
37. Holland J (1975) Adaptation in natural and artificial systems. University of Michigan Press, Oxford
38. Solati S, Habibagahi G (2006) A genetic approach for determining the generalized interslice forces and the critical non-circular slip surface. Iran J Sci Technol Trans B Eng 30(B1):1
39. McCombie P, Wilkinson P (2002) The use of the simple genetic algorithm in finding the critical factor of safety in slope stability analysis. Comput Geotech 29:699–714
40. Zolfaghari AR, Heath AC, McCombie PF (2005) Simple genetic algorithm search for critical non-circular failure surface in slope stability analysis. Comput Geotech 32(3):139–152
41. Kirkpatrick S (1984) Optimization by simulated annealing—quantitative studies. J Stat Phys 34(5–6):975–986
42. Metropolis N, Rosenbluth A, Rosenbluth M, Teller A, Teller E (1953) Equation of state calculations by fast computing machines. J Chem Phys 21(6):1087–1092
43. Blum C, Li X (2008) Swarm intelligence in optimization. Swarm intelligence: introduction and applications, 1st edn. Springer-Verlag, Berlin, pp 43–85
44. Soliman MM, Hassanien AE, Onsi HM (2014) Bio-inspiring techniques in watermarking medical images: a review. In: Hassanien AE, Kim TH, Kacprzyk J, Awad AI (eds) Bio-inspiring cyber security and cloud services: trends and innovations, vol 70, 1st edn. Springer, Berlin, pp 93–114. https://doi.org/10.1007/978-3-662-43616-5
45. Kennedy J, Eberhart R (1995) Particle swarm optimization. In: Proceeding of the IEEE international conference on neural networks, Perth, Australia. pp 1942–1948
46. Pant M, Thangaraj R (2007) Particle swarm optimization: performance tuning and empirical analysis. Stud Comput Intell 203:101–128
47. Kennedy J, Eberhart R, Shi Y (2001) Swarm intelligence. Morgan Kaufmann Academic Press, San Francisco, pp 1931–1938
48. Dorigo M, Maniezzo V, Colorni A (1996) The ant system optimization by a colony of cooperating agents. IEEE Trans Syst Man Cybern B 26(1):29–41
49. Dorigoa M, Stutzle T (2004) Ant colony optimization. MIT Press, Cambridge
50. Dorigoa M, Blum C (2005) Ant colony optimization theory: a survey. Theor Comput Sci 344 (2005):243–278

51. Blum C (2005) Ant colony optimization: introduction and recent trends. Phys Life Rev 2 (4):353–373
52. Geem ZW, Kim JH, Loganathan GV (2001) A new heuristic optimization algorithm: harmony search. Simulation 76:60–68
53. Yang XS (2009) Harmony search as a metaheuristic algorithm. In: Geem ZW (ed) Music-inspired harmony search algorithm. Springer-Verlag, Berlin., SCI 191, pp 1–14
54. Omran M, Mahdavi (2008) Global-best harmony search. Appl Math Comput 198:643–656
55. Fellenius W (1936) Calculation of the stability of earth dams. In: Proceedings of the 2nd Congress on large dams, International Commission on large dams of the World Power conference, vol 4, pp 445–462
56. Taylor DW (1948) Fundamentals of soil mechanics. Wiley, New York
57. Bishop AW (1955) The use of the slip circle in the stability analysis of earth slopes. Geotechnique 5(1):7–17
58. Morgenstern NR, Price VE (1965) The analysis of the stability of general slip surfaces. Geotechnique 15(1):79–93
59. Spencer E (1967) A method of analysis of the stability of embankments assuming parallel inter-slice forces. Geotechnique 17(1):11–26
60. Janbu N (1973) Slope stability computations. In: Embankment dam engineering. Wiley, New York
61. Sarma SK (1973) Stability analysis of embankments and slopes. Geotechnique 23(3):423–433
62. Chen Z, Ugai K (2008) Limit equilibrium and finite element analysis – a perspective of recent advances. Landslides and engineered slopes – Chen et al. Taylor & Francis Group, London, 25–38
63. Spall JC (2003) Introduction to stochastic search and optimization. Wiley, Hoboken. ISBN 0-471-33052-3
64. Haldar A, Mahadevan S (1999) Probability, reliability, and statistical methods in engineering design. Wiley, New York. ISBN: 978-0-471-33119-3
65. Coates DF (1981) Rock mechanics principles. Monograph 874. Canada Centre for Mineral and Energy Technology (CANMEeT, formerly Mines Branch, Energy, Mines and Resources Canada), Ottawa
66. Hasofer AM, Lind NC (1974) An exact and invariant first-order reliability format. J Eng Mech ASCE 100:111–121
67. Breitung K, Hohenbichler M (1989) Asymptotic approximations for multivariate integrals with an application to multinormal probabilities. J Multivar Anal 30:80–97. https://doi.org/10.1016/0047-259X(89)90089-4
68. Rubinstein RY, Kroese DP (2008) Simulation and the Monte Carlo method, 2nd edn. Wiley
69. Slide Version 6.0 (2010) User's guide. Rocscience, Toronto. Available from: https://www.rocscience.com/help/slide/webhelp/slide_interpret/probability/Probabilistic_Analysis_Overview.htm. Last accessed on 06.04.2017
70. Yang XL, Yin JH (2004) Slope stability analysis with nonlinear failure criterion. J Eng Mech 130(3):267–273. https://doi.org/10.1061/(ASCE)0733-9399(2004)130:3(267)
71. Baker R (1980) Determination of the critical slip surface in slope stability computations. Int J Numer Anal Methods Geomech 4:333–359
72. Chen ZY, Shao CM (1988) Evaluation of minimum factor of safety in slope stability analysis. Can Geotech J 25:735–748
73. Chen ZY, Morgenstern NR (1983) Extensions to the generalized method of slices for stability analysis. Can Geotech J 20:104–119
74. Nelder JA, Mead R (1965) A simplex method for function minimization. Comput J 7:308–313. https://doi.org/10.1093/comjnl/7.4.308
75. Arfken G (1985) The method of steepest descents. §7.4 in mathematical methods for physicists, 3rd edn. Academic, Orlando, pp 428–436
76. Davidon WC (1959) Variable metric method for minimization. A.E.C. Res. and Develop. Report ANL-5990 (Rev. TID-4500, 14th ed.)

77. Fletcher R, Powell MJD (1963) A rapidly convergent descent method for minimization. Comput J 6(2):163–168. https://doi.org/10.1093/comjnl/6.2.163
78. Bardet JP, Kapuskar MM (1989) A simplex analysis of slope stability. Comput Geotech 8:329–348
79. Goh A (1999) Genetic algorithm search for critical slip surface in multiple-wedge stability analysis. Can Geotech J 36(2):382–391
80. Donald IB, Giam PSK (1989) Improved comprehensive limit equilibrium stability analysis. Department of Civil Engineering Report No. 1/1989, Monash University, Melbourne, Australia
81. Cheng YM (2003) Location of critical failure surface and some further studies on slope stability analysis. Comput Geotech 30:255–267
82. Karaulov AM (2005) Statement and solution of the stability problem for slopes and embankments as a linear-programming problem. Soil Mech Found Eng 42(3)
83. Cheng YM, Li L, Chi SC, Wei WB (2007) Particle swarm optimization algorithm for the location of the critical non-circular failure surface in two-dimensional slope stability analysis. Comput Geotech 34:92–103
84. Sun J, Li J, Liu Q (2008) Search for critical slip surface in slope stability analysis by spline-based GA method. J Geotech Geoenviron 134(2):252–256
85. Cheng YM, Li L, Lansivaara T, Chi S, Sun Y (2008) An improved harmony search minimization algorithm using different slip surface generation methods for slope stability analysis. Eng Optim 40(2):95–115
86. Li L, Yu G, Chu X, Lu S (2009) The harmony search algorithm in combination with particle swarm optimization and its application in the slope stability analysis. In: International conference on computational intelligence and security. IEEE, pp 133–136
87. Kahatadeniya KS, Nanakorn P, Neaupane KM (2009) Determination of the critical failure surface for slope stability analysis using ant colony optimization. Eng Geol 108:133–141
88. Sabhahit N, Rao A (2011) Genetic algorithms in stability analysis of non-homogeneous slopes. Int J Geotech Eng 5(1):33–44
89. Gao W (2015) Determination of the noncircular critical slip surface in slope stability analysis by meeting ant colony optimization. J Comput Civ Eng 30(2):1–10
90. Kalatehjari R, Arefnia A, Rashid ASA, Ali N, Hajihassani M (2015) Determination of three-dimensional shape of failure in soil slopes. Can Geotech J 52(9):1283–1301. https://doi.org/10.1139/cgj-2014-0326
91. Yamagami T, Jiang JC (1997) A search for the critical slip surface in three dimensional slope stability analysis. Soils Found 37(3):1–16
92. Huang CC, Tsai CC (2000) New method for 3D and asymmetrical slope stability analysis. J Geotech Geoenviron 126(10):917–927
93. Regmi RK, Jung K (2016) Application of dynamic programming to locate the critical failure surface in a rainfall induced slope failure problem. KSCE J Civ Eng 20(1):452
94. Donald I, Chen ZY (1997) Slope stability analysis by the upper bound approach: fundamentals and methods. Can Geotech J 34(6):853–862. https://doi.org/10.1139/t97-061
95. Kim J, Salgado R, Yu HS (1999) Limit analysis of slopes subjected to pore-water pressures. J Geotech Geoenviron Eng 125(1):49–58
96. Kim J, Salgado R, Lee J (2002) Stability analysis of complex soil slope using limit analysis. J Geotech Geoenviron Eng 128(7):546–557
97. Pastor J, Thai TH, Francescato P (2003) Interior point optimization and limit analysis: an application. Commun Numer Method Eng 19:779–785. https://doi.org/10.1002/cnm.619
98. Liu FT, Fan YH, Yin JH (2011) The use of QP-free algorithm in the limit analysis of slope stability. J Comput Appl Math 235:3889–3897
99. Jia C, Huang Q, Xia B (2015) Stability analysis of soil slope using discontinuity layout optimization. Adv Mater Res 1065–1069:190–198. https://doi.org/10.4028/www.scientific.net/AMR.1065-1069.190

100. Rongfu X, Gaopeng T (2015) Slope stability limit analysis based on inclined slices technique. Electron J Geotech Eng 20:1813–1832
101. Lingxi Q, Xiong Z (1995) Rigid finite element and its applications in engineering. Acta Mech Sinica 11(l):44–50
102. Sloan SW (1988) Lower bound limit analysis using finite elements and linear programming. Int J Numer Anal Methods Geomech 12:61–77
103. Sloan SW (1989) Upper bound limit analysis using finite elements and linear programming. Int J Numer Anal Meth Geomech 13(3):263–282
104. Chen J, Yin JH, Lee CF (2005) The use of an SQP algorithm in slope stability analysis. Commun Numer Methods Eng 21:23–37. https://doi.org/10.1002/cnm.723
105. Liu F, Zhao J (2013) Limit analysis of slope stability by rigid finite-element method and linear programming considering rotational failure. Int J Geomech 13(6):827–839
106. Yamagami T, Ueta Y (1988) Search for noncircular slip surfaces by the Morgenstern-Price method. In: Proceedings of the 6th international conference on numerical methods in geomechanics, Innsbruck, Austria, 11–15 April. A.A. Balkema, Rotterdam, pp 1335–1340
107. Kim J, Lee S (1997) An improved search strategy for the critical slip surface using finite element stress fields. Comput Geotech 21(4):295–313
108. Fletcher R (1987) Practical methods of optimization, 2nd edn. Wiley, New York. ISBN 978-0-471-91547-8
109. Lyamin AV, Sloan SW (2002) Lower bound limit analysis using non-linear programming. Int J Numer Methods Eng 55:573–611. https://doi.org/10.1002/nme.511
110. Lyamin AV, Sloan SW (2002) Upper bound limit analysis using linear finite elements and non-linear programming. Int J Numer Anal Methods Geomech 26:181–216. https://doi.org/10.1002/nag.198
111. HTV P, Fredlund DG (2003) The application of dynamic programming to slope stability analysis. Can Geotech J 40(4):830–847
112. Yang Y, Xing H, Yang X, Zhou J (2016) Determining the critical slip surface of three-dimensional soil slopes from the stress fields solved using the finite element method. Mathematical problems in Engineering, Hindawi Publishing Corporation, vol 2016, Article ID 7895615, 11 pages. https://doi.org/10.1155/2016/7895615
113. Low B, Tang W (1997) Probabilistic slope analysis using Janbu's generalized procedure of slices. Comput Geotech 21(2):121–142
114. Bhattacharya G, Jana D, Ojha S, Chakraborty S (2003) Direct search for minimum reliability index of earth slopes. Comput Geotech 30(6):455–462
115. Greco VR (1996) Efficient Monte Carlo technique for locating critical slip surface. J Geotech Eng ASCE 122(7):517–525
116. Griffiths DV, Fenton GA (2004) Probabilistic slope stability analysis by finite elements. J Geotech Geoenviron 130(5):507–518
117. Xue J, Gavin K (2007) Simultaneous determination of critical slip surface and reliability index for slopes. J Geotech Geoenviron 133:878–886
118. Cho S (2007) Effects of spatial variability of soil properties on slope stability. Eng Geol 92 (3–4):97–109
119. Hong H, Roh G (2008) Reliability evaluation of earth slopes. J Geotech Geoenviron 134 (12):1700–1705
120. Schittkowski K (1986) NLPQL: a FORTRAN subroutine solving constrained nonlinear programming problems. Ann Oper Res 5(1):485–500
121. Tan X, Wang J (2009) Finite element reliability analysis of slope stability. J Zhejiang Univ Sci A 10(5):645–652
122. Khajehzadeh M, Taha M, El-Shafie A (2010) Harmony search algorithm for probabilistic analysis of earth slope. Electron J Geotech Eng 15:1647–1659
123. Khajehzadeh M, Taha M, El-Shafie A (2010) Modified particle swarm optimization for probabilistic slope stability analysis. Int J Phys Sci 5(15):2248–2258

124. Zhang H, Zhao Y (2010) Probabilistic slope stability analysis based on the upper bound theorem. In: International conference on E-Product, E-Service, and E-Entertainment (ICEEE), November 7–9. IEEE. https://doi.org/10.1109/ICEEE.2010.5660372
125. Farah K, Ltifi M, Hassis H (2011) Reliability analysis of slope stability using stochastic finite element method. Proc Eng 10:1402–1407
126. Celestino TB, Duncan JM (1981) Simplified search for non-circular slip surface. In: Proceedings of the 10th international conference on soil mechanics and foundation engineering, pp. 391–394
127. Zeng P, Jimenez R, Piña RJ (2015) System reliability analysis of layered soil slopes using fully specified slip surfaces and genetic algorithms. Eng Geol 193:106–117

Part V
Selected Case Studies

Chapter 12
Integration of Terrestrial Laser Scanning and GIS Analysis for Multi-temporal Landslide Monitoring: A Case Study of the Mont de La Saxe (Aosta Valley, NW Italy)

Gianpiero Amanzio, Ashwani Kumar Tiwari, Muriel Lavy, and Marina De Maio

Abstract In Italy, landslide phenomena and mass movements are very common, particularly along the Alps, the principal mountainous chains in the northern part of the country. In this study, we used a terrestrial laser scanner (TLS) to collect a multi-temporal dataset during 3 years (2012–2014) of observation, with the aim of accurately evaluating these phenomena in the Mont de la Saxe area (Aosta Valley region, Italy). Starting from the point clouds acquired with the TLS, we derived the digital surface models and we performed a multitemporal analysis in geographical information system (GIS) to identify the morphological features of the landslide and to delineate the displacement of the phenomena. The analysis allows quantifying the major elevation change occurred in the middle and the bottom side of landslide body during 2012–2014 with a high precision. The volume displaced in the second year (2013–2014) increased by 66% than the previous year (2012–2013), showing a progressive acceleration of the landslide phenomena. This result indicates that the volume estimation is crucial for planning future landslide emergency situations and to calibrate the early warning system, based on occurred phenomena.

Keywords Terrestrial laser scanning (TLS) · Landslide · Mont de La Saxe · Mass movement, DTM · GIS

G. Amanzio (✉) · M. Lavy
GENEGIS GI S.r.l., Turin Branch (TO), Turin, TO, Italy
e-mail: g.amanzio@genegis.net

A. K. Tiwari · M. De Maio
DIATI-Department of Environment, Land and Infrastructure Engineering
Politecnico di Torino, Torino, Italy

S. P. Pradhan et al. (eds.), *Landslides: Theory, Practice and Modelling*,
Advances in Natural and Technological Hazards Research 50,
https://doi.org/10.1007/978-3-319-77377-3_12

12.1 Introduction

A landslide is when rock, debris or earth move down a slope. Landslides occur when the materials which constitute the slope of a hill fail and the force of gravity intervenes. The failure of materials that produces a landslide can be the result of natural causes or human activity. Specifically, natural activities like heavy or long-lasting rainfall, melting snow, and rapid tectonic forcings, for example, volcanic activity or earthquakes, are trigger factors for landslides. Moreover, landslides occur due to human activities such as changes in the slope profile, excavation, irrigation deforestation and other factors [1]. Landslides are the cause of several adverse effects on many environmental elements such as a change in the topography surface; contamination of the quality of rivers and streams as well as the groundwater flow system; forest cover loss, and impact on the habitats of natural wildlife living on the earth's surface [2]. In recent time, several studies from different countries have used different methods to evaluate the landslides problems [3–11]. Landslide phenomena frequently occur in Italy, notably in the principal mountain chains like the Apennines or the Alps [12].

In fact, because of characteristic like its relief, lithology, and structure in Italy the risk of the landslide is very high. They occur throughout Italy and are the most frequent type of natural disaster. After earthquakes, landslides are responsible for the highest number of victims [13]. In the last 30 years, there have been disastrous landslides in Piemonte (1994), Sarno and Quindici (1998) and NW Italy (2000). Specifically, the Aosta Valley (NW Italy), where is located the case study discussed in the following, is a prevailing mountain territory particularly prone to arising from landslides phenomena. Among them, we can mention Becca France landslide [14], Beauregard landslide [15] and Bosmatto landslide [16].

In order to manage the risk deriving from landslide phenomena, geographical information system (GIS) spatial analysis is one of the most prominent effective techniques in order to map zones and assess risk regarding environmental health problems [17–19]. The handling and analyzing through spatial data tools of GIS software could make the application of quantitative analysis in landslide hazard assessment and mapping easier [20]. Some of these techniques monitor landslides using laser scanner instrumentation based on Light Detection and Ranging (LiDAR) technology. Geographical information system (GIS) represents a useful tool when quick decisions are required because graphical representation facilitates a policy decision making [21]. Several researchers have used GIS technique for the evaluation of the hazardous risk of landslides on the environmental parameters as well as mapping of landslides [22–31].

Furthermore, the use of Terrestrial Laser Scanner (TLS) represents an advanced approach to monitor instability phenomena and to detect the landslide displacement [32]. The intense use laser scanning technology, with automatic filter procedures allows a rapid high resolute Digital Terrain Models (DTM) production, with high accuracy and high automation which is crucial for a multitemporal monitoring system [33]. The landslide area is often characterized by a hard accessibility due to the high level of risk of moving around the dangerous area, for this reason, the use of long-range laser scanner represents a safety technique of monitoring without decrease the accuracy and the resolution during the survey [34]. Thus, the aim of this study was to identify the landslide phenomenon using TLS and GIS techniques in the Mont de La Saxe, Aosta Valley region. The information on the landslide phenomenon is very useful for the management of landslides in the area. It could also assist policy makers to tackle landslides in the area.

12.2 Study Area

The case study of the Mont de La Saxe landslide is situated in the Aosta Valley (NW Italy) in the Courmayeur municipality above the La Saxe hamlet (Fig. 12.1). In Italy, there are a lot of complex landslides and the one present on the Mont de La Saxe is one of these. The volume estimation raises about 8 million cubic meters [35], generating great risks for the small communities of La Palud and Entrèves. National and international viability is also affected, as this spot is crucial due to the nearby International Mont Blanc Tunnel. In the spring of 2013 a sudden acceleration occurred in the landslide body, so the local administration had to evacuate the area, closing any access to the Ferret Valley [36]. Mapping extant slope failures are thought to be the principal tool for assessing landslide phenomenon [37]. In landslide studies, high-resolution Digital Terrain Models (DTMs) allows analysis of the morphological features of the landslide slope, which are necessary to understand the space-and time-dependent processes. LiDAR (light detection and ranging)-derived DTMs are extensively used to study the landslide, focusing on estimating the volume, boundary and topographic change of landslides [38–40]. So, data collected by long-range terrestrial laser scanner are processed in a GIS with the aim to extract the landslide morphological features of rapid movement and to generate high resolute DTM.

12.2.1 Climate and Geological

The Aosta Valley is a mountainous region in northwestern Italy and it is intersected by glacial tributaries, creating eighteen minor valleys, mainly along the North-South axis and converging towards the Aosta Valley. The Dora Baltea River is the principal water course, flowing West to East.

A typical alpine climate dominates in the Aosta Valley region with very low temperature in winters and cool summers. The Aosta Valley has a significant amount of rainfall during the year. Specifically, it is highest peaks in autumn and spring, and lowest in winter and summer. In fact, the peak mean precipitation is approximately 140 mm of rainfall in a month, and the minimum value is 30 mm [41].

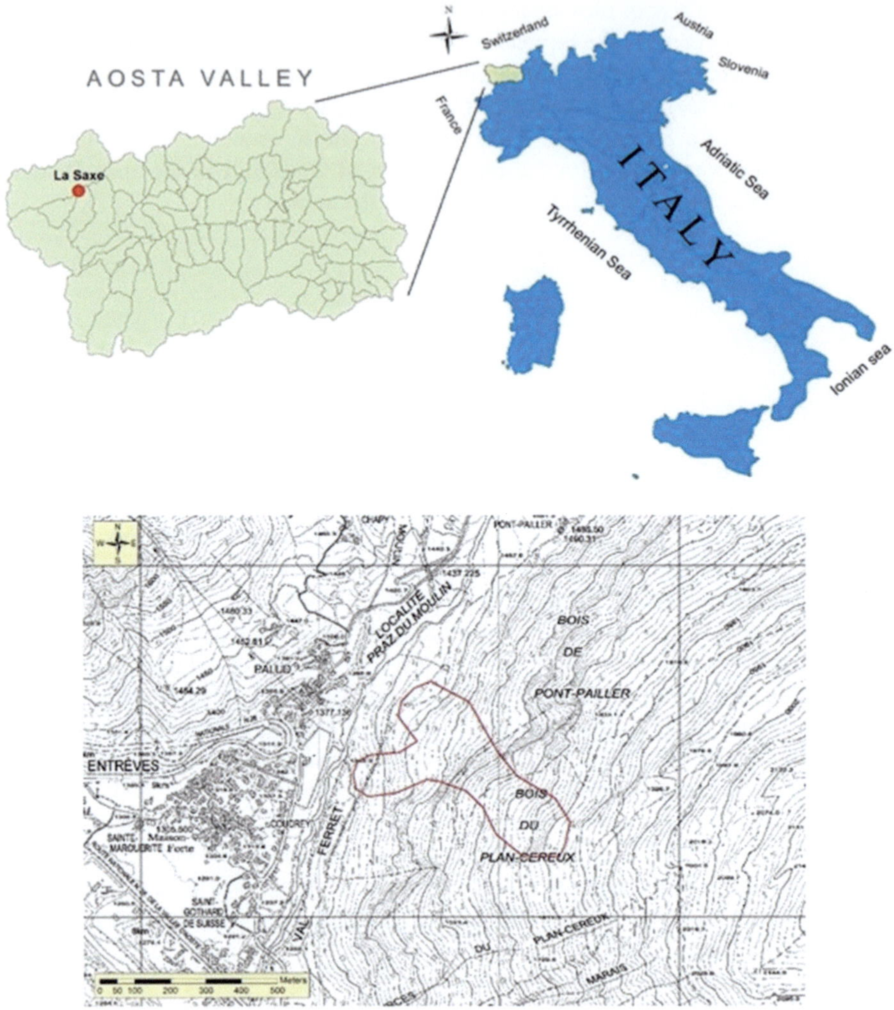

Fig. 12.1 Test sites location (red polygon) at the Mont de La Saxe of the Aosta Valley region, Italy

The alpine range comprises a number of crystalline and metamorphic tectonic domains. It is from the external to the internal structural side [42, 43] (Fig. 12.2).

The Austroalpine is the uppermost super unit, widely exposed in the Eastern Alps. This composite nappe system originated from the ocean-facing (distal) part of the Adriatic passive continental margin, which principally developed in the Cretaceous (Eoalpine) orogeny. The Sesia–Lanzo Zone (eclogitic schists and gneiss) and the Dent Blanche nappe (schists, gabbro, Mesozoic cover, and metagranites) are present in this system.

The Piedmont Zone is characterized by a part of terrigenous flysch-type metasediments (quartzites, calcareous schists and marbles) and one of the prevalent

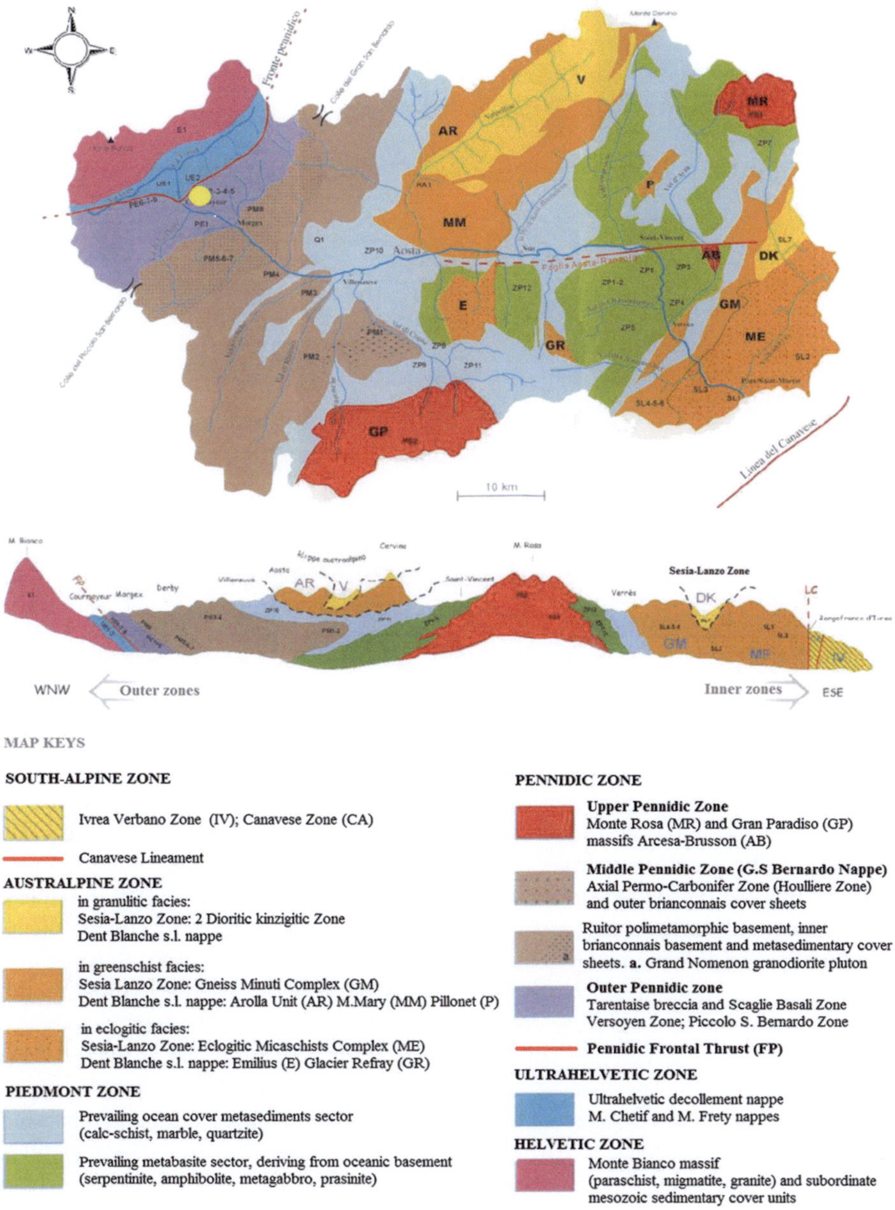

Fig. 12.2 Geological map of the Aosta Valley [42]. The yellow circle marks the case study location

metabasites derived from the oceanic substrate (leucocratic gneisses, prasinites, metagabbros and serpentinites).

The Pennidic zone has an Upper Pennidic Zone (metaconglomerates, schists, metagranites and gneiss), a Middle Pennidic Zone (dolomites quartzites, gneiss,

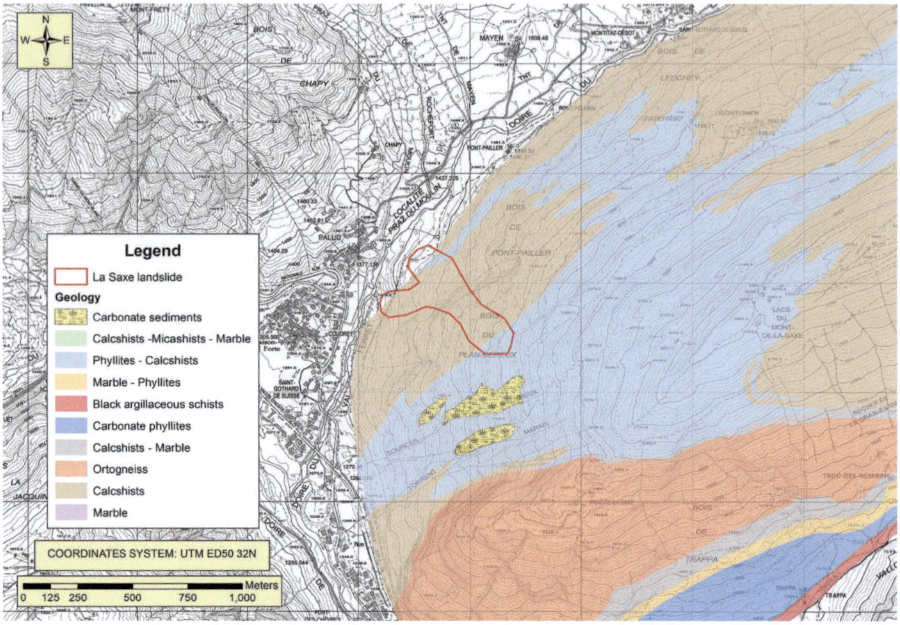

Fig. 12.3 Geological map of the Mont de La Saxe landslide. The red polygon mark the landslide area

calcareous schists, metagranites and gypsum) and an Outer Pennidic Zone (quartzites, prasinites, flysch, serpentinites, calcareous schists and schists).

Our case study, the Mont de la Saxe landslide is situated in the Ultrahelvetic Zone, consisting of, the Mont Chetif nappes (carniole, limestones, calcareous schists, and porfiroids), the Ultraelvetic nappes and the Mont Frety nappes (Fig. 12.3). The Helvetic Zone consists of the Mont Blanc Massif (Granites, migmatites and paraschists) and, subordinately, by the Mesozoic covers sheets (Regione Valle d'Aosta and Università degli Studi di Torino, 2005). Specifically, Mont de la Saxe is situated to the south of the Mont Blanc crystalline Massif, in the western Italian Alps, which originated in the Cretaceous period through the subduction of a Mesozoic ocean and collision between the European and Adriatic continental margins [44]. The study area is located in deformed Middle Jurassic meta-sedimentary sequences, consisting mainly of tightly foliated carbonate-bearing argillaceous schists, black schists and arenaceous limestones with quartz arenites levels [45, 46]. South of the study landslide area, penninic units are found, which comprise the low-grade meta-sedimentary deposits of Courmayeur zone [44]. This unit, dominated by black argillaceous schists, dips towards SE, with angles from 20 to 60°, forming an imbricated structure. The landslide (of about 8×106 m^3) extends from 1400 to 1870 m a.s.l., covering an area of approximately 150,000 m^2. Its maximum horizontal length is around 550 m, maximum width about 420 m, and mean slope gradient 37°. The upper scarp, approximately 200 m wide, is characterized by a steep rock wall some tens of meters high, along subvertical schistosity planes [47] (Fig. 12.3).

12.3 Materials and Methods

The landslide analysis is based on multi-temporal higher-resolute datasets collected with a long-range terrestrial laser scanner (TLS) which uses LiDAR technology. The principle is based on the time-of-flight distance measurements using an infrared sensor (near-infrared laser with a wavelength of 1550 nm, a measurement range of 4000 m, an accuracy of 15 mm, a precision of 10 mm) [48].

The general workflow of the study, from the data acquisition to the final elaboration is illustrated in the following.

12.3.1 Data Acquisition

During 3 years (from April 2012 to April 2014) a TLS survey (once per year) was executed with the aim of detecting the geomorphological variations (elevation and volume) that characterize the landslide phenomena of the Mont de La Saxe.

Specifically, the TLS was used to acquire and generate a high-resolute 3D point cloud of the investigated area. The scans have been designed to obtain a final ground resolution of 10 cm, which was deemed suitable to capture both large-scale structure and micro-topography [49]. To effectively scan the majority of the landslide zone, scanning has been performed from an elevation higher than the area of interest in order to obtain an oblique point of view from above the landslide. This approach also allows the laser pulse to penetrate deeply in the topographic depression [50, 51]. For this reason, the scan position chosen is located near the Rifugio Pavillon, on the Mont Blanc Massif, at an altitude of 2020 m a.s.l. From this position, a complete survey can be undertaken of the Mont de La Saxe landslide with a long-range laser scanner.

12.3.2 Data Elaboration

The first processing step of the 3D point cloud data involves the removal of the isolated points and additional noises data (such as aerosols and water droplets) which may lie around the scan position. Furthermore, a coarse error removal elaboration was implemented on the point cloud. After this procedure, it is necessary to apply an ICP (Iterative Closest Point) algorithm, which is an efficient algorithm for robust rigid registration of 3D data [52] to align multiple scans. The implementations of the ICP allow reducing the standard deviation of the point clouds alignment. In this way, it is possible to merge point cloud from different scan position using an overlap between the scans. The algorithm works at best with planar surfaces in RiSCAN PRO, in this study area we obtained an average 3D error of about 0.02 m.

The third step aims to separate the ground points from the non-ground-point (vegetation). Therefore, a terrain filter tool, compatible with RiSCAN PRO, was used. This filter works in a hierarchic manner with several levels of details, which is based on a grid representation of the data at each level [53]. Since the method may lead to misclassification of a steep area and edges as vegetation, manual checking and correction were performed at the end of the process.

The fourth step uses the registration tool of RISCAN PRO that allows associating a proper coordinate system to the point cloud using tie point targets (reflectors). Specifically, for the first survey (April 2012), the recording of the coordinates of individual reflectors was undertaken by using a global navigation satellite system (GNSS) receiver. The GNSS data were refined by a post-processing differential correction based on close to the permanent station (Aosta area, 20 km away). One of the scan should at least contain three reflector targets for carrying out the registration by tie points with common point configuration algorithm in RiSCAN PRO software. The other two TLS surveys (April 2013 and April 2014) were aligned using the ICP algorithm based on the first georeferenced survey (April 2012).

Finally, the last step consists of the generation of three DTMs using ArcGIS 10.2 software. The quality of the DTM largely depends on the accuracy of individual survey points and the method of interpolation [54]. Various factors can produce errors in the digital elevation model (DEM), such as topographic complexity, survey point quality, surface composition, sampling strategy and interpolation methods [55–58]. The quality of the raw survey data is of vital importance.

We evaluated the optimal pixel size resolution using the Nyquist-Shannon sampling theorem [59] and generating a density map on LAS-dataset (LAS file is an industry-standard binary format for storing LIDAR data) to analyze the point cloud distribution before evaluating the better resolution to DTM.

According to Lichti et al. 2006, in order to avoid blurring effects (the resolution can be much lower than expected if the beamwidth is larger than the sampling interval), the final DTMs were generated with a resolution of 0.25 m. This operation was carried out by a Natural Neighbor interpolator tool with minimum cell assignment in ArcGIS (10.2) [8, 60]. Natural Neighbor interpolator was chosen as it can leave a coarser morphology avoiding smoothing effects [61].

12.4 Results and Discussion

In order to the estimate elevation change and change of volume during 3 years (2012–2014), the elevation difference maps and the erosion and deposition maps were created with TLS-derived DTMs in ArcMap, as shown in Figs. 12.4, 12.5, and 12.6.

Table 12.1 summarizes the change of volume, the volume balance between the erosion and deposition phenomena in the different time intervals.

From 2012 to 2013 (Fig. 12.4), the change of surface elevation was mainly in the middle and up side of the landslide body. Figure 12.5 shows how the change of surface elevation was mainly in the middle and bottom side of the Mont de La Saxe.

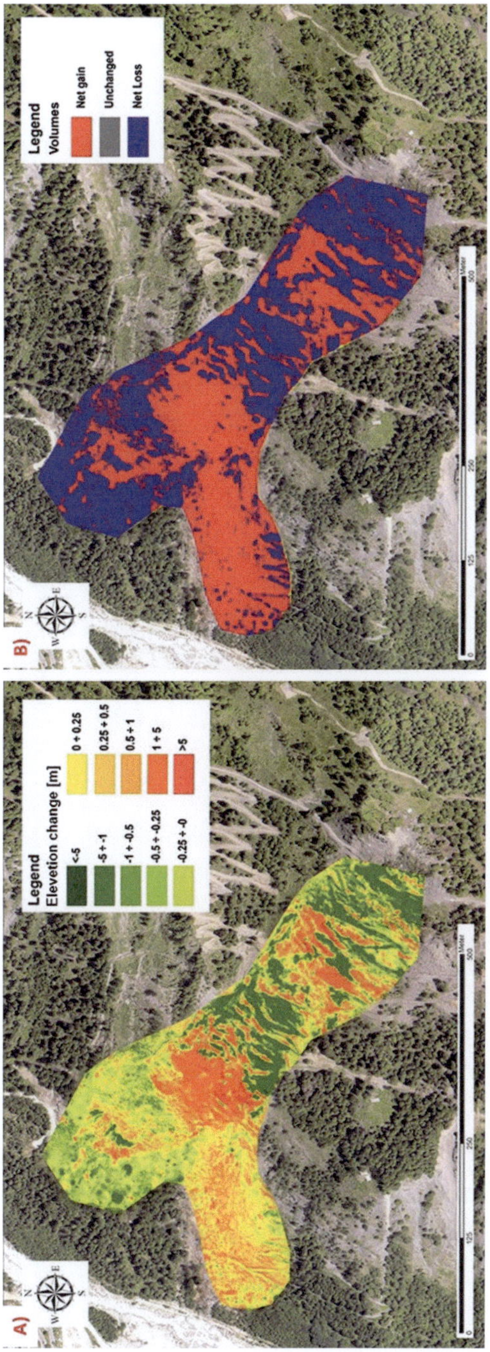

Fig. 12.4 Elevation and volume change maps of the landslide generated from the comparison between DTMs (2012–2013). (**a**) Map of elevation change; (**b**) Map of volume change

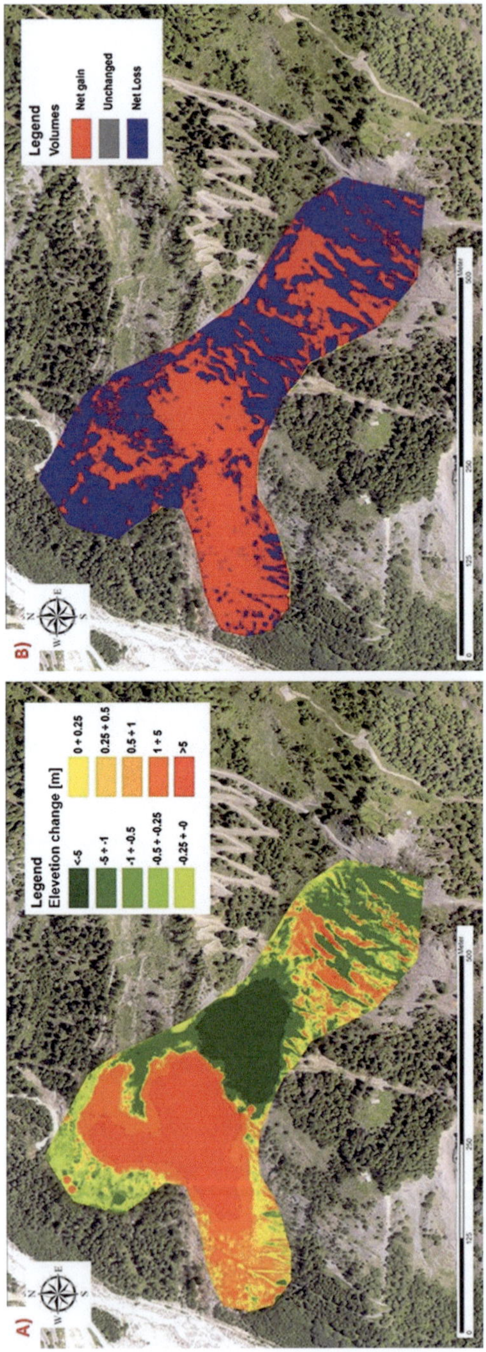

Fig. 12.5 Elevation and volume change maps of the landslide generated from the comparison between DTMs (2013–2014). (**a**) Map of elevation change; (**b**) Map of volume change

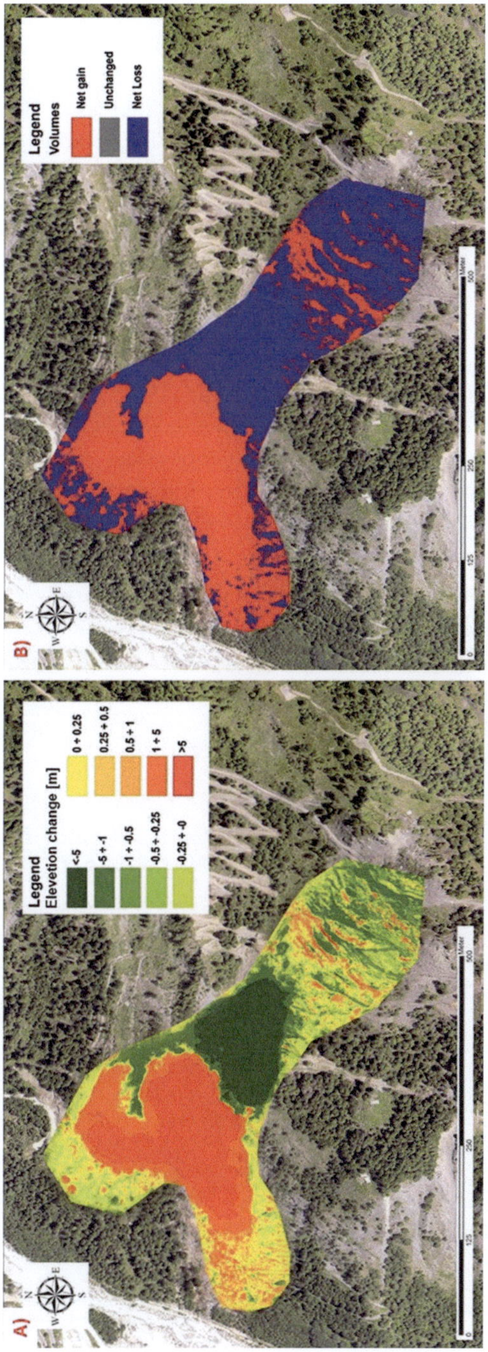

Fig. 12.6 Elevation and volume change maps of the landslide generated from the comparison between DTMs (2012–2014). (**a**) Map of elevation change; (**b**) Map of volume change

Figure 12.6 illustrates the evolution of the landslide between 2012 and 2014, from which it was noticed that the major elevation change occurred in the middle and bottom side of landslide body.

The local uplift reached up to 5 m from 2013 to April 2014, the large erosion happened in the middle zone and the mass movement increased the elevation of the bottom part of the landslide up to 5 m. The acquisitions elaborated from 2012 to 2014, well presents the evolution process of the mass movement phenomena occurred in 2 years. During this period the elevation of the topographic surface in the southwest crown zone decreased significantly. The change primarily consisted in a lowering of the topographic surface in the lower part of the crown zone. The mass flow separated in two directions in the toe zone. The surface elevation in the south part of the landslide toe abruptly changed, producing a deposit with local elevation increased to 10 m. The volume change computed for different time intervals are shown in Table 12.1. These data indicated that the period of 2013–2014 has major changes in volume than the previous year.

During the period 2009–2012 [62] several geophysical surveys were performed on the Mont de La Saxe landslide. Specifically, five seismic tomography profiles were acquired along the landslide slope in 2010 (Fig. 12.7).

Table 12.1 Change in volume computed for different time intervals based on DTMs

Time interval	Erosion (m^3)	Deposition (m^3)	Volume balance (m^3)
2012–2013	44,238	41,341	2897
2013–2014	246,318	240,602	5716
2012–2014	271,586	261,845	9741

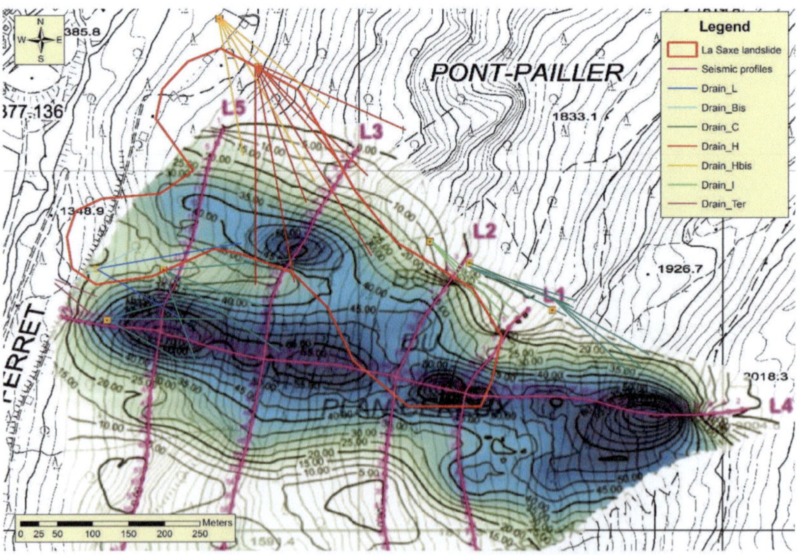

Fig. 12.7 Seismic profiles map and drain networks in the landslide area

These surveys show that either relaxed or fractured rock masses reaching depths up to 70–90 m are present (blue area in Fig. 12.7). Under these rocks, some relatively sound or slightly relaxed rock was detected. This thickened material tends to decrease in outer direction respect to the geophysical area; however, it maintains an important thickness in the landslide area (40–50 m of depth). Specifically, the areas where the highest values are detected corresponding to the landslide part where the major elevation change occurred on the landslide body.

Moreover, from 2010, several underground water drainage works were installed to reduce the groundwater pressure behind the landslide body (Fig. 12.7). The monitoring of discharge yielded from drainpipes shows that the ones located in NW part of the geophysics area return major values respect to the drainage work located in SW part of that area. So, we can hypothesize that the NW part of the geophysical area is highly influenced by the groundwater flow which affects the stability of the thickened material. This hypothesize seems to be confirmed by the elevation change and the volume variation that identify a major displacement in the middle and the bottom side of landslide body during 2012–2014.

12.5 Conclusions

Large landslides can produce significant hazard emergency, so that understanding their mechanism and quantifying precisely the volume displacements require acquisition of accurate datasets. The Mont de La Saxe case study was analyzed through the use of a long-range terrestrial laser scanner (TLS). This instrumentation allowed collecting several point-cloud datasets of the topographic surface of the landslide during 3 years (from 2012 to 2014). These provide a robust multitemporal dataset to study the landslide evolution. The digital surface models, derived from the point clouds, allowed to reconstruct the 3D deformation of the mass movement and to calculate the elevation change and the volumetric changes from 2012 to 2014. The analysis showed how the landslide increased the velocity which consequently increased the displaced volume (66% of the total volume displaced occurred during the 2nd year). The 85% of the total area affected by the landslide underwent a change, showing the size of the phenomena.

Furthermore, monitoring datasets allow the subdivision of the landslide in different zones with different behaviors in terms of elevation and volume changes. The geophysical analysis performed in 2010 showed the presence of thickened material extending to depths up to 40–50 m in the landslide area. Furthermore, drainage work installed in the same period of time indicates that the discharge collected in NW part of the area return major values with respect to the drainage works located in the SW part of that area. So, it seems that groundwater flow affects the stability of the thickened material located in NW part of the area causing landslide movements.

This study shows that how the integration between laser scanner datasets and GIS allows performing a multi-temporal analysis with important results that support the landslides risk assessment and helps to improve the environmental management system for hazard emergency. The future investigation aims to integrate highly precise topographical measurement with hydrological analysis, with the aim of defining the influence of the hydrogeological system on the local movement of the landslide body.

Acknowledgments We thank the Regione Autonoma Valle d'Aosta, Assessorato Opere pubbliche e difesa del suolo – Dipartimento difesa del suolo e risorse idriche for support and data sharing. We grateful to the anonymous reviewers and Editors for their precious suggestions.

References

1. Guzzetti F, Reichenbach P, Cardinali M et al (2005) Probabilistic landslide hazard assessment at the basin scale. Geomorphology 72(1):272–299
2. Geertsema M, Highland L, Vaugeouis L (2009) Environmental impact of landslides. In: Landslides–disaster risk reduction. Springer, Berlin, pp 589–607. doi:https://doi.org/10.1007/978-3-540-69970-5_31
3. Refice A, Capolongo D (2002) Probabilistic modeling of uncertainties in earthquake-induced landslide hazard assessment. Comput Geosci 28(6):735–749
4. Liu JG, Mason PJ, Clerici N et al (2004) Landslide hazard assessment in the Three Gorges area of the Yangtze river using ASTER imagery: Zigui–Badong. Geomorphology 61(1):171–187
5. Pourghasemi HR, Moradi HR, Aghda SF et al (2014) GIS-based landslide susceptibility mapping with probabilistic likelihood ratio and spatial multi-criteria evaluation models (North of Tehran, Iran). Arab J Geosci 7(5):1857–1878
6. Nichol JE, Shaker A, Wong MS (2006) Application of high-resolution stereo satellite images to detailed landslide hazard assessment. Geomorphology 76(1):68–75
7. Hong Y, Adler R, Huffman G (2006) Evaluation of the potential of NASA multi-satellite precipitation analysis in global landslide hazard assessment. Geophys Res Lett 33(22):1–5
8. Prokop A, Panholzer H (2009) Assessing the capability of terrestrial laser scanning for monitoring slow moving landslides. Nat Hazards Earth Syst Sci 9(6):1921–1928
9. Abellán A, Calvet J, Vilaplana JM et al (2010) Detection and spatial prediction of rockfalls by means of terrestrial laser scanner monitoring. Geomorphology 119(3):162–171
10. Niethammer U, James MR, Rothmund S et al (2012) UAV-based remote sensing of the Super-Sauze landslide: evaluation and results. Eng Geol 128:2–11
11. Jebur MN, Pradhan B, Tehrany MS (2014) Optimization of landslide conditioning factors using very high-resolution airborne laser scanning (LiDAR) data at catchment scale. Remote Sens Environ 152:150–165
12. Catani F, Casagli N, Ermini L et al (2005) Landslide hazard and risk mapping at catchment scale in the Arno River basin. Landslides 2(4):329–342
13. Landslides in Italy – Special Report 2008 (2008) ISPRA (Italian National Institute for Environmental Protection and Research. Rapporti 83/2008, p 38. ISBN 978-88-448-0355-1
14. Forno MG, Gattiglio M, Gianotti F (2012) Geological context of the Becca France historical landslide (Aosta Valley, NW Italy). Alpine Medit Quat 25(2):125–140
15. Barla G, Antolini F, Barla M et al (2010) Monitoring of the Beauregard landslide (Aosta Valley, Italy) using advanced and conventional techniques. Eng Geol 116:218–235
16. Luino F (2005) Sequence of instability processes triggered by heavy rainfall in the northern Italy. Geomorphology 66:13–39
17. Ghosh A, Tiwari AK, Das S (2015) A GIS based DRASTIC model for assessing groundwater vulnerability of Katri Watershed, Dhanbad, India. Model Earth Syst Environ 1(3):1–14
18. Tiwari AK, Singh PK, De Maio M (2016) Evaluation of aquifer vulnerability in a coal mining of India by using GIS-based DRASTIC model. Arab J Geosci 9(6):1–15
19. Tiwari AK, Singh AK, Mahato MK (2017) GIS based evaluation of fluoride contamination and assessment of fluoride exposure dose in groundwater of a district in Uttar Pradesh, India. Hum Ecol Risk Assess 23(1):56–66

20. Guzzetti F, Carrara A, Cardinali M et al (1999) Landslide hazard evaluation: a review of current techniques and their application in a multi-scale study, Central Italy. Geomorphology 31 (1):181–216
21. Singh PK, Tiwari AK, Panigarhy BP, Mahato MK (2013) Water quality indices used for water resources vulnerability assessment using GIS technique: a review. Int J Earth Sci Eng 6 (6–1):1594–1160
22. Gupta RP, Joshi BC (1990) Landslide hazard zoning using the GIS approach—a case study from the Ramganga catchment, Himalayas. Eng Geol 28(1–2):119–131
23. Aleotti P, Chowdhury R (1999) Landslide hazard assessment: summary review and new perspectives. Bull Eng Geol Environ 58(1):21–44
24. Rautela P, Lakhera RC (2000) Landslide risk analysis between Giri and Tons rivers in Himachal Himalaya (India). Int J Appl Earth Obs Geoinf 2(3):153–160
25. Temesgen B, Mohammed MU, Korme T (2001) Natural hazard assessment using GIS and remote sensing methods, with particular reference to the landslides in the Wondogenet area, Ethiopia. Phys Chem Earth Pt C 26(9):665–675
26. Dai FC, Lee CF (2002) Landslide characteristics and slope instability modeling using GIS, Lantau Island, Hong Kong. Geomorphology 42(3):213–228
27. Ayalew L, Yamagishi H (2005) The application of GIS-based logistic regression for landslide susceptibility mapping in the Kakuda-Yahiko Mountains, Central Japan. Geomorphology 65 (1):15–31
28. Yalcin A (2008) GIS-based landslide susceptibility mapping using analytical hierarchy process and bivariate statistics in Ardesen (Turkey): comparisons of results and confirmations. Catena 72(1):1–12
29. Yilmaz C, Topal T, Süzen ML (2012) GIS-based landslide susceptibility mapping using bivariate statistical analysis in Devrek (Zonguldak-Turkey). Environ Earth Sci 65 (7):2161–2178
30. Akgun A, Kıncal C, Pradhan B (2012) Application of remote sensing data and GIS for landslide risk assessment as an environmental threat to Izmir city (west Turkey). Environ Monit Assess 184(9):5453–5470
31. Perotto-Baldiviezo HL, Thurow TL et al (2004) GIS-based spatial analysis and modeling for landslide hazard assessment in steeplands, southern Honduras. Agric Ecosyst Environ 103 (1):165–176
32. Tesa G, Galgaro N, Zaltron N et al (2007) Terrestrial laser scanner to detect landslide displacement fields: a new approach. Int J Remote Sens 28:3425–3446
33. Bitelli G, Dubbini M, Zanutta A (2004) Terrestrial laser scanning and digital photogrammetry techniques to monitor landslide bodies. Int Arch Photogramm Remote Sens Spat Infor Sci 35 (B5):246–251
34. Corsini A, Castagnetti C, Bertacchini E et al (2013) Integrating airborne and multi-temporal long-range terrestrial laser scanning with total station measurements for mapping and monitoring a compound slow moving rock slide. Earth Surf Process Landforms 38(11):1330–1338
35. Crosta G, Cancelli P, Tamburini A et al (2012) Chasing a complete understanding of a rapid moving rock-slide: the La Saxe landslide. EGU General Assembly 2012. Vienna, Austria
36. Roncella R, Forlani G, Fornari M et al (2014) Landslide monitoring by fixed-base terrestrial stereo-photogrammetry. ISPRS Ann Photogramm Remote Sens Spat Inf Sci II-5:297–304. https://doi.org/10.5194/isprsannals-II-5-297-2014
37. McKeana J, Roering J (2004) Objective landslide detection and surface morphology mapping using high-resolution airborne laser altimetry. Geomorphology 57:331–351
38. Joyce KE, Samsonov S, Manville V, Jongens R et al (2009) Remote sensing data types and techniques for lahar path detection: a case study at Mt Ruapehu, New Zealand. Remote Sens Environ 113:1778–1786
39. Kasai M, Ikeda M, Asahina T et al (2009) LiDAR-derived DEM evaluation of deep-seated landslides in a steep and rocky region of Japan. Geomorphology 113:57–69

40. Ventura G, Vilardo G, Terranova C et al (2011) Tracking and evolution of complex active landslides by multi-temporal airborne LiDAR data: the Montaguto landslide (Southern Italy). Remote Sens Environ 115:3237–3248
41. Mercalli L, Cat Berro D, Montuschi S (2003) Atlante climatico della Valle d'Aosta. Società Meteorologica Subalpina Torino, p 405
42. Bonetto F, Gianotti F (1998) Il Giardino delle Rocce di Pollein. Comune di Pollein, pieghevole
43. Lo Russo S, Amanzio G, Ghione R et al (2015) Recession hydrographs and time series analysis of springs monitoring data: application on porous and shallow aquifers in mountain areas (Aosta Valley). Environ Earth Sci 73(11):7415–7434
44. Dal Piaz GV, Bistacchi A, Massironi M (2003) Geological outline of the Alps. Episodes 26:175–180
45. Perello P, Piana F, Martinotti G (1999) Neoalpine structural features at the boundary between the Penninic ahd Helvetic domains. Eclogae Geol Helv 92:347–359
46. Hölbling D, Füreder P, Antolini A et al (2012) A semi-automated object-based approach for landslide detection validated by persistent scatterer interferometry measures and landslide inventories. Remote Sens 4:1310–1336
47. Crosta GB, Prisco C, Frattini P et al (2014) Chasing a complete understanding of the triggering mechanisms of a large rapidly evolving rockslide. Landslides 11:747–764
48. Crepaldi S, Zhao Y, Lavy M et al (2015) Landslide analysis by multitemporal terrestrial laser scanning (TLS) data: the Mont de la Saxe landslide. Rend Online Soc Geol It 35:92–95
49. Dunning SA, Massey CI, Rosser NJ (2009) Structural and geomorphological features of landslides in the Bhutan Himalaya derived from terrestrial laser scanning. Geomorphology 103(1):17–29
50. Schürch P, Densmore AL, Rosser NJ et al (2011) Dynamic controls on erosion and deposition on debris-flow fans. Geo Soc America 39(9):827–830
51. Blasone G, Cavalli M, Marchi L et al (2014) Monitoring sediment sources areas in a debris-flow catchment using terrestrial laser scanning. Catena 123:23–36
52. Almhdie A, Léger C, Deriche M et al (2007) 3D registration using a new implementation of the ICP algorithm based on a comprehensive lookup matrix: application to medical imaging. Pattern Recog Letters 28(12):1523–1533
53. Riegl (2013) RiSCAN PRO operating & processing software for RIEGL 3D laser scanners
54. Heritage GL, Milan DJ, Large ARG et al (2009) Influence of survey strategy and interpolation model on DEM quality. Geomorphology 112(3–4):334–344
55. Lane SN (1998) The use of digital terrain modelling in the understanding of dynamic river channel systems. In: Lane SN, Richards K, Chandler J (eds) Landform monitoring, modelling and analysis. Wiley, Chichester, pp 311–342
56. Wise SM (1998) The effect of GIS interpolation errors on the use of digital elevation models in geomorphology. In: Lane SN, Richards K, Chandler J (eds) Landform monitoring, modelling and analysis. Wiley, Chichester, pp 139–164
57. Wechsler S, Kroll C (2006) Quantifying DEM uncertainty and its effects on topographic parameters. Photogramm Eng Remote Sens 72:108–1090
58. Krzysztof S (2009) Evaluation of digital terrain models generated from laser scanning data under forest conditions. MSc thesis submitted in the framework of and according to the requirements of the UNIGIS Master of Science programme (Geographical Information Science & Systems). Jagiellonian University, Kraków, Paris Lodron University of Salzburg
59. Shannon CE (1949) Communication in the presence of noise. Proc Inst Radio Eng 37:10–21
60. Bater CW, Coops NC (2009) Evaluating error associated with lidar-derived DEM interpolation. Comput Geosci 35(2):289–300
61. Tarolli P, Sofia G, Dalla Fontana G (2012) Geomorphic features extraction from high resolution topography: landslide crowns and bank erosion. Nat Hazards 61:65–83
62. Progeo Srl (2010) Indagini geognostiche e geofisiche al di caratterizzare la natura geologico-tecnica del versante in frana in località Coudrey, Courmayer (AO), frana del Mont de la Saxe

Chapter 13
Machine Learning Techniques in Landslide Susceptibility Mapping: A Survey and a Case Study

Taskin Kavzoglu, Ismail Colkesen, and Emrehan Kutlug Sahin

Abstract Machine learning techniques have been increasingly employed for solving many scientific and engineering problems. These data driven methods have been lately utilized with great success to produce landslide susceptibility maps. They give promising results particularly for mapping large landslide prone areas with limited geotechnical data. This chapter surveys their use in landslide susceptibility analysis and presents a case study investigating their effectiveness with regard to a conventional statistical method, namely logistic regression. It starts with the importance of spatial prediction of future landslides from past and present ones and discusses the requirement of advanced techniques for landslide susceptibility mapping. A critical literature survey is given under five main categories including core algorithms and their ensembles together with their hybrid forms. An application is presented for machine learning application using bagging, random forest, rotation forest and support vector machines with their optimal settings.

13.1 Introduction

Having a primary role in the establishment and development of residential settlements, natural disasters have been a major research topic particularly for geoscientist and engineering professionals. Rapid population growth increases pressure on natural resources and the natural environment, and raises the consequent risk associated with human activities [1]. Prediction and risk assessment of natural disasters, which can be classified into two broad groups as hydro-meteorological and geophysical disaster, are vital for planning and mitigation studies that reduce the number of human and economic losses. It is a fact that the frequency and the scale of natural disasters have increased considerably, mainly as a result of the climate change and uncontrolled human-induced changes in the landscape

T. Kavzoglu (✉) · I. Colkesen · E. K. Sahin
Department of Geomatics, Gebze Technical University, Gebze-Kocaeli, Turkey
e-mail: kavzoglu@gtu.edu.tr

S. P. Pradhan et al. (eds.), *Landslides: Theory, Practice and Modelling*,
Advances in Natural and Technological Hazards Research 50,
https://doi.org/10.1007/978-3-319-77377-3_13

(e.g. deforestation and road construction), which weakens the soil stability. As a result, the cost related to the disasters has been an increase over time. According to the Centre for Research on the Epidemiology of Disasters, at least 17% of all fatalities from natural hazards are due to the landslides [2].

Landslides, a type of natural disaster causing severe human loses and property damage, are geological phenomena related to ground movements of rock fall, and debris flow. A landslide can be described as the movement of a mass of rock, debris, or earth down a slope, under the influence of gravity [3]. Landslides are the sudden onset disasters that are usually triggered by several factors, such as intense rainfall, snow melting, earthquakes, volcanic eruptions and land use changes undermining slope instability. It should be also mentioned that climate change resulting from the global warming is one of the driving forces for landslide since it causes increased temperature, higher intensity and frequency for rain events, and lower summer precipitations. Therefore, attention should be paid to disaster mitigation and contingency planning studies by the land-use planners and policy makers, pertaining to sustainable development and reducing the risk from potential landslide events. As underlined by [4], prevention from landslides may only possible to a limited degree. However, improved understanding of the causes may help to stop or limit the human actions that increase ground instability. Although more emphasis has been given to the investigation of possible landslide locations, the increasing trend is expected to continue for three major reasons: increased and uncontrolled urbanization, continued deforestation and increased precipitation caused by climate change [5].

Determining the location of landslide prone areas is of crucial importance for hazard management studies, which is generally conducted through landslide susceptibility analysis considering various meteorological and geo-environmental parameters. It is well-known that conventional methods of ground geotechnical survey are costly both in terms of time and money, also impractical for large regions although they produce more reliable landslide hazard maps. Landslide susceptibility mapping is based on a basic modelling concept that new landslides are most likely to take place at lands having similar geological, geophysical and environmental characteristics of the previous landslide locations. The susceptibility modelling approach, compared to the conventional ones, is rapid and cost-effective with no limitation for the size of the study area. In general, landslides are complicated geophysical processes related to geology, geomorphology and hydrogeology of the ground. They occur due to the existence of various factors that show site-to-site variation. In fact, susceptibility analysis includes a variety of uncertainties that make its modelling a difficult task requiring improved knowledge about the study area characteristics and advanced techniques to model inherent relationships from a complicated structure represented by various data types. The study of factors or conditions that cause slope instability and the triggering factors or processes is of primary importance in the analysis of landslide susceptibility [6].

Prior to any conceptualizing and modelling, dealing with the landslide phenomenology requires a profound understanding of the triggering and conditioning factors that are in control of the landslide process [7]. In the estimation of landslide susceptibility, causative (i.e. conditioning or preparatory) factors are usually considered in modelling or learning the characteristics of the problem rather than triggering factors that are usually sudden hydrological or geotechnical changes. Causative factors are mainly related to geology, geomorphology, soil structure, road network, land cover type generally derived from remotely sensed images, weathering condition and hydrogeological condition of the study area. Determination of optimum number of conditioning factors for a study area is one of the most important and difficult task in landslide susceptibility assessment. Up to now, there are not any guidelines agreed universally for the determination of case-specific conditioning factors [8]. It should be pointed out that landslide conditioning factors showed variation with respect to the study area and its geographical locations. Thus, every study area has its own particular set of factors causing landslides [9]. In other words, a specific factor can be a causative one for one region but not contributing one in another region. Therefore, selection of causative factors for a particular landslide problem is a difficult task that is usually performed by the user's experience and availability of the data. In the literature, a large number of factors have been considered as causative factors in susceptibility analysis with varying degree of usage [10, 11]. Since the number of causative factors has recently increased, some data analysis techniques are certainly required to identify the directly related factors [12]. Increasing the number of causative factors appears beneficial to improve modelling quality, but it may reduce the prediction accuracy due to the involvement of highly correlated, redundant, sometimes irrelevant factors. This issue is likely to be a major problem for future studies having large number of factors at hand. Several solutions to this problem including the use of genetic algorithm, a priori ranking of factors, using certainty factor, testing various combinations of the factors have been lately investigated by researchers [8, 13, 14].

Since the 1970s, many scientists have proposed approaches to produce susceptibility maps showing the location of possible landslide locations. Methods used in landslide susceptibility mapping are diverse and numerous [15, 16]. They can be categorized into two groups as qualitative and quantitative methods. Qualitative methods, simple methods mainly applied in 1970s, are used with expert judgments and experience considering direct field measurements. Quantitative methods, on the other hand, are based on mathematically and statistically rigorous objective methodologies. They are intended to reduce the subjectivity of landslide susceptibility evaluation by incorporating statistical and geotechnical models. These methods have become popular due to their simple expressions of the dependent (i.e. landslides) and independent (i.e. conditioning factors) variables [17]. Quantitative methods can also be categorized into several subgroups: statistical, geotechnical and heuristic methods. While the statistical methods estimate the relationship between the causative factors and past landslides using bivariate and multivariate methods, deterministic methods also called geotechnical methods are case-specific ones using some factors measured on site, ignoring climate and human-induced factors. Logistic regression, weight of evidence and analytical hierarchy process are well-known statistical methods that are usually applied as benchmark methods when a new method or approach is proposed. Heuristic methods also known as machine learning techniques employ advanced algorithms to model the inherent complex relationship through the analysis of causative factors for landslide and non-landslide locations. They introduce nonlinearity and do not assume any distribution for the dataset. Machine learning is an automated model building approach for data analysis that learns the underlying relationships or hidden insights in the data to construct analytical models. Thus, they can be used to produce accurate and repeatable results through iterative learning, despite not being explicitly programmed to do so. Up to now, many machine learning methods have been proposed and practiced in landslide susceptibility assessment, but the most popular ones have been the support vector machines, decision trees, artificial neural networks, and ensemble methods including bagging, random forest and rotation forest. Machine learning methods with some selected studies are listed in Table 13.1, showing that kernel and tree-based methods together with their hybrid versions are now popularly practiced in the current literature. While first application of machine learning methods were applied using the core algorithms of neural networks, decision trees and support vector machines, current studies have largely focused on their improved versions using their derivatives, hybrids or ensemble forms. These algorithms have been not only used for susceptibility assessment but also used for some other applications including detection of landslide locations [18], selection of relevant conditioning factors [8] and landslide displacement prediction [19].

Table 13.1 Review of machine learning methods applied in modelling landslide susceptibility

Category	Method	Citations
Neural networks	Adaptive neuro-fuzzy inference system	[14, 20–22]
	Back-propagation neural network	[23–29]
	Extreme machine learning	[30]
	Learning vector quantization	[31]
	Multivariate adaptive regression splines	[32–34]
	Radial basis function neural networks	[26]
	Self-organizing map	[35]
Fuzzy-based	Fuzzy clustering	[36, 37]
Hybrid	k-means with particle swam opt.	[35]
	SVM-particle swarm optimization	[38]
	Wavelet packet-statistical models	[39]
Kernel-based	Gaussian process	[40]
	Kernel logistic regression	[26, 41]
	Support vector machines (SVM)	[7, 13, 14, 25, 26, 29, 42–44]
Tree-based	Bagging	[15, 45, 46]
	Boosting	[45, 46]
	Chi-squared automatic interaction detection	[47, 48]
	Decision trees	[8, 14, 28, 32, 49, 50]
	Functional tree	[46]
	Logistic model tree	[26]
	Random forest	[43, 51–53]
	Rotation forest	[54]

Machine learning methods have become increasingly ubiquitous throughout the hazard management and mitigation studies, especially in landslide susceptibility zonation. A comparative analysis of machine learning methods for the production of landslide susceptibility maps of Macka district of Trabzon in Turkey was carried out using available eight conditioning factors. For this purpose, the most popular machine learning algorithms, namely bagging, random forest (RF), rotation forest (RotFor) and support vector machines (SVM), used in susceptibility assessment. Moreover, their performances were compared with the conventional method, i.e. logistic regression (LR). For the evaluation of predictive ability of the bagging, RF, RotFor, SVR and LR models the root mean squared error (RMSE) and mean absolute error (MAE) were estimated from the differences between the predicted susceptibility index values derived from the models and known values of the test samples to determine the precision and bias of the predictions, respectively. Three common statistical measures, namely overall accuracy, receiver operating characteristic (ROC) curve and value of area under the ROC curve (AUC), was also calculated to compare their performances. The differences in model performances were analysed using Wilcoxon's signed rank test.

13.2 Study Area

This study was conducted on Mackaregion of Trabzon, Turkey (Fig. 13.1). The study area covers approximately 855 km^2 rugged terrain, situated between 39° 19′ and 39° 47′ longitudes, and 40° 55′ and 40° 36′ latitudes. Due to its physiographic

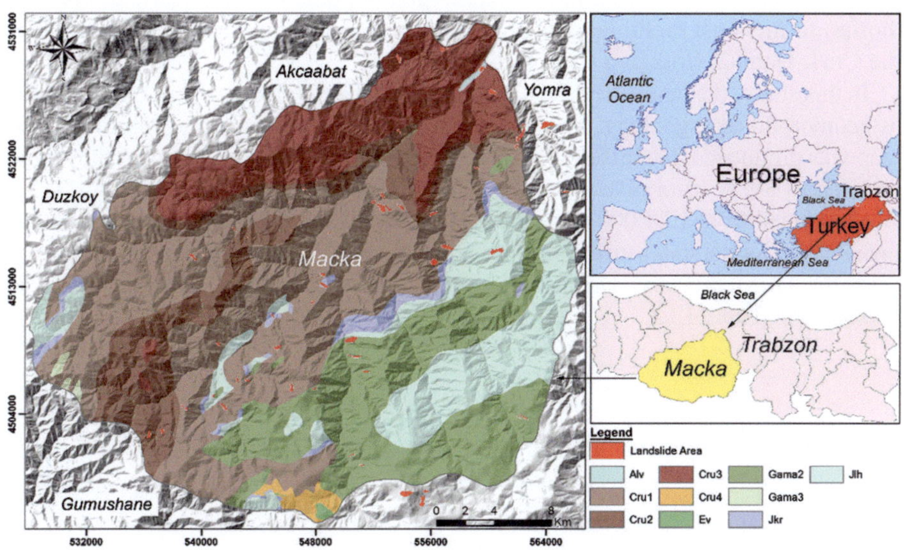

Fig. 13.1 Geological map of the study area and landslide inventory

Table 13.2 Detailed information related to factor maps

Major factors	Sub-factors	Sub-classes
Geology	Lithology	Jlh, Jcr, Cru1, Cru2, Cru3, Cru4, Ev, Gama2, Gama3, Alv
Geomorphology	Elevation (m)	80–576, 576–888, 888–1182, 1182–1.468, 1468–1742, 1742–2005, 2005–2266, 2266–2820
	Slope (°)	0–10.16, 10.16–16.41, 16.41–22.14, 22.14–27.09, 27.09–31.52, 31.52–35.95, 35.95–41.42, 41.42–66.42
	Aspect	Eight principal directions (N, NE, E, SE, etc.) and flat areas (−1°)
	Plan curvature	Concave, flat, convex
Hydrology	TWI	−0.83–1.07, 1.07–1.93, 1.93–2.78, 2.78–3.80, 3.80–5.21, 5.21–7.22, 7.22–10.07, 10.07–17.94
Land cover	Land use/cover	Urban, water, green tea, hazelnut, agriculture, deciduous, pasture, coniferous, soil/rock
	NDVI	−0.24–0.16, 0.16–0.26, 0.26–0.35, 0.35–0.42, 0.42–0.49, 0.49–0.55, 0.55–0.62, 0.62–0.76

conditions, land characteristics and climatic conditions such as above average rainfall and soil structure, the region has witnessed many landslides at different scales. The mean annual precipitation over the study area is approximately 200 mm and many of the recent landslides in the study area are triggered by heavy rainfalls. Elevations range from 80 to higher than 2800 m, and the slope angles reach 66°. The lithology map (Fig. 13.1) produced by the General Directorate of Mineral Research and Exploration covering ten types of geological formations was utilized. The study area is covered by Cru1 (basalt, andesite, lava and pyroclastic), Cru3 (basalt, andesite, lava and pyroclastic) and Gama2 (granite, granodiorite, quartz diorite and diorite) formations. Preliminary analysis shows that most of landslides occur under the Cru1, Gama2, Cru2 and Cru3 formations.

In this study, the landslide causative factors can be divided into four major groups as geomorphology, geology, hydrology and land cover. Since each causative factor map was produced from different sources, they were at different scales, so they were rescaled to 30 × 30 m pixel resolution. Moreover, digital elevation model (DEM) was produced from 1:25,000 scale topographic maps through digitization of contour lines. Elevation, slope, aspect, plan curvature and TWI factor maps were extracted from the DEM imagery. Detailed information related to the causative factors is presented in Table 13.2.

13.3 Landslide Inventory

Preparing a representative landslide inventory map is of crucial importance in all versions susceptibility mapping models. According to the basic assumption that future landslides will most likely happen in similar physiographic settings of the past and present landslides [55]. Therefore, it is highly important to designate the location the past and current landslides correctly. In the literature, there is not any consensus on how landslide inventory maps should be prepared. Some sampling strategies have been proposed by researchers. In the present case study, a widely-used strategy based on the use of polygons representing the spatial location of a landslide was employed in the preparation of inventory map. In this study, landslide locations were taken from the map produced with "Turkish Landslide Inventory Mapping Project" by MTA Institute, Turkey. Totally 54 landslide (5144 pixels) and 23 non-landslide (1188 pixels) polygons were taken into consideration. Location analysis of the past landslides revealed that the north-east and south-west directions of the study area witnessed substantially more landslides. Spatial distribution of the landslides in the inventory map was statistically analysed, and found that the average landslide size was 85,908 m^2 and their acreage ranged from 6017 to 291,924 m^2. The minimum and maximum lengths of landslide are 117 m and 1092 m, respectively. Sixty percent of the landslides range from 120 to 500 m in length. Their width ranged between 65 and 451 m.

13.4 Methodology

Figure 13.2 shows the process adopted in this study for landslide susceptibility modelling. The process begins with the preparation of landslide inventory and causative factor maps. Subsequently, eight landslide conditioning factors (lithology, elevation, slope, aspect, plan curvature, topographical wetness index, land use/cover and NDVI) were considered for susceptibility mapping process using bagging decision tree, random forest (RF), rotation forest (RotFor) and support vector machine (SVM) methods. In addition, the logistic regression (LR) was utilized for comparison purpose.

13.4.1 Bagging Algorithm

Introduced by [56], bagging (or bootstrap aggregating) has been successfully applied to many classification and regression problems. Bagging algorithm aims to select a training sample using a bootstrap aggregating (a sample collected with replacement) from the original input training set and build a learning model. By means of bootstrap aggregating technique, it is possible to generate different training data sets and hence construct diverse learning algorithms in ensemble model. In other words, bagging decreases the overall prediction error or stabilizes individual weak learners (i.e. decision tree) by reducing variance. The resulting ensemble model created by bagging combines the predictions of multiple learning algorithms to make a final decision. Consider a given an input training data set containing n number of training examples, a sample of n training examples is generated by sampling with

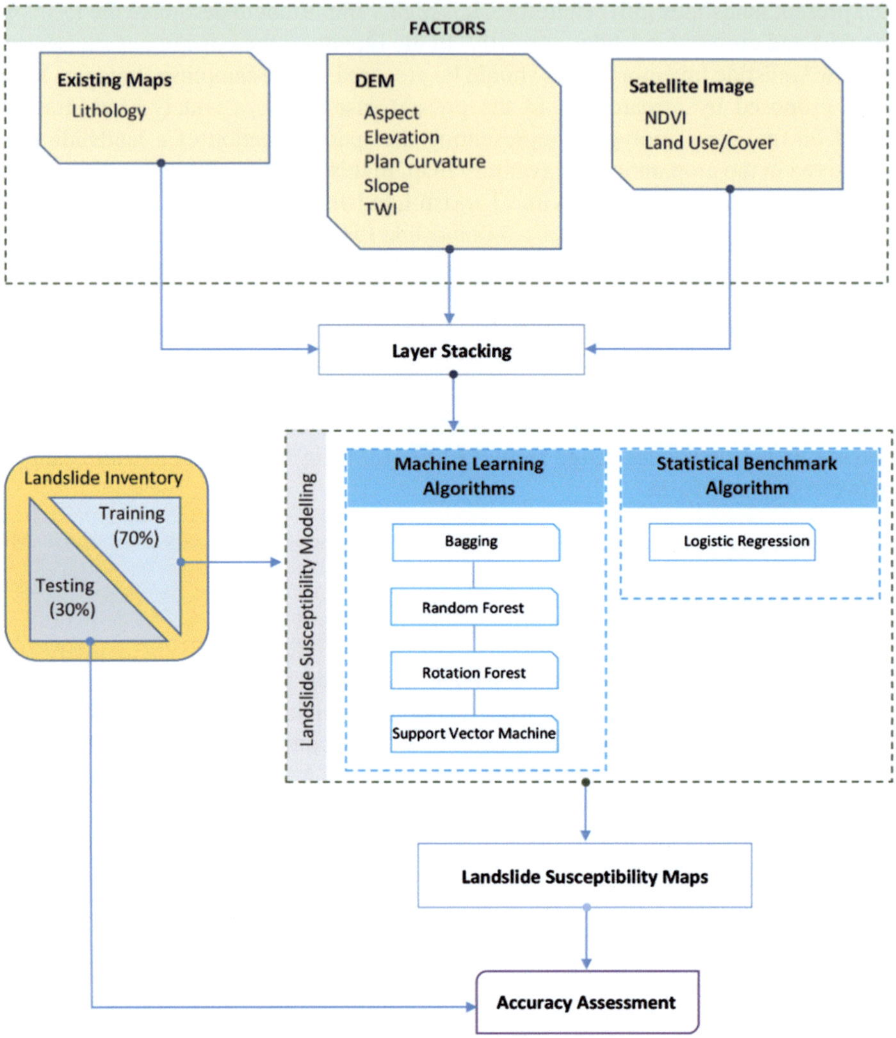

Fig. 13.2 Flowchart of the methodology used in this study

replacement. For training of each individual learning algorithm in ensemble model, this process is employed iteratively. Final prediction of a test sample is performed by combining a vote of the predictions of each individual learner (i.e. majority voting procedure). Although bagging is used to reduce the model variance, it is not successful in reducing the model bias. Thus, the trees in the ensemble model become correlated, limiting the level of error reduction. Therefore, it is advisable to determine components of the ensemble model to minimize the bias at the possible expense of variance [57].

13.4.2 Random Forest Algorithm

Random forest (RF) developed by [58] is widely-used ensemble learning algorithms successfully applied for classification, regression and feature selection purposes. RF is based on the idea that builds a set of decision trees, using randomly selected training samples through bootstrap aggregating strategy to make final a prediction. About two thirds of the selected samples known as in-bag samples are used for training of the decision tree with the remaining one third known as out-of-bag samples are used in an internal cross-validation to estimate the predictive accuracy of the constructed tree model. The output is decided by a majority voting. The underlying philosophy of RF is that the 'strength' of the trees is maintained while reducing the correlation between the trees in the forest. For the implementation of RF algorithm, two parameters (the number of trees and the number of variables) have to be set by the analyst. In order to construct a random forest ensemble model, two randomization processes are employed. First, training samples for each individual tree are randomly selected by applying bootstrap sampling strategy. Second, instead of selecting the best split, the tree inducer randomly samples a subset of the attributes and chooses the best one [59]. For this reason, RF can be viewed as an enhanced or generalized version of the bagging method that builds a randomized decision tree at each iteration.

13.4.3 Rotation Forest Algorithm

Rotation forest (RotFor) is an advanced ensemble learning algorithm, used to generate accurate and diverse classifiers [60]. RotFor applies a linear transformation method, principal component analysis (PCA), to the original feature subsets to project data in to a new feature space for each individual classifier in the ensemble model [61]. In the each iteration of the ensemble model construction process, the input features are randomly divided into k subsets. Then, PCA is applied to the each subset to extract the principal components of rational features. As a result, k sets of principal components are used to training of the each individual classifier of the ensemble model. To increase diversity, the bootstrap sampling strategy applied to the data created in each circle before the principal components transformations are applied.

13.4.4 Support Vector Machine Algorithm

Survey of literature reveals that support vector machine (SVM) has been one of the most popular kernel-based supervised learning algorithms, successfully applied to various fields. The main idea behind the SVM is to seek an optimal hyperplane that provides maximum separation between linearly separable two classes. For non-separable cases, the data set is moved to a higher dimensional space using a kernel function to find the linear separation. When the SVM is used to model the complex non-linear problems including function approximation and regression estimation, the algorithm is often referred to as support vector regression (SVR) [62]. SVR algorithm assumes that each set of input features (i.e. landslide conditioning factors) has unique relation to its target variable (i.e. landslide susceptibility index). Thus, the SVR algorithm identifies the rules to estimate the target values of unknown test data samples from a set of inputs [63].

13.4.5 Logistic Regression Algorithm

Logistic regression (LR) is the most commonly applied multivariate analysis for producing landslide susceptibility maps. The LR method seeks relationship between a dependent variable (the presence or absence of landslides) and independent variables (i.e. conditioning factors). A linear fitting model is estimated describing the relationship between the dependent and independent variables. The LR process is associated with the probability of landslide phenomena to the "logit" Z (where $-\infty < Z < 0$ for higher odds of non-occurrence and $0 < Z < \infty$ for higher odds of occurrence) [64]. The LR function *Logit(p)* equation is as follows:

$$\text{Logit}(p) = \log\left[\frac{p}{1-p}\right] \tag{13.1}$$

In this equation, p is the probability that the dependent variable ranging from 0 to 1, and $(p/(1-p))$is the so-called odds or likelihood ratio. Using the logit transformations, the multiple linear regression equation can be written as:

$$\text{Logit(pi)} = \beta_0 + \sum_{i=1}^{n} \beta_i x_i \tag{13.2}$$

Where β_0 is the intercept, β_{i} indicates the coefficients measuring the contribution of independent variables x_{i}, and n show the number of independent variables.

13.5 Results

This study investigates the performance of the machine learning algorithms in comparison to logistic regression method for a study area in Turkey. For building landslide susceptibility models, training and test datasets including landslide and non-landslide samples were randomly selected from a landslide inventory map. The inventory data were randomly divided into training and testing datasets considering

70:30 sampling ratio. The same datasets were used to build regression models of the algorithms considered in this study. The Weka software (v.3.8) was utilized for bagging, RF, SVR algorithms, and SPSS (v.22) software was chosen for implementation of the LR method. Susceptibility index maps produced by the algorithms were reclassified into five common susceptibility levels by applying equal interval approach. For a successful susceptibility analysis using any parameter-based technique, it is crucially important to find and set optimal parameter values. Parameterization of machine learning algorithms considered here is explained as follows.

For the construction of bagging ensemble prediction model, decision tree algorithm was used as the base learner. The number of iteration is a critical user-defined parameter for the implementation of the bagging algorithm. A cross-validation strategy (i.e. fivefold) was applied to determine the optimum iteration number using a training dataset. The cross-validation result showed that optimum iteration number of bagging ensemble model was 30 for this study. In order to evaluate the predictive power of bagging, two standard statistical metrics namely, RMSE and mean absolute error (MAE) were also calculated from the test dataset, and the RMSE and MAE values of 0.303 and 0.186 were estimated respectively.

For the application of RF algorithm, the number of trees (n) and the number of input variables considered in each node split (k) are to be set by the user. The input data set consisted of eight landslide conditioning factors, hence the number of input variables (m) was set to be 3 (i.e. $k = \sqrt{m}$ variables at each split). On the other hand, out-of-bag (OOB) error results of RF ensemble model were used to determine the number of trees parameter. For this purpose, input data set was firstly classified using a large number of trees (i.e. 500 trees) to estimate changes in OOB error with increasing number of trees. The resulting graph showing the relations between OOB error and the number of trees (n) was given in Fig. 13.3.

It was observed that there was a sharp decline in OOB error from 0.183 to less than 0.05 as number of tree increased from 1 to 50. After that, OOB error continued to decrease slightly until the number of trees takes value 200. From this critical point to larger tree sizes, OOB error stays stable. For this reason, the number of trees (n) was set to be 200 for the current study. The predictive accuracy of RF model constructed with the user-defined parameters was tested using RMSE and MAE statistics, estimated as 0.290 and 0.193, respectively.

Two parameters of RotFor method had to be determined. In order to determine the optimum number of iterations, a cross-validation strategy was applied considering the training dataset and 70 iterations were estimated optimal. In the search for optimal number of splits (K), it was observed that changes in the parameter value have no effect on the prediction results. Consequently, it was set to 3 for the model building. RMSE and MEA were calculated as 0.329 and 0.256, respectively.

Radial basis function (RBF) kernel function was chosen in the implementation of the SVR algorithm. In SVR application, three parameters are needed to define from user-side. Meta-parameters of regularization parameter C, threshold value ε, and kernel width γ were determined by grid search method. As a result, ε value of 0.001, C value of 1.250, and γ value of 0.1 were determined as optimal. RMSE and MEA were calculated as 0.308 and 0.173, respectively.

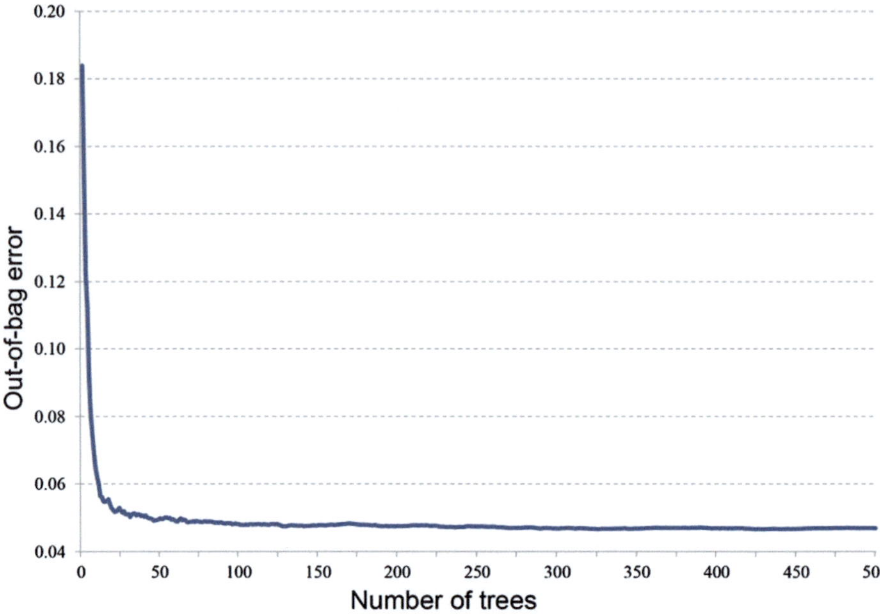

Fig. 13.3 Out-of-bag (OOB) error graph for random forest estimation

In performing the LR method, independent variables including the eight causative factors (lithology, elevation, slope, aspect, plan curvature, TWI, LULC and NDVI) and the dependent variable as landslide areas were used. The LR method was utilized to model the spatial relationship between the landslides and causative factors. The standard errors (SE), regression coefficients (β), Wald test statistic and associated p-values were estimated. Among all factors, slope was the most contributing factor since it had the highest coefficient value. Other effective factors were lithology and elevation. On the other hand, the coefficients estimated for NDVI and TWI were close to 0, indicating the minor impacts or weak relation to landslide occurrence. All causative factors had p-values lower than 0.1, indicating statistical significance between factors and the susceptibility to landslide at the 90% confidence level.

All methods with the above-mentioned parameter settings were applied to the multi-layer dataset to yield susceptibility maps. Histogram values were categorized into five susceptibility classes using quantile approach to obtain susceptibility maps (Fig. 13.4). The predictive powers of the landslide susceptibility models were measured using overall accuracies calculated using the test dataset. It should be noted that two susceptibility classes as very high and high level of the susceptibility map were considered as potential landslide occurrence and the rest (i.e., moderate, low and very low) were considered as non-landslide in accuracy assessment process. Overall accuracies for bagging, RF, RotFor, SVR and LR methods were estimated as 83.08%, 87.23%, 85.31%, 84.85% and 78.46%, respectively. Results revealed that the highest accuracy was estimated with the RF ensemble model (87.23%), whereas

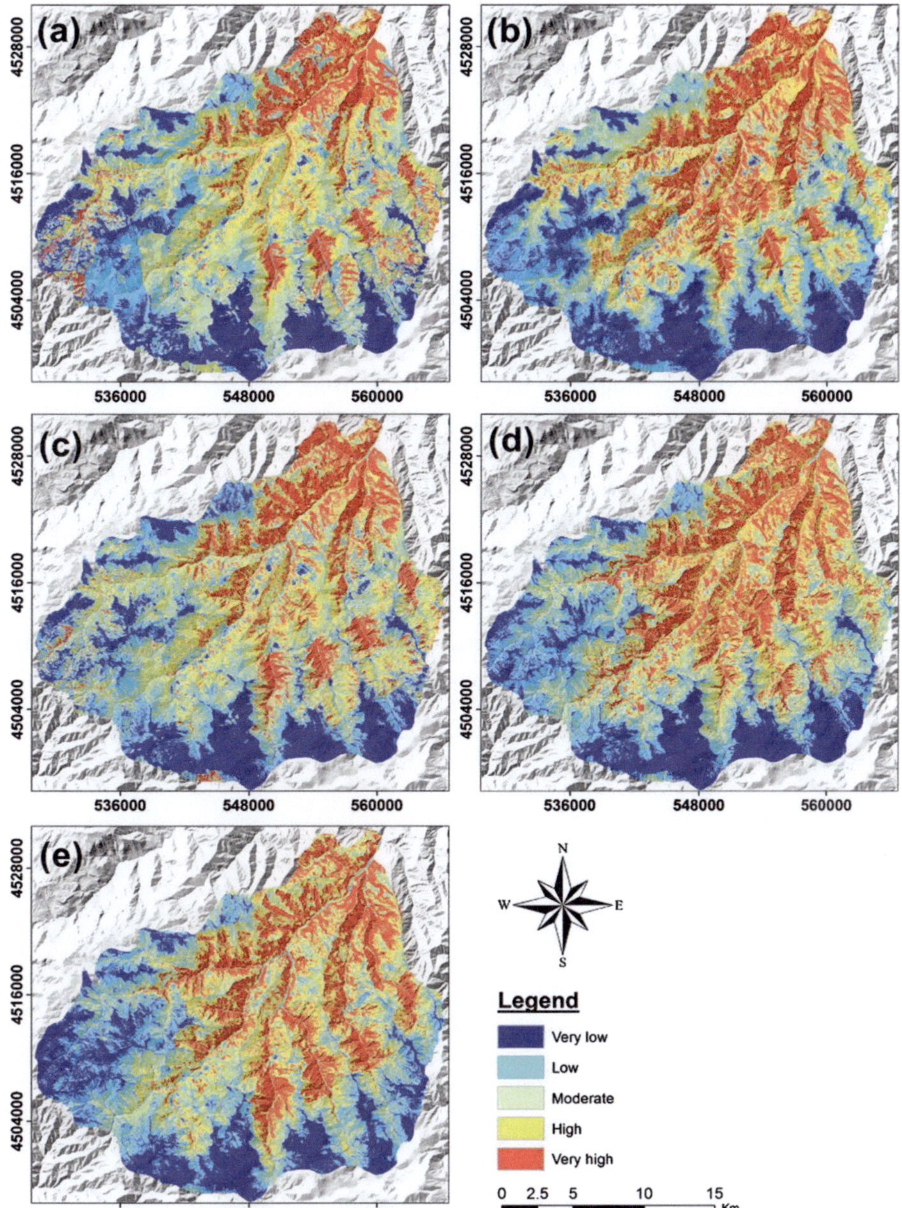

Fig. 13.4 Landslide susceptibility maps produced by (**a**) bagging, (**b**) random forest, (**c**) rotation forest, (**d**) support vector machine and (**e**) logistic regression

the lowest accuracy was calculated with traditional LR model (78.46%). RotFor and SVR algorithms showed similar performances and calculated accuracies of their predictive models were 85.31% and 84.85%, respectively. Results clearly showed that the machine learning algorithms produced more accurate results in comparison with the traditional LR method, and they improved the prediction accuracy up to 9% in terms of overall accuracy. This finding supported the results of some previous studies [65–68] whilst there are limited studies showed the effectiveness of LR method compared to machine learning algorithms [28, 41, 69].

In order to further evaluate the performances of the obtained susceptibility models, ROC curve and AUC value were applied. ROC curves are constructed by correctly classified pixels (sensitivity) and incorrectly identified pixels (1-specificity).

The estimated AUC values for bagging, RF, RotFor, SVR and LR models were 0.931, 0.963, 0.959, 0.955, and 0.868, respectively (Fig. 13.5). AUC values confirm the overall accuracy results for the method performances. From plotted the ROC curves, it was revealed that all susceptibility models produced acceptable results. When the estimated AUC values were analysed, the machine learning algorithms were much more effective for landslide susceptibility assessment, and the RF method produced the highest AUC value of 0.963, followed by RotFor (AUC = 0.959), SVR (AUC = 0.955) and bagging (AUC = 0.931). It is clear from the obtained results that the machine learning algorithms outperformed the standard LR model with higher AUC values.

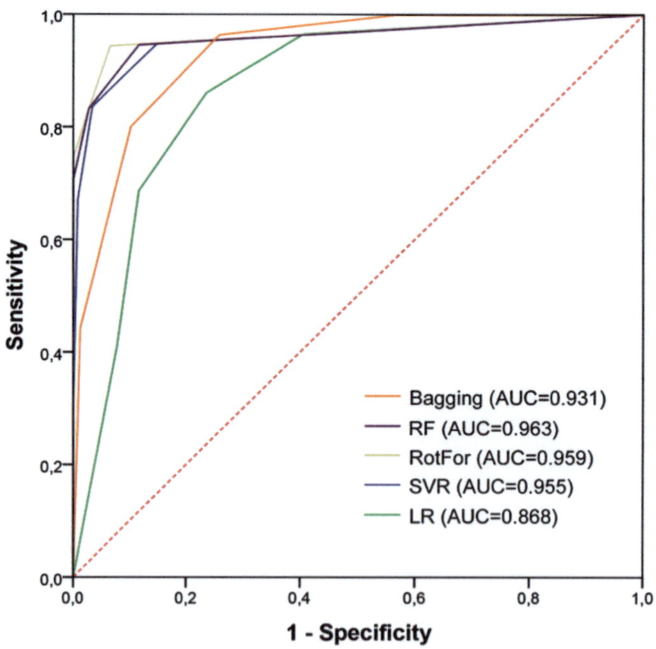

Fig. 13.5 Results of ROC curves and AUC values for the different methods

Table 13.3 Wilcoxon's signed-rank test statistic for landslide susceptibility models. Note that estimated values greater than the table value ($Z_{1,\ 0.05} = 1.96$) indicates statistical significance

	Bagging	RF	RotFor	SVR	LR
Bagging	–	8.464	5.104	2.954	9.913
RF		–	6.755	7.073	18.961
RotFor			–	2.860	11.824
SVR				–	10.741

In addition to the assessment of five susceptibility model performances using overall accuracy and AUC values, Wilcoxon's test was also employed to validate the significance of differences statistically. If the estimated statistic value is larger than critical table value ($Z_{1,\ 0.05} = 1.96$), the null hypothesis can be rejected with 95% confidence level. Calculated statistical test results were given as a matrix in Table 13.3. From the table, all estimated statistic values obtained by pairwise comparisons were greater than the critical table value. Therefore, it can be concluded that differences in the model performances was found to be statistically significant. In other words, performance of the RF method was statistically better than the other machine learning methods. This could be related to the RF characteristics of being non-parametric, capable of using continuous and categorical data, easy to parametrize, robust against overfitting, and not being sensitive to noise in the dataset.

13.6 Conclusions

Producing accurate and reliable landslide susceptibility maps representing the areas prone to landslides has been one of the most concentrated topics in hazard management. These maps are used as a base map in many global and regional studies on hazard management and planning. Therefore, reliability and accuracy of the landslide susceptibility maps has crucial importance in order to achieving the desired goals of the management plans. Landslide susceptibility mapping consists of complex and multi-stage steps including preparation of landslide inventories, selection of landslide causative factors, determining an appropriate prediction algorithm and accuracy assessment. Up to now, many methods or frameworks have been proposed so as to increase prediction accuracy of landslide models, and their performances have been investigated for susceptibility of landslides. In this study, four well-known machine learning algorithms, namely ensemble based bagging, random forest, rotation forest and kernel-based support vector machines, were employed in the process of susceptibility assessment. Their performances were compared to that of the LR method, which can be regarded as a conventional statistical approach.

Result of this study revealed some important findings. Firstly, when the estimated overall accuracies were analysed, the machine learning methods clearly outperformed conventional logistic regression methods (up to 9% improvement). ROC curves and related AUC statistics also supported the above finding. In addition, performance differences were concluded as statistically significant based on Wilcoxon's test. Secondly, the results produced in this study also revealed that RF algorithm produced the best performance (overall accuracy of 87.23% and AUC value of 0.963) among the machine learning algorithms and the difference in their performances was statistically significant at 95% confidence interval. Thirdly, among the ensemble models based on bootstrap aggregating strategy considering accuracy results, the RF and RotFor algorithms was superior to bagging algorithm. This is an expected result since both the RF and RotFor methods are improved versions of the bagging. However, the processing time required to form an ensemble model is longer, particularly for RotFor method including PCA processing behind. From the findings of the study and the literature review, it was induced that the machine learning methods are invaluable tools for landslide susceptibility assessment, and they should be favoured over statistical methods particularly for cases including a large number causative factors and limited landslide locations at hand. On the other hand, literature survey reveals that instead of single usage of the methods, hybrid and ensembles of machine learning methods will play an important role in future studies for improvement in predictive power of landslide susceptibility evaluation.

References

1. Kreimer A, Munasinghe M (1991) Managing environmental degradation and natural disasters: an overview. In: Kreimer A, Munasinghe M (eds) Managing natural disasters and the environment. World Bank, Washington, DC, pp 3–6
2. Lacasse S, Nadim F (2009) Landslide risk assessment and mitigation strategy. In: Sassa K, Canuti P (eds) Landslides – disaster risk reduction. Springer, Berlin
3. Varnes DJ (1978) Slope movement types and processes. In: Schuster RL, Krizek RJ (eds) Landslides: analysis and control. Transportation Research Board, National Academy of Sciences, Washington, DC
4. Marsh SH (2000) Landslide hazard mapping: summary report. British Geological Survey Technical Report WC/00/11, 25p
5. Schuster RL (1996) Socioeconomic significance of landslides. In: Turner AK, Schuster RL (eds) Landslides: investigation and mitigation, Special Report 247. Transportation Research Board, National Research Council, National Academy Press, Washington, DC, pp 12–35
6. Dai FC, Lee CF, Ngai YY (2002) Landslide risk assessment and management: an overview. Eng Geol 64:65–87
7. Marjanović M, Kovacevic M, Bajat B et al (2011) Landslide assessment of the starcˇa basin (Croatia) using machine learning algorithms. Acta Geotech Slov 2011:45–55
8. Kavzoglu T, Sahin EK, Colkesen I (2015) An assessment of multivariate and bivariate approaches in landslide susceptibility mapping: a case study of Duzkoy district. Nat Hazards 76:471–496
9. van Westen CJ, Rengers N, Soeters R (2003) Use of geomorphological information in indirect landslide susceptibility assessment. Nat Hazards 30:399–419
10. Hasekiogullari GD, Ercanoglu M (2012) A new approach to use AHP in landslide susceptibility mapping: a case study at Yenice (Karabuk, NW Turkey). Nat Hazards 63:1157–1179
11. Suzen ML, Kaya BS (2012) Evaluation of environmental parameters in logistic regression models for landslide susceptibility mapping. Int J Digit Earth 5:338–355

12. Smola A, Vishwanathan SVN (2008) Introduction to machine learning. Cambridge University Press, Cambridge, UK
13. Costanzo D, Rotigliano E, Irigaray C et al (2012) Factors selection in landslide susceptibility modelling on large scale following the GIS matrix method: application to the river Beiro basin (Spain). Nat Hazards Earth Syst 12:327–340
14. Pradhan B (2013) A comparative study on the predictive ability of the decision tree, support vector machine and neuro-fuzzy models in landslide susceptibility mapping using GIS. Comput Geosci 51:350–365
15. Brenning A (2005) Spatial prediction models for landslide hazards: review, comparison and evaluation. Nat Hazards Earth Syst 5:853–862
16. Marrapu BM, Jakka RS (2014) Landslide hazard zonation methods: a critical review. Int J Civ Eng Res 5:215–220
17. Bai SB, Wang J, Thiebes B et al (2014) Susceptibility assessments of the Wenchuan earthquake-triggered landslides in Longnan using logistic regression. Environ Earth Sci 71:731–743
18. Wan S, Yen JY, Lin CY et al (2015) Construction of knowledge-based spatial decision support system for landslide mapping using fuzzy clustering and KPSO analysis. Arab J Geosci 8:1041–1055
19. Lian C, Zeng ZG, Yao W et al (2013) Displacement prediction model of landslide based on a modified ensemble empirical mode decomposition and extreme learning machine. Nat Hazards 66:759–771
20. Pradhan B, Sezer EA, Gokceoglu C et al (2010) Landslide susceptibility mapping by neuro-fuzzy approach in a landslide-prone area (Cameron Highlands, Malaysia). IEEE T Geosci Remote 48:4164–4177
21. Choi J, Lee YK, Lee M et al (2011) Landslide susceptibility mapping by using an adaptive neuro-fuzzy inference system (ANFIS). In: 2011 I.E. international geoscience and remote sensing symposium (IGARSS):1989–1992
22. Aghdam IN, Varzandeh MHM, Pradhan B (2016) Landslide susceptibility mapping using an ensemble statistical index (Wi) and adaptive neuro-fuzzy inference system (ANFIS) model at Alborz Mountains (Iran). Environ Earth Sci 75:553
23. Lee S, Ryu JH, Lee MJ et al (2003) Use of an artificial neural network for analysis of the susceptibility to landslides at Boun, Korea. Environ Geol 44:820–833
24. Gómez H, Kavzoglu T (2005) Assessment of shallow landslide susceptibility using artificial neural networks in Jabonosa River Basin, Venezuela. Eng Geol 78:11–27
25. Yilmaz I (2010) Comparison of landslide susceptibility mapping methodologies for Koyulhisar, Turkey: conditional probability, logistic regression, artificial neural networks, and support vector machine. Environ Earth Sci 61:821–836
26. Bui DT, Tuan TA, Klempe H et al (2016) Spatial prediction models for shallow landslide hazards: a comparative assessment of the efficacy of support vector machines, artificial neural networks, kernel logistic regression, and logistic model tree. Landslides 13:361–378
27. Gorsevski PV, Brown MK, Panter K et al (2016) Landslide detection and susceptibility mapping using LiDAR and an artificial neural network approach: a case study in the Cuyahoga Valley National Park, Ohio. Landslides 13:467–484
28. Wang LJ, Guo M, Sawada K et al (2016) A comparative study of landslide susceptibility maps using logistic regression, frequency ratio, decision tree, weights of evidence and artificial neural network. Geosci J 20:117–136
29. Xu C, Shen LL, Wang GL (2016) Soft computing in assessment of earthquake-triggered landslide susceptibility. Environ Earth Sci 75:767
30. Vasu NN, Lee SR (2016) A hybrid feature selection algorithm integrating an extreme learning machine for landslide susceptibility modeling of Mt. Woomyeon, South Korea. Geomorphology 263:50–70
31. Pavel M, Nelson JD, Fannin RJ (2011) An analysis of landslide susceptibility zonation using a subjective geomorphic mapping and existing landslides. Comput Geosci 37:554–566

32. Felicisimo A, Cuartero A, Remondo J et al (2013) Mapping landslide susceptibility with logistic regression, multiple adaptive regression splines, classification and regression trees, and maximum entropy methods: a comparative study. Landslides 10:175–189
33. Wang LJ, Guo M, Sawada K et al (2015) Landslide susceptibility mapping in Mizunami City, Japan: a comparison between logistic regression, bivariate statistical analysis and multivariate adaptive regression spline models. Catena 135:271–282
34. Conoscenti C, Rotigliano E, Cama M et al (2016) Exploring the effect of absence selection on landslide susceptibility models: a case study in Sicily, Italy. Geomorphology 261:222–235
35. Wan SA (2013) Entropy-based particle swarm optimization with clustering analysis on landslide susceptibility mapping. Environ Earth Sci 68:1349–1366
36. Wang LJ, Sawada K, Moriguchi S (2013) Landslide susceptibility analysis with logistic regression model based on FCM sampling strategy. Comput Geosci 57:81–92
37. Alimohammadlou Y, Najafi A, Gokceoglu C (2014) Estimation of rainfall-induced landslides using ANN and fuzzy clustering methods: a case study in Saeen Slope, Azerbaijan province, Iran. Catena 120:149–162
38. Yu XY, Wang Y, Niu RQ et al (2016) A combination of geographically weighted regression, particle swarm optimization and support vector machine for landslide susceptibility mapping: a case study at Wanzhou in the three gorges area, China. Int J Environ Res Publ Health 13:487. https://doi.org/10.3390/ijerph13050487
39. Moosavi V, Niazi Y (2016) Development of hybrid wavelet packet-statistical models (WP-SM) for landslide susceptibility mapping. Landslides 13:97–114
40. Colkesen I, Sahin EK, Kavzoglu T (2016) Susceptibility mapping of shallow landslides using kernel-based Gaussian process, support vector machines and logistic regression. J Afr Earth Sci 118:53–64
41. Hong HY, Pradhan B, Xu C et al (2015) Spatial prediction of landslide hazard at the Yihuang area (China) using two-class kernel logistic regression, alternating decision tree and support vector machines. Catena 133:266–281
42. Yao X, Tham LG, Dai FC (2008) Landslide susceptibility mapping based on support vector machine: a case study on natural slopes of Hong Kong, China. Geomorphology 101:572–582
43. Goetz JN, Brenning A, Petschko H et al (2015) Evaluating machine learning and statistical prediction techniques for landslide susceptibility modeling. Comput Geosci 81:1–11
44. Hong HY, Pradhan B, Jebur MN et al (2016) Spatial prediction of landslide hazard at the Luxi area (China) using support vector machines. Environ Earth Sci 75:40
45. Bui DT, Ho CT, Revhaug I et al (2014) Landslide susceptibility mapping along the national road 32 of Vietnam using GIS-based J48 decision tree classifier and its ensembles. In: Buchroithner M, Prechtel N, Burghardt D (eds) Cartography from pole to pole: selected contributions to the XXVIth international conference of the ICA, Dresden 2013. Springer, Berlin/Heidelberg, pp 303–317. https://doi.org/10.1007/978-3-642-32618-9_22
46. Bui DT, Ho TC, Pradhan B et al (2016) GIS-based modeling of rainfall-induced landslides using data mining-based functional trees classifier with Ada Boost, Bagging, and Multi Boost ensemble frameworks. Environ Earth Sci 75:1101
47. Althuwaynee OF, Pradhan B, Park HJ et al (2014) A novel ensemble bivariate statistical evidential belief function with knowledge-based analytical hierarchy process and multivariate statistical logistic regression for landslide susceptibility mapping. Catena 114:21–36
48. Althuwaynee OF, Pradhan B, Lee S (2016) A novel integrated model for assessing landslide susceptibility mapping using CHAID and AHP pair-wise comparison. Int J Remote Sens 37:1190–1209
49. Saito H, Nakayama D, Matsuyama H (2009) Comparison of landslide susceptibility based on a decision-tree model and actual landslide occurrence: the Akaishi Mountains, Japan. Geomorphology 109:108–121
50. Yeon YK, Han JG, Ryu KH (2010) Landslide susceptibility mapping in Injae, Korea, using a decision tree. Eng Geol 116:274–283

51. Catani F, Lagomarsino D, Segoni S et al (2013) Landslide susceptibility estimation by random forests technique: sensitivity and scaling issues. Nat Hazards Earth Syst 13:2815–2831
52. Trigila A, Iadanza C, Esposito C et al (2015) Comparison of logistic regression and random forests techniques for shallow landslide susceptibility assessment in Giampilieri (NE Sicily, Italy). Geomorphology 249:119–136
53. Were K, Bui DT, Dick OB et al (2015) A comparative assessment of support vector regression, artificial neural networks, and random forests for predicting and mapping soil organic carbon stocks across an Afromontane landscape. Ecol Indic 52:394–403. https://doi.org/10.1016/j.ecolind.2014.12.028
54. Pham BT, Bui DT, Prakash I et al (2016) Rotation forest fuzzy rule-based classifier ensemble for spatial prediction of landslides using GIS. Nat Hazards 83:97–127
55. Klose M (2015) Landslide databases as tools for integrated assessment of landslide risk. Springer Theses, Springer, Berlin
56. Breiman L (1996) Bagging predictors. Mach Learn 24:123–140
57. Aggarwal CC (2015) Data classification: advanced concepts. In: Data mining: the textbook. Springer, Cham, pp 345–387
58. Breiman L (2001) Random forests. Mach Learn 45:5–32
59. Rokach L (2016) Decision forest: twenty years of research. Inform Fusion 27:111–125
60. Rodriguez JJ, Kuncheva LI (2006) Rotation forest: a new classifier ensemble method. IEEE T Pattern Anal 28:1619–1630
61. Rokach L (2010) Ensemble-based classifiers. Artif Intell Rev 33:1–39
62. Schölkopf B, Smola AJ (2002) Learning with kernels: support vector machines, regularization, optimization and beyond. MIT Press, Cambridge, MA
63. Gleason CJ, Im J (2012) Forest biomass estimation from airborne LiDAR data using machine learning approaches. Remote Sens Environ 125:80–91
64. Mancini F, Ceppi C, Ritrovato G (2010) GIS and statistical analysis for landslide susceptibility mapping in the Daunia area, Italy. Nat Hazards Earth Syst 10:1851–1864
65. Kavzoglu T, Sahin EK, Colkesen I (2014) Landslide susceptibility mapping using GIS-based multi-criteria decision analysis, support vector machines, and logistic regression. Landslides 11:425–439
66. Wang YT, Seijmonsbergen AC, Bouten W et al (2015) Using statistical learning algorithms in regional landslide susceptibility zonation with limited landslide field data. J Mt Sci-Engl 12:268–288
67. Feng HJ, Yu JJ, Zheng JL et al (2016) Evaluation of different models in rainfall-triggered landslide susceptibility mapping: a case study in Chunan, southeast China. Environ Earth Sci 75 (21):1–15
68. Pham BT, Pradhan B, Bui DT et al (2016) A comparative study of different machine learning methods for landslide susceptibility assessment: a case study of Uttarakhand area (India). Environ Model Softw 84:240–250
69. Yilmaz I (2009) Landslide susceptibility mapping using frequency ratio, logistic regression, artificial neural networks and their comparison: a case study from Kat landslides (Tokat-Turkey). Comput Geosci 35:1125–1138

Chapter 14
Landslides in Permafrost Zone of Russia

Stanilovskaya Julia

Abstract Recently the intensification of cryogenic processes at high rate is widespread in Russian cryolithozone. Landslides in permafrost area are characterised by frozen bottom of sliding and known as thaw slumps or mudslides. Landslide body consists of frozen soils, which thaws and forms thermo cirques and depressions. Long warm summers and snowy winters of last decade trigger cryogenic landslides over Siberia. The situation with existing linear infrastructure (roads, railways and pipelines) is changing due to permafrost thawing that has not been taken into account during design 5–30 years ago. Apart from deformations to infrastructure cryogenic landslides provoke gas release from permafrost, change of water content in lakes and rivers, and disturbance of lands (tundra and taiga) for rendering. Diversity of landslides in permafrost area and nine study cases are presented in this paper. The mitigation measures are not applicable for cryogenic landslides at high rates. There is a call for innovative solutions in this area.

14.1 Introduction

This paper aims to show the diversity of landslides in permafrost area and some hazardous case studies in Russia. This paper demonstrates demand for further investigation to prepare the mitigation measures with innovative solutions. The instrumental research of cryogenic landslides in Russia is limited to team work on the Kara sea region [1].

The international classification of landslides, including mudflows, rock falls etc., was used. A genesis of cryogenic landslides is specific to thawing or freezing processes on gently slopes. Cryogenic landslide is characterized by thaw blocks and flows. Apart from heavy precipitations and high seismicity, this kind of landslides is developed due to permafrost thawing. The main trigger parameters are increase in soil temperature and depth of thaw layer. Permafrost thawing with

S. Julia (✉)
Sergeev Institute of Environmental Geoscience, Moscow, Russia

S. P. Pradhan et al. (eds.), *Landslides: Theory, Practice and Modelling*,
Advances in Natural and Technological Hazards Research 50,
https://doi.org/10.1007/978-3-319-77377-3_14

formation of thermo cirques and depressions is linked with the climate change and surface disturbance near infrastructure or after fire. The development of cryogenic land slide including coastal erosion is modeled like slope process during thawing of frozen soils or freezing of thaw soils (in case of Mirnyy below). In addition, widespread coastal erosion and shoreline erosion of lakes at high rates, in the Arctic is essentially accompanied by a thaw blocks/flows displacement of high banks. But it is not discussed in this paper.

14.2 Case Studies

In recent years, winters became shorter and warmer with a prolonged period of autumns and summers with abnormal high air temperatures. The surface layer does not freeze deep enough. These minor factors have strongly influenced the permafrost state. During long warm summers the seasonal thaw layer ("protective layer") becomes deeper contributing to sliding on the ice screen of frozen soils. As a result, underground ice laid below the thaw depth (2–3 m) have already begun to melt.

Landslides in Russian permafrost are unique and at least nine types presented in this paper (Fig. 14.1). The main triggering factor of cryogenic landslides is permafrost thawing on gentle slopes. Sliding of thawed soils on frozen table and solifluction are typical for permafrost. These processes have been accelerated during last decade near Bovanenkovo gas field on the Yamal peninsula; an anthropogenic landslide near Mirnyy caused by freezing of thaw tailing dump during temperature fluctuations in autumn; mud and rock flows are appeared in Chara-China and Baikal-Amur Main Line after heavy rains and earthquakes. Landslide near fault and kurums in Chulman area and paleo cryogenic forms on Olkhon Island (Baikal Lake) are relict and stable.

Nine landslides presented in Table 14.1 are characterised by formation date (from relict to suddenly appeared), seismicity (from 5 to 10), hazard (from potential to high), susceptible infrastructure (roads, pipelines), mitigation measures (from non to thermo insulation and monitoring).

The development of cryogenic land sliding including coastal erosion is modeled like slope process during thawing of frozen soils or freezing of thaw soils (in case of Mirnyy below).

14.2.1 Bovanenkovo

Cryogenic landslides of two types are monitored in Yamal peninsula [1]: mud flows (retrogressive thaw slumps due to thawing of massive ground ice) and translational landslides (active-layer detachments due to thawing of ice lenses in the active-layer base). A large group of cryogenic landslides occurred around Bovanenkovo area in 1980 observed from time series satellite imagery 1963–2013. In 2016 specialists at

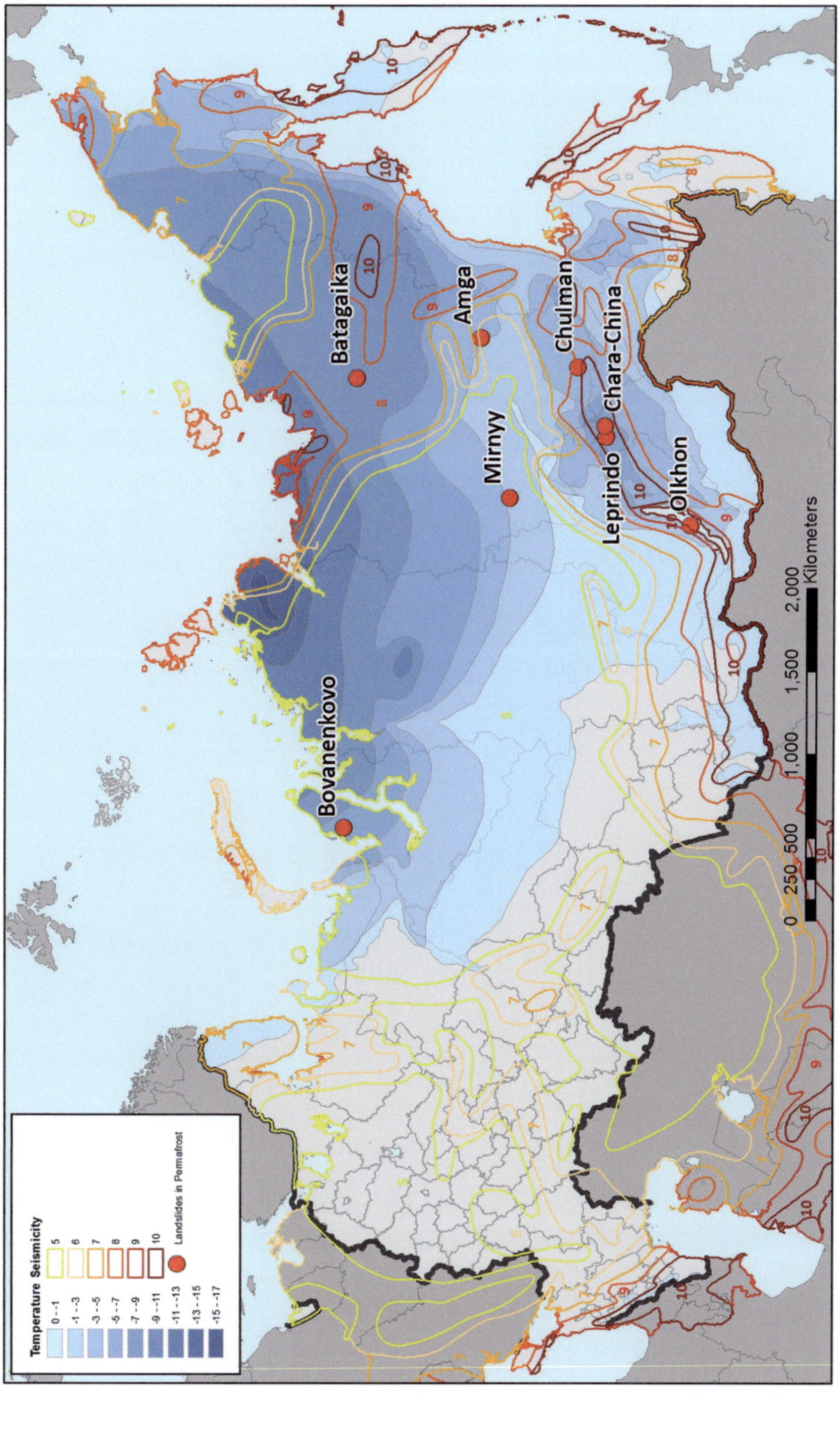

Fig. 14.1 Landslides locations, permafrost temperature and seismicity. (Design by D. Volkov)

Table 14.1 Landslide in permafrost zone of Russia

Landslide type	Location	Formation date	Seismicity	Hazard	Infrastructure	Mitigation
Retrogressive thaw slump (mudslides)	Bovanenkovo	Before 1980	5	High	Road, electrical power lines	Thermo insulation layers, monitoring
Cryogenic landslide (anthropogenic)	Mirnyy	2015	5	High	Road	None
Mudflow	Amga	2013	7	High	Road	None
Thaw slump (depression)	Batagaika	1960-now	8	Potential	None	None
Landslide near fault	Chulman	Before 1980	9	High	Pipeline	Special design, monitoring
Kurums (coarse debris)	Chulman	Before 1980	9	Potential	Pipeline	Special design, monitoring
Cryogenic landslide	Olkhon	Before 1980	10	Potential	Road	None
Polygenetic landslides	Chara–China	After 2000	10	High	Railway	Galleries
Mudflow	Leprindo	Before 1980	10	High	Railway	Trenches

Fig. 14.2 Landslides directed to the road near Bovanenkovo. (Picture of 2016: Vladimir Olenchenko)

company Gazprom Dobycha Nadym had identified a lot of hazardous landslides to linear infrastructure (Fig. 14.2). Protecting measures like thermo insulation layers near infrastructure was installed and regular monitoring were carried out. The seismicity of the region is low and equal to 5.

Retrogressive thaw slump is typical geo hazard for existing and future infrastructure on the developing area of the Yamal peninsula with enormous gas/oil fields.

14.2.2 Mirnyy

The anthropogenic landslide was happened on the highway "Mirnyy-Lensk" due to the pressure in thaw layer between frozen soils in November of 2015 [2]. A landslide blocked the road and power line was damaged (Fig. 14.3). The landslide moved from the tailing dump of diamonds unexploited since 2004. The landslide consists of sand and clay, which remained after washing the soils and extraction of diamonds from alluvial deposits. First, the tailings pond was drained, and thaw soils started freezing from the bottom for several years. Then, in autumn of 2015 soil began freezing from the top forming excess pressure that squeezed the thaw and wet soils down. As the result the extruded raw diamond-bearing sands were moved away in form of landslide.

This kind of land sliding is unique and related to cryogenic process. The major triggering factor was freezing of thaw soils.

Fig. 14.3 Catastrophic landslide from diamond tailing dump caused by excess pressure in freezing soils near Mirnyy. (Photo of November 2015 [2])

14.2.3 Amga

In July of 2013 as a result of prolonged rainfall mudflow was occurred in several parts of highway near Amga [3]. The road Yakutsk–Amga was closed for couple days. Active mudflow covered the highway with a mixture of mud, fallen trees and rocks. About 3 km of road were completely washed away. Fourteen large and small mudflows up to 100 m in width were recorded (Fig. 14.4). The watershed is covered by burning larch forests of 2011. Apart from heavy rains probably fire was a trigger for thaw layer increase and sliding mass preparation.

14.2.4 Batagaika

The Batagaika depression in Yakutia began to form in the 1960s after deforestation and continuous development [4]. This form belongs to mega thaw slump according huge sizes of the thermo cirque: 850 m in length and width and up to 80 m deep (Fig. 14.5). Retreat rate is 15–30 m/year. Permafrost at this site is ice-rich with ground ice and thermal denudation process plays leading role in the Batagaika slump formation.

Fig. 14.4 Catastrophic mudflow after heavy rains closed the highway Yakutsk-Amga. (Photo of July 2013 [3])

14.2.5 Chulman Landslide

The landslide on the route of buried pipelines is situated near Chulmakan fault. The landslide occupies part of the watershed and the steep slope of 20–35°. The difference in height reaches 180 m alluvial sediments of the watershed are under thawed condition. Middle and bottom parts of slope are frozen. The slope has a stepped profile that indicates the horizontal bedding of rocks (Fig. 14.6). Special design for the pipeline and real time monitoring are suggested. The landslide is in stable condition and Intensification is only possible during earthquakes.

This typical landslide as presented one below is located in the southern part of cryolithozone with high seismicity magnitude.

14.2.6 Chulman Kurums

Kurum is debris coarse grained with icy layers inside. Kurums in Chulman area on flat and medium-sized slopes with a length of 2 km and a width of 20–100 m are characterized by very slow displacement and now they are under preserved conditions (Fig. 14.7).

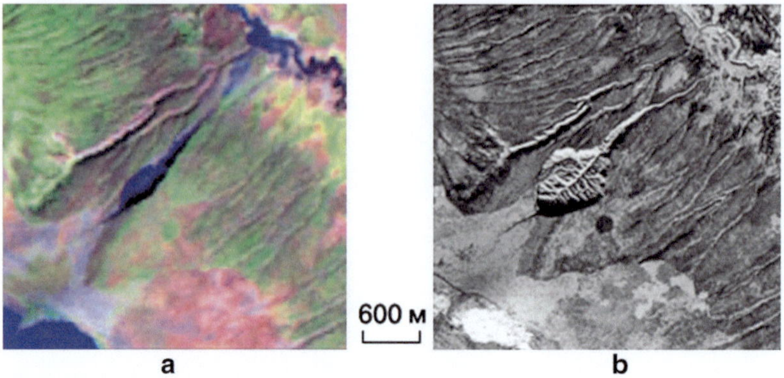

Fig. 14.5 Dynamics of the Batagaika mega thaw slump [4]: (**a**) 1991, (**b**) 2016

Fig. 14.6 Landslide on a steep slope along pipeline in Chulman area, 2010

14.2.7 Olkhon

Cryogenic landslides were investigated on very arid steppes in the northern coast of Olkhon Island (Lake Baikal) with permafrost patches (Fig. 14.8). The assumed mechanism is sliding of thawed soils on the seasonal frozen table.

14.2.8 Chara - China Railway

The precise date of the series of polygenetic landslides in the Udokan Ridge along the Chara-China Railroad is unknown (approximately between 2001 and 2015). The

Fig. 14.7 Kurum along pipeline near Chulman area, 2010

Fig. 14.8 Cryogenic landslides on Olkhon Island (Lake Baikal), March 2014

common force of rock falls, debris flow, snow-stone avalanches, and kurum-shift leaded to dramatic damage of the structures (Fig. 14.9).

14.2.9 Leprindo

Numerous ancient and fresh mudflows cones descend directly to the Lake Leprindo in Transbaikalia (Fig. 14.10). A catastrophic mudflow stopped the transportation on

Fig. 14.9 Catastrophic landslide – kurum-shift in the mountain part of Chara-China Railroad, 2015. (Photo by D.O.Sergeev)

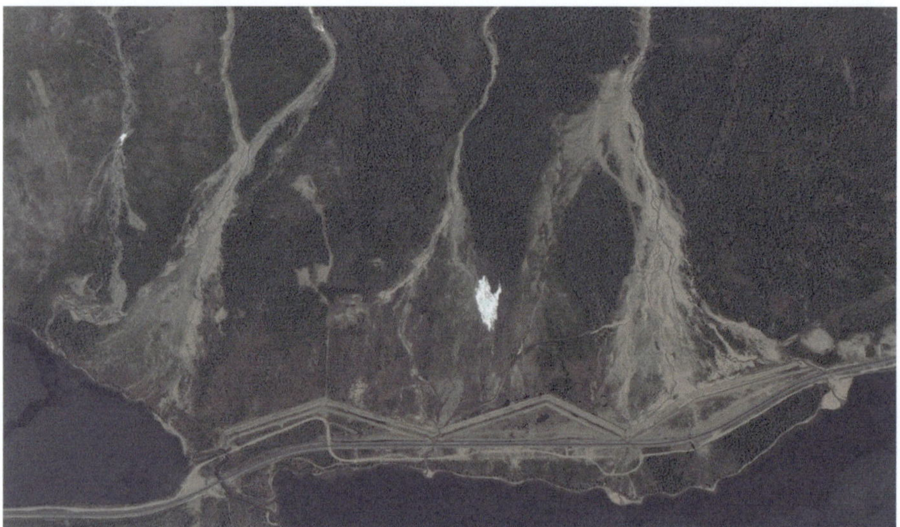

Fig. 14.10 Protecting mudflows trenches along the Baikal – Amur Main Line. (Satellite image from Bing maps, 2017)

the Baikal – Amur Main Line for couple days in July of 2001. The source of mud material was glacial and colluvial deposits. Active tectonic crushing of bedrocks plays major role in the development of a powerful mass of unconsolidated rocks. Heavy rainfall triggered this mudflow. The mitigating trenches were constructed in 2008.

14.3 Conclusions

Many cryogenic landslides are widespread in remote tundra and taiga area, and only several have catastrophic consequences along roads to block it for several days. Clearly visible trend of increase in the number of natural disasters caused by the permafrost degradation due to climate change and industrial operation. This leads to adaptation of existing infrastructure to accelerated cryogenic landslides and creation of new design for future structures, probably the re-routing or relocation should be planned.

A technogenic landslide on the diamond tailing during freezing in autumn was happened despite geotechnical monitoring. This kind of land sliding is unique and related to cryogenic process. The major triggering factor was freezing of thaw soils. It means that thaw and frozen soils are very sensitive to little changes in operation. There is no typical monitoring system to be able to predict acceleration of cryogenic landslides near infrastructure.

The Law on the Permafrost Protection is under review of the Russian Legislation in 2017. It is highly recommended to consider cryogenic landslides in this law to mitigate and adapt to this rare process in permafrost.

Acknowledgments The paper was prepared within the framework of the Presidium of the Russian Academy of Sciences 2016–2018: "Changes in permafrost zone of Russia caused by global warming: natural hazards and contemporary problems of permafrost".

References

1. Leibman M, Khomutov A, Gubarkov A (2012) Cryogenic landslides in the Arctic plains of Russia. In: Proceedings of the first symposium on landslides in cold region, Harbin, China. pp 37–48
2. http://news.iltumen.ru/komentarii/2015/12/opolzen-razbor-poleta
3. http://old.sakha.gov.ru/node/123042
4. Kunitsky VV, Syromyatnikov II, Schirrmeister L, Skachkov YB, Grosse G, Wetterich S, Grigoriev MN (2013) Ice-rich permafrost and thermal denudation in the Batagay area (Yana Upland, East Siberia). Cryosphere Earth XVII(1):56–68
5. Khimenkov A, Sergeev D, Vlasov A, Kozireva E, Rybchenko A, Svetlakov A (2015) Contemporary and paleo-cryogenic formations on Olkhon island. Cryosphere Earth XIX(4): 54–63

MIX
Papier aus verantwortungsvollen Quellen
Paper from responsible sources
FSC® C105338

If you have any concerns about our products, you can contact us on
ProductSafety@springernature.com

In case Publisher is established outside the EU, the EU authorized representative is:
Springer Nature Customer Service Center GmbH
Europaplatz 3, 69115 Heidelberg, Germany

Printed by Libri Plureos GmbH
in Hamburg, Germany